MALT
WHISKY
YEARBOOK
2016

First published in Great Britain in 2015 by
MagDig Media Limited

© MagDig Media Limited 2015

ISBN 978-0-9576553-2-4

MagDig Media Limited
1 Brassey Road
Old Potts Way, Shrewsbury
Shropshire SY3 7FA
ENGLAND

E-mail: info@maltwhiskyyearbook.com
www.maltwhiskyyearbook.com

Previous editions

Contents

Introduction

Those who claim to know me well, often say I'm a person who doesn't like changes. I admit I enjoy routine in some parts of my daily life. On the other hand change, development and innovation in the world of whisky is what gives me the reason to produce the Malt Whisky Yearbook year after year. Change is what ignites the passion in me to persevere.

One change this year, which you will hopefully discover if you have bought previous editions, refers to the book itself. It has a new layout. For ten years it has remained more or less unchanged, but after many discussions with a good friend (thanks Kristoffer), I was persuaded to do something new. I like it – and I hope that you do as well. By the way, I wish to add one more novelty. Over the years, some readers have wished for guidance on how to pronounce the names of the various distilleries. This year I have added a phonetic script for all the Scottish distilleries. I'm sure that some of you will have some comments and you are more than welcome to share these with me. Please remember though, that phonetics is not an absolute science. There can be many different ways to pronounce a word or a name.

My excellent team of whisky writers have excelled themselves again this year and have contributed with some fascinating articles;

For many whisky lovers, peated whisky is what counts. Jonny McCormick widens the perspective and investigates how countries outside of Scotland produce the smoky liquid.

How does the environment impact on the taste of a whisky? Neil Ridley is struggling to become a believer when it comes to the question of whisky and terroir.

Martine Nouet tempts us with delicious combinations of whisky and food and serves great tips on how to do it yourself.

Is the customer always right? Gavin D Smith wants to know if the whisky producers listen to what their core consumers have to say.

The cask is often described as the most important factor defining the flavour of a whisky. Charles MacLean explains why.

Craft has become a word of fashion in the whisky world. Ian Buxton feels the marketers are abusing the expression.

The interest in Japanese whisky has exploded and the category is now at a crossroad. Nicholas Coldicott and Stefan Van Eycken sum up the present and look ahead into the future.

In Malt Whisky Yearbook 2016 you will also find the unique, detailed and much appreciated section on Scottish malt whisky distilleries. It has been thoroughly revised and updated, not just in text, but also including numerous, new pictures, new distilleries and tasting notes for all the core brands. The chapter on Japanese whisky is completely revised and the presentation of distilleries from the rest of the world has been expanded. You will also find a list of more than 150 of the best whisky shops in the world with their full details and suggestions where to find more information on the internet and through books and magazines. The Whisky Year That Was provides a summary of all the signficant events during the year. A new feature is Milestones in Scotch Whisky history covering seven signifi-cant innovations and occasions that helped shape the whisky industry we see today. Finally, the very latest statistics gives you all the answers to your questions on production and consumption.

Thank you for buying Malt Whisky Yearbook 2016. I hope that you will have many enjoyable moments reading it and I can assure you that I will be back with a new edition in 2017.

Malt Whisky Yearbook 2017 will be published in October 2016.
To make sure you will be able to order it directly, please register at
www.maltwhiskyyearbook.com.

If you need any of the previous ten volumes of Malt Whisky Yearbook,
some of them are available for purchase (in limited numbers) from the website
www.maltwhiskyyearbook.com

Acknowledgments

First of all I wish to thank the writers who have shared their great specialist knowledge on the subject in a brilliant and entertaining way – Ian Buxton, Nicholas Coldicott, Stefan Van Eycken, Charles MacLean, Jonny McCormick, Martine Nouet, Neil Ridley and Gavin D. Smith.

A special thanks goes to Gavin who put in a lot of effort nosing, tasting and writing notes for more than 100 different whiskies. Thanks also to Gavin for the tasting notes for independent bottlings and to Nicholas and Stefan for the Japanese notes. I am also grateful to Philippe Jugé for his input on French distilleries.

The following persons have also made important photographic or editorial contributions and I am grateful to all of them:

Alistair Abbott, Iain Allan, Alasdair Anderson, Russel Anderson, Paul Aston, Duncan Baldwin, Andrew Ballantyne, Jan Beckers, Becky Bell, Barry Bernstein, Jodi Best, Matt Blaum, Bartley Blume, Etienne Bouillon, Keith Brian, Ross Bremner, Stephen Bremner, Stephanie Bridge, Katrin Broger, Andrew Brown, Graham Brown, James Brown, Gordon Bruce, Nicolas Le Brun, Alexander Buchholz, Simon Buley, Neil Cameron, Peter Campbell, Jim Casey, Bert Cason, Walter Catton, Ian Chang, Ian Chapman, Yuseff Cherney, Oliver Chilton, Ashok Chokalingam, Stewart Christine, Casper Chuang, Gordon Clark, Suzanne Clark, Margaret Mary Clarke, Hilary Cocalis, Willie Cochrane, Francis Conlon, Graham Coull, Georgie Crawford, Andrew Crook, Gloria Cummins, Francis Cuthbert, Kirsty Dagnan, Susie Davidson, Paul Dempsey, Scott Dickson, David Doig, Jean Donnay, Ludo Ducrocq, Jonas Ebensperger, Lenny Eckstein, Winston Edwards, Bryan Ensall, Simon Erlanger, Graham Eunson, Karen Farber, Hannah Fisher, Andy Fiske, Robert Fleming, Callum Fraser, Shane Fraser, Simon Fried, Rosemary Gallagher, Hannah Gardner, James Geddes, Ewan George, Jonathan Gibson, Mark Armin Giesler, Nir Gilat, Colin Gordon, Kenny Grant, Jasmin Haider, Jim Harrelson, Stuart Harrington, Steve Hawley, Ralph Haynes, Mickey Heads, Joe Heller, Holger Henrich, Gemma Hertwig, Paul Hletko, Séamus Holohan, Liam Hughes, Robbie Hughes, Peter Hunt, David Hynes, Caryn Inglis, Jill Inglis, Helen Jagger, Don Jennings, Valero Jimenez, Michael John, Jenny Karlsson, Collin Keegan, Andrew Laing, Bill Lark, Chad Larrabee, Roar Larsen, Claudia Liebl, Allan Logan, Ian Logan, Alistair Longwell, Bill Lumsden, Horst Lüning, Iain MacAllister, Des McCagherty, Alan McConnochie, Alistair McDonald, Andy Macdonald, John MacDonald, Lynne McEwan, Joseph McGirr, Tim McGuire, William McHenry, Sandy Macintyre, Doug McIvor, Alistair Mackenzie, Bruce Mackenzie, Doug MacKenzie, Jaclyn McKie, Ian MacMillan, Sarah McNaught, Grant MacPherson, Ian McWilliam, Dennis Malcolm, Graham Manson, Stephen Marshall, Jennifer Masson, Victor Matthews, Kwanele Mdluli, Lee Medoff, Ian Millar, Ann Miller, Leah Miller, Andrew Milne, Carol More, Nick Morgan, Jari Mämmi, Andrew Nelstrop, Ingemar Nordblom, Tim Obert, Jason O'Donnell, Hans Offringa, Ewan Ogilvie, Jane Overeem, Chieh-Chang Pan, John Pastor, Stephen Paul, Richard Pelletier, David Perkins, John Peterson, Sean Phillips, Alexandra Piciu, Don Poffenroth, Mike Rasmussen, Carl Reavey, Tony Reeman-Clark, Malcolm Rennie, Bryan Ricard, Nicol van Rijbroek, Patrick Roberts, Jackie Robertson, James Robertson, Stuart Robertson, Brian Robinson, Darren Rook, Colin Ross, David Roussier, Hans Rubens, Iain Russell, Caroline Rylance, Joseph Sammons, Kelly Sanders, Ernst Scheiner, Daric Schlesselman, Tyler Schramm, Heather Shade, Andrew Shand, Nicholas Sikorski, Sam Simmons, Daniel Smith, Alison Spowart, Richard Stabile, Mark Steven, Karen Stewart, Rick Stillwagon, Reto Stoeckli, Duncan Tait, Elizabeth Teape, Jack Teeling, Marcel Telser, Marie Tetzlaff, David Thomson, Laura Thomson, Roselyn Thomson, Haraldur Thorkelsson, Jarrett Tomal, Louise Towers, Jon Tregenna, Zac Triemert, Stuart Urquhart, Laura Tolmie, Sean Venus, Lee Walker, Heather Wall, Jennifer Watson, Ranald Watson, Mark Watt, Andy Watts, Iain Weir, Nick White, Ronald Whiteford, Anthony Wills, Alan Winchester, David Wood, Lee Wood, Sandy Wood, Stephen Woodcock, Nick Yoder, Allison B Young, Patrick van Zuidam.

Finally, to my wife Pernilla and our daughter Alice, thank you for your patience and your love and to Vilda, the lab, my faithful companion during long working hours.

Ingvar Ronde
Editor
Malt Whisky Yearbook

Turf's Up

by Jonny McCormick

"Look, I really love my Islay malts," argues Bill Lark.
"I've become accustomed to that iodine flavour. I describe it as a downright,
dirty mongrel of a peaty whisky, but I love it to bits!"
Acquiescing, the Tasmanian distilling legend slowly surrenders
his upturned palms, and who can blame him?

No different to you or I, Islay's peatiest malts fascinate him. Bill Lark is effusive, generous, and enthusiastic company, as we sit ensconced in the corner of a bar in one of London's finer hotel establishments. "I love peaty whiskies, but I didn't initially. I couldn't stand them, but I considered that there would be a lot of consumers just like me. When you're learning to enjoy malt whiskies, peat straight up can be something that can put you off." Periodically, we decline the waitress's polite interruptions for further libations, allowing her to wheel away and blend back into the afternoon hubbub of genteel chatter and tinkling piano keys. Bill continues, moving on to the subject of Lark Distillery, "So in making our whiskies, I knew we needed to have peat in the whisky."

There is a thirst for peated whiskies. Stop for a moment and contemplate just how special the last two years have been for new peaty releases; Ardbeg Supernova 2014 landed on planet earth, Jim McEwan unmuzzled Octomore 6.3 Islay Barley - a mighty whisky hewn from malt peated to warp factor 258ppm, John Campbell engineered his retro vision using cold smoked, floor-malted barley for his Laphroaig Cairdeas 200th Anniversary edition, Lagavulin reached back to 1991 for their Fèis Ìle 2015 edition and Compass Box brought out the unstoppable force of The Peat Monster Cask Strength Magnum which is drenched in peaty notes. Demonstrably, Islay peat is as vitally important to whisky drinkers as the Nürburgring is to weekend petrolheads. Yet, this is emphatically not an Islay story. This is how the potency of popular peated whiskies spurred whisky innovators around the world to devise novel methods of satisfying the need for peat.

Kavalan Distillery in Taiwan

Where the peats have no name
King Car Distillery, Taiwan

On the other side of the world, I am being driven through the stop-start traffic filling the Hsuehshan Tunnel, leaving Taipei behind and heading for the King Car Distillery. The 13km long tunnel pierces deep under Snow Mountain, emerging into Yilan County and rewarding the traveller with a panoramic vista stretching over the flooded rice fields to Guishan Island (known colloquially as Turtle Island because of its striking profile). Once rested, Master Distiller Ian Chang leads a distillery tour to explain how they produce Kavalan single malt whisky, accompanied by their distillery consultant Dr. Jim Swan.

Dr. Swan's specialist knowledge has seen him assuredly adapt the distillation and maturation process to cope with local climates remarkably different from his native Scotland. For example, at King Car distillery, integral cooling coils within the double-skinned stainless steel fermenters are essential to controlling the process, but it's not a feature required

in a tunroom on Islay. Like an alembic éminence grise, Dr. Swan has been instrumental in commissioning more distilleries around the world than he cares to remember. Each one seems to have been blessed by his golden touch. Ian explains how the distillery is set up to follow Dr. Swan's exact instructions for a single recipe.

Kavalan's secret weapon is its warehouse, where the Taiwanese spirit interacts intensely with top quality wood in the humid atmosphere. The cask types enable many different styles of Kavalan whisky, the climate permitting full maturation within 3-4 years. Understandably, there had never been a need to experiment with production, for example, by altering the mashing in their semi-lauter mash tun, varying fermentation times, or amending the spirit cuts from their Forsyth copper pot stills.

A few months later, I run into Ian Chang in Chicago. He is excited to show me some fresh samples of peated Kavalan whisky and begins to explain how they recognised the demand for peated whiskies coming from domestic and international whisky

Kavalan's secret weapon – the warehouse

drinkers. "When it comes to peated Kavalan, it has always been our aim not to make it too peated," Ian reassures. "To begin with, we have to make sure that the malted barley has the right amount of peatiness. By using the right mashing procedure, we can obtain the peatiness from the malted barley. During distillation, we need to make the right cuts to select the phenolic contents we desire at 7-8ppm, making sure that we don't collect all of it."

He's talking about phenolic parts per million (ppm) in the final liquid of course, a more meaningful number than the original ppm of the malt that many producers talk about. The ppm of malt is not strictly comparable, i.e. 50ppm malt will produce a different ppm in the bottle in different distilleries and this final figure is dependent on significant variables such as mashing, fermentation, and spirit runs. This will be the first time a second Kavalan recipe has been enacted, which involves a completely different set-up of the production equipment by Dr. Swan, as Ian anticipates,

"It's very exciting. We can't wait to taste the results after maturation in four years' time." It's Kavalan

remixed: Peats by Dr. J. Existing whiskies such as the peated Kavalan Distillery Reserve have been achieved using the original recipe and an old trick of David Stewart's deployed on The Balvenie 17 year old Islay Cask. "Before we started making our own peaty new make, we used to buy ex-Islay casks," shares Ian. These are used for full maturation, however, not finishing, "Over 4-5 years' maturation, the heat drives most of the peatiness out of the wood to infuse into the spirit." Blenders will tell you that this process creates a superficial peatiness, good on the nose and palate, but lacking in some of the oils derived from the peated malt in the real thing. It's common knowledge that it's getting increasingly difficult to buy casks, not to mention peaty casks.

"For future sustainability, we recognised that it was time to make our own peaty new make. We can use the peaty spirit to increase the complexity and create a new category of Kavalan expressions." Given their high performance warehouse, these whiskies should be ready for drinking within 3-5 years. You can sense Ian's high hopes, "We are very much looking forward to that."

A peat car famed for fire

Mackmyra Distillery, Sweden

Stumbling around the spooky darkness of a Swedish forest on a chilly winter's night, we gravitate, moth-like, through the pines towards the suffusion of milky light emitting from a towering obelisk. Our beacon of hope turns out to be Mackmyra distillery. Before the tour begins, we're beckoned to one side to see where they peat smoke their malt. No pretty pagoda roofs here, not even one of those large, mechanised industrial malting facilities seen in Scotland. There is no need to trade off some manufactured romanticism around disused architectural heritage: functional practicality trumps bygone aesthetics here. Mackmyra do things their own way. Production was a steady 180,000 lpa until 2011, when it tripled with the move to the new Mackmyra distillery, and they have been making peated whisky since 2004. One of the double doors is cranked opened on a sea freight shipping container: its contents comprise the entire Mackmyra kilning operations. Smoke is drawn from the fire through to the main body of the corrugated steel box and over the shelves of malt laid out within.

They started with a little smoking wagon on wheels when Mackmyra distillery was situated in the old, working village. The landowners were less appreciative of their endeavours and confined their smoking activities to just six weeks a year. "That meant we had very restricted amounts of smoky new make," says Master Blender Angela D'Orazio. She admits that most of it went into their private reserve casks, as there was no shortage of whisky enthusiasts queuing up to buy it. "We've still got a few casks from 2004 but they're not very smoky anymore."

White moss peat is gathered from two local peat bogs, the Big Moss and Karinmossen. "We began to experiment to see what kind of smokiness we could make, trying different woods such as pinewood, trying pure Swedish peat, but then we saw what happened with juniper," says Angela. "We built our stack of peat and put two juniper twigs on top." It's a protected species, so Mackmyra can only source it from very specific locations with permission.

"Juniper is such a dominant flavouring that we can use it without having a lot of it," she says reassuringly. "Even though it's less than 1%, the fact that people say they can sense the juniper just confirms it. It's peat, but there's something else in there: something Swedish and distinctive. It's not like we are using juniper wood that might produce a sauna-like type of smokiness. The twigs are the most delicate parts of it, so that's the ones we use."

No trip to Mackmyra is complete without a journey deep into the Bodås mine, a former iron ore facility, then mushroom farm, now used to store the distillery's casks over 50 metres underground. Each warehouse room has its own microclimate. The temperature sits around a cool 9°C, and though the atmosphere is moist and heady with vapours, the stability and reduced airflow diminish the angel's share when compared with a warehouse above ground in Scotland. Dunnage and palletised casks sleep quietly in their chambers. Down there, you'll find an abundance of cask types from new sherry, ex-Jack Daniels casks from Lynchburg, Swedish oak, Sauternes casks, quarter casks, compact 30 litre private reserve casks, and other experimental batches. Angela explains how she uses all the variables at her disposal to deploy their peaty whisky to full effect,

"We call our original recipe R1, the 'elegant' recipe, and we do a 'smoky' recipe called R2 whilst R3 is our 'extra smoky' one. As we had very little stock when we created the recipe, we cut it when we realised it was a very peaty malt. We created a recipe that was two third 'elegant' and one third 'smoked' and that became the R2. A lot of our customers love smoke, and they requested more smoke. R2 is the range between 15-30ppm, usually around 25 ppm [in the bottle]. Then we created the extra smoky R3 and it's lovely. With that, we cut it two thirds 'smoked', one third 'elegant'. However, it takes longer to mature because of the phenols. If you compare an R2 recipe in first-fill bourbon and an R3 recipe in first-fill bourbon, then you see that you get less oak extraction with R3, as it takes longer to mature."

"When I make Mackmyra Svensk Rök (Swedish smoke), I want to create a complex product, so I use a range of four to nine year old whisky," she says, sharing some of the secrets of her blending. "Some 70% of the whisky comes from smoky casks, and 30% are older, first-fill bourbon casks. You get all these dimensions from the different sizes of bourbon casks we use in the recipe; the 200L, the 100L, and the 30L plus a few casks of Swedish oak, and a few sherry to give it a bit of roundness. The youngest four year old whisky is the stuff that gives it the oomph," she gestures. However unconventional their smoker may look from the outside, the ends more than justify the means.

Drying the malted barley - Mackmyra Distillery in Sweden

Peat smoking at Lark Distillery, Tasmania

Thank you for smoking

Lark Distillery, Tasmania

As the Founder of Lark Distillery, Bill Lark can recall the Eureka moment that led him to his earliest peat experiments with malt. Naturally, a pub was involved and plenty of beer. "We had that dilemma: how do we introduce peat smoke into our whisky? We sat in the pub one night and drank lots of beer and talked about if it would be possible to bubble smoke through the wash. We discussed it until we had drunk enough beer that we knew it was time to go home." This was in the days when you could still smoke in pubs. Outside, Bill's wife remarked about the reek of other peoples' cigarette smoke coming from Bill's jumper. "We looked at each other and I wondered, 'What if we build a smokehouse and use sterile cotton wool to collect the peat smoke and put it in the still, like a gin still? Would that work?" It worked beautifully. Later, they went one step further and built a peat furnace to refine the technique. "Our first peat smoker was a tall chimneystack that we

used to load with malt on a steel shelf three metres above the fire," says Bill with pride. "We would dampen the malt already created and keep it high enough above the peat fire so there was just enough warmth to dry that dampness out, but not enough to change the character of the malt. We figured as the malt dried, it drew in the peat phenolics."

Lark eschewed the option of bringing in peated malt from the U.K. like other Tasmanian distilleries have done. "For us, it was always very important that we stuck to our routes of origin, and that our ingredients are Tasmanian where possible. When we bought our malt from Cascade Brewery, they didn't want peat anywhere near the place because they make beer. In fact, we were able to secure our own peat bog, which has been an amazing opportunity."

"We peat smoked only 16% of our malt to make our first peated whisky," Bill goes on to explain. "I wanted people to taste and experience all of the elements that go into making our whiskies first; the floral note of the barley, the effect on the front of the palate from the wood whether it was port, or sherry or bourbon, the rich, oily malt in the middle and just a hint of peat smoke on the back of the palate, so that balance was very important to us."

Lark developed a strong following for their whiskies and sensed that their customers were ready to explore more heavily peated malt locally, nationally and internationally. Bill commissioned something akin to an updraft pottery kiln in the next phase, with three large shelves where the peat smoke travels over and under the malt. "We found that works an absolute treat. In fact, we sent some samples of our peated whiskies over to Glenfiddich for testing, and we were getting 8-11ppm in our final bottled whiskies. We thought that was perfect, because we were getting that sweetness and depth of character." For the last six to eight years, they have peat smoked 50% of their malt and they are now doing some runs where they are peat smoking 100%. "When we start releasing our 100% post-malt peat smoked whiskies, in terms of peat phenolics, we could be up around 25ppm in the bottle."

With ten working distilleries producing whisky on the island and up to another five in the pipeline, Bill has been helping other Tasmanian distillers to get deeper into peat. "We've always made the wash for Overeem. They chose their own yeast regime, but they wanted the peating done our way." Redlands distillery will be using Tasmanian peat in the same way as Lark distillery. Situated in the beautiful Derwent Valley, Redlands Estate is a paddock to bottle distiller, where they grow their own barley and use their own floor maltings. Think of it like the Tasmanian Kilchoman.

Peat smoke is not necessarily the only source of smoky character in your whisky, as wood-derived phenols can be produced in more mature whiskies. Bill remembers when he was starting out and struggling to tell the difference, "One of the most important things I discovered when I started going to whisky tastings was when I was forced to stand up and describe whiskies: I was getting it wrong time and time again. There was this confusion between oak and peat. It wasn't until we actually smoked

our own peat and you were there smelling the peat smoke with the barley, when all of a sudden, my eyes opened and I went, 'That's what it is'. These days, we love to take people up to our peat bog and cook them Tasmanian ocean trout over a peat fire. As much as it's a very pleasant experience eating and drinking whisky up there, they can instantly relate to what effect peat is having as opposed to anything else."

Many whisky connoisseurs detect the gradual waning of peaty character from bottles over time. This may be down to further chemical interactions occurring over time. Is it really changing in the bottle? "It does and it's funny, when I first started drinking whisky, people said to me, you must tell people that whisky doesn't change in the bottle," says Bill. "You know, once you've taken it out of the barrel and put it in the bottle, it's never going to change. That's not right, and if you ask distillers and they're honest, once you put whisky or any other spirit in the bottle, it does soften and it does change. I mean it's still noticeably the same whisky, but the balance is different."

Finally, I quiz Bill about the plant constituents of Tasmanian peat, and whether the biodiversity delivers a unique flavour profile? "I'd love to be able to say there is, but we had a visit from 40 peat bog experts from all around the world and they told me, 'You know what? Your peat is almost 100% the same as the peat in the Scottish Highlands.'" Regretfully, Bill says, "I used to think that maybe some of the dry sclerophyll forest material would have washed down into the peat bog but it hasn't happened." It turns out Tasmania has deposits of blanket bog formed almost entirely from compacted, decomposed sphagnum moss, just like Scotland.

Ultimately, I've been spun around the globe and come full circle to land back on Scottish peat. Kavalan, Mackmyra, and Lark are breaking new ground in their own unique ways to bring us enticing, smoky temptations. Their spirit is a good maxim of how to live life: March to the dram of your own peat. Well, something like that anyway.

Whisky writer, author, and photographer Jonny McCormick has written hundreds of whisky articles for publications including Whisky Advocate, Whisky Magazine and the Malt Whisky Yearbook. McCormick created the Whisky Magazine Index and the Whisky Advocate Auction Index to track trading and value in the secondary market. He is a Keeper of the Quaich and he has led presentations about whisky in Europe, Asia, and North America.

Reign of Terroir

by Neil Ridley

'I Want To Believe'. Those of you with a penchant for 1990's American Sci-fi TV will perhaps remember this as the seminal phrase from The X Files, boldly emblazoned on a poster situated in the office of lead character – the complex Fox Mulder. It relates to Mulder's steadfast belief in extra-terrestrial activity, in the face of mass opposition – or indeed, any conclusive proof that We. Are. Not. Alone.

Fast-forward to a surprisingly warm spring day this year on Islay, when, as my car pulls through the gates of the Bruichladdich distillery, the phrase makes an unexpected re-entry into my consciousness. The reason for this recall concerns the current theories that surround the very fabric of what makes whisky such a beguiling and – largely misunderstood subject: Terroir. As opinions go, nothing seems to be so divisive amongst the whisky community right now (well… perhaps the concept of No Age Statement whiskies, but let's not go there.)

I am perhaps stuck somewhat in the precarious middle ground: with one half of me thinking there's more to this than meets the eye, the other half more cynically feeling that a lot of marketing guff and misinformation on Terroir is doing the business no favours in the long run.

Terroir is of course a very well-feathered subject in the world of French wine making. In fact, it's the same when it comes to French spirits. From Cognac to Armagnac, the concept of Terroir is fairly well documented and seemingly unchallenged. When you look at the roots of what Terroir means, the interaction between the earth and the grape vines which

Examining the barley grain

produce the aforementioned fine spirits, it all makes perfect sense: some grapes grow better than others in different soil types, ergo you're going to get a different style of product from region to region – and looking a little more closely, from field to field.

Take the regions of Cognac for a second: the vines that come from the Grand Champagne region grow exceptionally well in the more coveted chalky soil type. Move further out from this region and the soil type changes, becoming more adulterated with clay, flint, limestone and sand – less conducive for growing the best grapes, so less desirable for the wine makers and distillers. All clear so far. One can even adequately apply the term to the spirits produced in Mexico. The Agave plants that provide the raw materials for mezcal and Tequila grow in some startlingly different climates and soil types, consequently giving the distiller a range of different flavours to play with, from sweeter, fruity distillates in the Highlands (thanks to the lower temperatures and mineral-rich soils) to much more herbaceous, vegetal flavours of the hotter lowlands and its largely volcanic soil.

So Terroir is not a complicated subject – anywhere that is, unless you're a distiller in Scotland perhaps. Can the term adequately apply to whisky making? And if so, how far reaching can it apply? Quite a lot – and quite far, if you're from Bruichladdich, it seems.

Ignoring terroir is lazy

"The effects of Terroir haven't been explored in the world of Scotch whisky and at Bruichladdich we believe that this is lazy and wrong," thinks Carl Reavey, Media Manager for the distillery. "We believe that the Scotch whisky industry has hidden behind the smokescreen of brand image because that is easier than getting a grip on the esoteric micro-influences that come to play on the development of its fundamental raw material, barley, – the most flavour complex cereal in the world."

So what has the distillery been doing to explore the concept further? Conceptually, the idea of different strains of barley growing in different soil types should have some influence on flavour – but, as I have always been led to believe by good, old fashioned science, the effects are purely on alcohol yield.

"Terroir varies according to place," continues Reavey. "Terroir varies from one field to another, from harvest to harvest and from one vintage to the next. Its effect will inevitably vary from plant species to plant species and from crop to crop. We believe that it will inevitably impart subtle nuances and variety. It will have an effect on any food or drink. The more complex the flavours inherent in that food or drink, the more profound that effect. And single malt Scotch whisky is the most flavour complex spirit in the world…"

Professor Otto Hermelin in a Bruichladdich warehouse

Bruichladdich's optimism and passion for such a concept is clear to all and its experiments – from distilling different varieties of barley, including Optic, Concerto, Chalice and even ancient Bere Barley are likely to help open up the discussion to a wider audience. But until anything conclusive comes out, there will always be sceptics – including me, it has to be said.

Just hang on a minute though. The distillery might just have science on their side – from the point of a different kind of Terroir. Otto Hermelin, Associate Professor in the Department of Geological Sciences at the University of Stockholm has been working with Bruichladdich (as well as Ardbeg, Kilchoman and Bunnahabhain) to look into the effects of maturation, from a temperature and humidity perspective. I'm intrigued to find out what the professor is hoping to achieve with his project, which is on track to last ten years.

"The reason why I am doing this is that when I first saw new make spirit running through the spirit safe I noticed there was a huge amount of green copper oxide produced," he explains "so I was intrigued to see how much made it into a finished bottle of whisky. I initially analysed samples of new make against whisky matured for 10 years from the same distilleries and it was quite reduced, which demonstrated how maturation takes out the copper. This was the beginning and the more I read, the more I realised how little research there was into the maturation process."

So how does this relate to Terroir - if at all?

"Well, the study has only been running for four years, but what I have noticed is that alongside the copper content going down in the samples, I have noticed the level of sodium going up – and I believe this is directly from the sea spray, as the warehouses I am monitoring are close to the sea.'

NOW we have an emerging mystery to solve! The Fox Mulder in me has suddenly pricked up his ears. So there is evidence to suggest that whisky matured closer to the sea has a higher salt content directly because of the environment it is matured in?

"Comparing the samples against Mackmyra here in Sweden, which is matured nowhere near the sea," he continues, "there is almost nowhere near the same level of sodium content. I believe that if you took an Islay new make and matured it on the mainland, it would taste different."

So the long held belief of the influence of the elements actually has some scientific basis – albeit in its embryonic stages. Whilst this won't come as much of a shock to anyone who has visited Islay (given its landscape has a visceral effect on pretty much every visitor) the professor's research has bought up a profoundly different type of 'Terroir', which is

Would a Bowmore taste different had it been matured on the mainland and not in Warehouse No. 1 by the shores of Loch Indaal

unique to location. To put this another way: can the same spirit be replicated exactly in a different location? Or do the elements – and in effect – a different type of Terroir come into their own here?

Peter Mackie´s trials

I remember an informative talk I attended several years ago in which an impassioned Jim McEwan steadfastly flew the Terroir flag, whilst several other distillery group representatives had to bite their tongues before making their cases. What should have ended up in a bar room brawl actually balanced itself out, thanks in part to some fascinating historical insights from Diageo's Dr Nick Morgan, so I'm keen to revisit some of his findings.

"My memory from the discussion we had is that in the conventional understanding of the word 'Terroir', there is, at first sight, no equivalent in Scotch whisky production. The reason for this," he continues "is that even if you look at the supply routes back in the late 19th century, distillers used malt from all over the place. Ironically, we probably use more Scottish malt now than we ever have done. There's also no real way to tie vintages down in the same way that the French wine producers do, which relates directly to the Terroir of the land."

But what Dr Morgan concedes is that there is a contradiction to his viewpoint when it comes to the distinctiveness about where whisky is made.

"The classic example of this was Peter Mackie's attempts to make Speyside whisky in Campbeltown after the First World War," he notes.

"Mackie owned the Hazelburn distillery [alongside founding Craigellachie and inheriting Lagavulin] and the reputation of Campbeltown whiskies had declined enormously in the late 19th and early 20th centuries for two reasons: firstly because when they were well made they were considered too pungent and heavy to be used in blends and then later, it was likely that the distillers were using poor quality barley from Eastern Europe. But Mackie was a control freak in the sense that he believed one could apply science and knowledge to do anything."

"He bought in distillers and raw materials from Speyside, re-engineered the distillery to mirror the exact set ups in Speyside, but still couldn't get any where near the same whisky produced there."

So travelling forward into the 21st century, what does Dr Morgan put this down to? Immeasurable and incremental changes in the whisky making process, or just simply Location Location Location?

"One of the things that has to be borne in mind is that although the fermentation of the wash is lead by yeast which is added to the washbacks, there is of course the idea of secondary fermentation by indigenous yeast," he continues. "One imagines that given the lack of control in the process over 100 years ago that the impact of those indigenous yeasts would have been far greater than today."

We're closing in on the truth, Scully, methinks. Let's throw the net wider, this time to whiskies made in locations other than Scotland. Japan, Tasmania and India have all been hugely influenced by the Scots when it comes to whisky-making practices – and raw materials – but yet produce markedly different styles of spirit.

I start my international investigation in the shadow of Japan's mist-shrouded Mount Asama, once the home to the Karuizawa distillery, which officially closed in 2011. One thing I couldn't quite understand is just why some of the vintage bottlings released several years ago had such insanely high ABVs, (well into the 60+% territory) despite resting in cask for upwards of 30 years. Did Location Terroir have something to do with this strange, unexplained phenomenon?

Marcin Miller, the man behind Number One Drinks and former owner of the remaining Karuizawa stock takes up the story.

"The altitude of the distillery site is 850 metres above sea level and it is located on Mount Asama, a volcano that recently showed signs of activity," he explains. "The average temperature is 10 degrees and the average humidity is 80%. This results in the area being shrouded in mist perhaps three times a week. My understanding is that there are no hard and fast rules and, further, that each distillery enjoys a unique microclimate based on a variety of factors including those listed above. However," he deduces, "it appears that at Karuizawa, water evaporated before alcohol resulting, not only in the high strength whiskies to which you refer, but also in extraordinary richness and concentration of aroma and flavour, which, to me, is what sets Karuizawa apart."

Similarly, Ichiro Akuto, owner of the Chichibu distillery has a few things to back up this Location concept. "I'm not really sure what Terroir means for us in the conventional sense, because our barley comes from UK and the oak for our casks was not grown in Japan. But we think the Terroir for whisky making is the environment for the maturation. It is one of the most important elements for flavour, we believe."

Does Akuto-san think he will ever see a bigger swing towards using domestically grown barley in Japanese whisky making? If so, would the flavours radically change?

"To be honest, we think using domestically grown barley won't make a big swing in Japan," he points

Maturing casks in Karuizawa warehouse

Chichibu Distillery in Japan

out. "So the flavour of Japanese whisky won't change immediately either. We used our local floor malted barley mixed with the imported one experimentally so it's not 100% locally made. That's why we are not sure about the flavour yet. We'll see it soon."

Over on Tasmania, Bill Lark, owner of the Lark Distillery – and many would say the founding father of modern day distillation on the island – has a few interesting views on Terroir from the perspective of the other building block of whisky making: water source. Certainly the PH balance can influence the style of wash produced, but science tells us that distillation effectively separates alcohol from water, so how can it really have an effect on spirit flavour?

"I agree wholeheartedly with that view regarding water for the preparation of wash and subsequent distillation and have expressed that similar opinion on many occasions," explains Lark. "In fact I have sourced water from a number of locations around Tasmania as a matter of interest and I think it is fair to say that no one at our distillery could discern any real difference in the character of the new make spirit."

"But," he continues, "I think I would also have to agree that the water used for breaking down the mature spirit for bottling is vitally important to the character of the whisky. We have all seen the effect the minerals in the water have on producing 'flocking', [essentially the opaqueness a whisky takes on when heavier proteins and oils are released from suspension] it is then hard to imagine that water from

different sources doesn't play a part in the character of the whisky at that stage."

I also pose the Terroir question to the Amrut distillery in Bangalore, where the heat and humidity have a hugely profound and unique effect on maturation. But what about the raw materials themselves?

"Terroir certainly has a meaning from two defining movements as far as Amrut is concerned," thinks Ashok Chokalingam, brand ambassador for the distillery. "The first one is that we use a 6-row barley, which is an admix variety, rather than any specific strain of barley and is grown at the foothills of the Himalayas. The geology of that particular region makes a great contribution to this."

"Secondly," he continues, "the topography of Bangalore is so unique in such a way that we have great contrast between day and night with temperature and humidity, coupled with the altitude of 3000 ft above the sea level. Both these two points makes a unique whisky in Bangalore."

So in one respect, perhaps we see Terroir in the conventional sense (a unique barley cultivation style) and from a location perspective. What does Ashok think about a Terroir concerning the water source?

"Right from the very beginning I am a strong advocate on this point," he interjects.

"A water source does not define the character of the whisky at all and I would go on to say it is irrelevant to a greater extent when it comes to its contribution in defining the flavour of the whisky. This 'marketing gimmick' has started somewhere a long time ago and been implanted into consumers minds in a strong way."

From field to flask

The phrase 'From Farm to Fork' did wonders for the British meat industry a few years ago. In fact, you couldn't so much as read a menu, without a waiter eagerly thrusting the name of the cow at you, the type of grass it regularly consumed and the number of bowel movements it probably had in its happy, fun-filled life. All this extra info meant a supposedly more knowledgeable customer and a cash register ringing excitedly in the background. Whisky has been relatively slow off the mark here but clearly now, times are a-changin'.

"We are aware that casual whisky drinkers have, on the whole, not previously considered Terroir, just as it is not high on the list of concerns for casual wine drinkers", explains Carl Reavey. "This is at least partly because industrial producers of both whisky and wine would rather that was the case and it is why they concentrate on promoting image, rather than espousing more challenging concepts. At Bruichladdich we produce some of the most thought-provoking spirits in the world. They will not be for everyone. That's OK…"

Yes, this is indeed 'OK', but as Reavey freely admits, "the nature of maturation means that the results would always take time to emerge." At some point, the burden of proof is going to weigh in particularly heavily here and as we have been witnessing in recent months (see the epic 'marketing' fail that tripped up Templeton Rye and its more-than-erroneous 'provenance' claims) there's nothing folk like more than to pick holes in a marketing concept – then sue the pants off you.

One aspect of Terroir that I have so far neglected to mention has actually been staring me in the face all along. In fact, in the literal sense, maybe the earth itself has created a Terroir for whisky making – of the peated variety.

In 2012, The Scotch Whisky Research Institute introduced a thesis by scientist Barry Harrison into the 'Impact of Peat Source on the Flavour of Scotch Malt Whisky'. Conducting lab distillation tests using malt peated from various peating sources (Islay, Orkney, St Fergus and Tomintoul) demonstrated pronounced differences in guaiacols and phenols, meaning that the make up of different peat (moss, grass, heather, woody plants/trees) would indeed inevitably lead to different flavours. The conclusion here was pretty clear: change your peated malt

Ashok Chokalingam – Amrut's brand ambassador

source and you're probably going to change your whisky's flavour.

Transport yourself to any of the locations mentioned above and…well, they're all very different landscapes. The treeless, gusty Orkney is covered in aromatic heather. Islay was once home to densely forested areas. Apparently, Skye once had a preponderance of hazelnut trees (although I can find no conclusive evidence to back this up). All that plant matter clearly 'matters' and has helped to fundamentally shape the modern peat bogs over millennia. Floral, woody and pronounced nuttiness are all smoky notes we associate with the different whiskies produced in these locations, so throw in a little romance and there's certainly a lot to believe in.

There's that phrase again. I Want To Believe. Perhaps there is more to Terroir than first meets the eye. Perhaps some disciples have asserted their beliefs more robustly than others, making the whole concept a little fuzzier for the rest of us. Either way, the jury's still out. But in true Fox Mulder fashion, albeit adapted slightly:

I'm Starting To Believe.

Neil Ridley writes regularly about whisky for a wide range of drinks and lifestyle publications such as Whisky Magazine, Square Meal, Imbibe, The Mayfair Magazine (where he is Food & Drink Editor) The Spirits Business, as well as daily and luxury publications including The Evening Standard, City AM, Aston Martin, Bentley and Sunseeker International. He is also one of the authors (together with Gavin D Smith) of Let Me Tell You About Whisky and in 2014, his latest book, Distilled (written together with Joel Harrison) was published.

Whisky & food
strike the good match

by Martine Nouet

I remember some twenty years ago, when I started experimenting with food and whisky pairings, I was looked at as an eccentric if not a lunatic. At this time, it seemed a heresy to match a first class single malt with a dish. Now it has become almost a classic, which I am both delighted with and proud that my determination stimulated emulation.

Choosing the appropriate whiskies to match dishes requires exactly the same efforts as with wine. You need to have some knowledge about the drink. The rest is practise and a question of taste and sensitivity as the technique is entirely based on sensory evaluation. I know that some people have tried a scientific approach, basing their pairings on chemistry (similar molecular profiles) but I am not convinced. For me, the appreciation of flavours goes far beyond chemistry. You can't leave out personal tastes nor cultural environment.

Let's have a look at the whiskies first. The fantastic wheel of aromas displayed by single malts, pot-still whiskeys, bourbons or rye whiskeys generates just as many descriptives as wine does (maybe more!). It makes sense to offer one's dram a wider source of enjoyment than just aperitif or after-dinner. Tasters' notes can often be read as a full menu. A lot of sensations, whether at nosing or at tasting, are related to food. What about "kippers", "oxtail broth", "peppered mackerel", "bacon", "crème brûlée", "lemon meringue pie"? A rich whisky reveals itself as a "liquid" dish ! All these obvious connections between food and whisky naturally ask for matching.

Martine Nouet demonstrating cooking with whisky

Photo: Alex Mitchell

Taken individually, a whisky and a dish have their own richness and character. When well matched, they create additional character, bringing in new flavours. One plus one equals three, not two, an odd mathematics formula I admit but so true. Serving whisky with a meal is not just having your favourite dram with your favourite dish. If you drink a highly peated single malt, say an Octomore, along with a red fruit pavlova, you will not appreciate either of them. It is like wearing a red dress with an orange scarf and purple shoes. Not the best demonstration of elegance and taste.

Matching food and drinks requires simple but basic rules. It is important to note the key aromas in the whisky profile. Then you look for the "bridge" which will make the two have a conversation. It does not have to be the main flavour. It will rather be a herb, a spice, a fruit... Then, you think of dishes which could echo or complement the aromatic palette. The importance is to keep the balance so that one does not overwhelm the other. You can also play contrast: sweet against bitter or sour, smooth against crisp. That last suggestion implies to consider the texture. A satin-like feel will pair with a creamy sauce if you want to put the stress on smoothness. But you can

also contrast that satin-like texture with a crunchy dish to make it better stand out (raw vegetables, thin slices of toasted bread).

The choice will also depend upon the season. Rich, creamy, heavily aromatic whiskies with a sherry influence will be outstanding in autumnal recipes with such ingredients as duck, beef, foie gras, parsnips, leeks, raisins, apples, figs, ginger, cinnamon. But certainly not light salads, red fruit, scallops or veal. Whereas younger, lighter-bodied single malts will marry with springtime and summertime products such as shellfish, fish (salmon, sea bass), herbs (basil and coriander), broad beans, artichoke, spinach, fennel, rhubarb, red and black berries. These are the ideal combination with a single malt matured in ex-bourbon or refill casks. Many malt lovers tend to associate smoked food to peaty single malts. Not an inspired choice as one smokiness conflicts with the other and overpowers the dish. It is better to go for a honeyed and malty whisky.

So let's review some easy pairings. For a better understanding purpose, we will examine different types of food and stick to Scottish single malts.

Parma prosciutto, gingerbread and orange

Fish and shellfish

The key-word is freshness. This is why young,
vibrant, bourbon-cask matured whiskies will feature
best. For fish, unpeated whiskies are better, for
instance a Speysider like Glenfiddich 12 or Glenlivet
Nàdurra will match white fish in a creamy sauce.
Non peated Highlanders such as Old Pulteney 12
or Oban 14 will go well with smoked fish or oily
fishes. Smoky island whiskies will pair with oysters,
mussels or lobsters. The perfect tip: a few drops of
Laphroaig 10 on an oyster. Caol Ila, with its vanilla
profile, pairs well with langoustines or lobster. As
with scallops, it depends on how they are prepared:
Bruichladdich Islay Barley with a "surf and turf"
combination such as scallops and black pudding, Isle
of Jura 16 if scallops are pan-fried with mushrooms,
For scallops served with butter, ginger and lemon
sauce, any bourbon matured whisky will shine
through (Auchentoshan American wood, The Laddie
Ten, Glenmorangie 10...)

Meat

There is a large spectrum of meat dishes. To sum
up, beef and venison will pair well with sherried

Glenlivet Nàdurra and scallops

matured whiskies. Oaky notes will go well with spicy and wine sauces. Dried fruit in the sauce or the stuffing can be a good bridge. Fruit like oranges and duck are perfect too and then, peated whiskies will find a good connection with citrus fruit. For a roasted breast of duck and an orange sauce, go for a sherry cask matured whisky such as Glenfarclas 15, Mortlach 15 (G & McP) or Glenrothes 1988. But depending upon the type of meat, bourbon cask matured malts will be chosen for poultry or lamb cooked with dried fruit. For instance, Glenmorangie Nectar d'Or will make a scrumptious match with an apricot and almond lamb tajine, Benromach 10 with roasted pork fillet and prunes. Spicy dishes such as a steak au poivre marry well with Talisker Storm for example.

Vegetables

The "king" vegetable with whisky is a root vegetable – parsnips, carrots, Jerusalem artichokes, neeps, celeriac – because of its sweetness. Chestnuts used as vegetables (a soup or a purée) feature very well too. Beetroots with their earthy flavours marry with sherried whiskies. Squash, with their rich smooth texture and their sweet taste allow interesting combination, especially when seasoned with exotic spices or with vanilla (cream of butternut squash and coconut milk paired with a bourbon cask matured malt). Potatoes are a "base" and thus, rather neutral. If they are seasoned with herbs, spices, fruit, they offer a wider range of possibilities. Raw crunchy vegetables (carrots, fennel, radishes, celery) are interesting texture wise but also taste wise (fennel and bourbon matured whiskies make a refreshing combination, agreeing on aniseed flavours).

Cheese

Cheese is maybe the easiest pairing to start with. The most important is to work with one type of cheese only. forget about serving an assortment of cheeses unless you are ready to open as many bottles of whisky as the number of cheeses you offer! It is impossible to find a consensual whisky. So here are some suggestions with one type only. With blue cheeses, go for a peated whisky (Lagavulin and roquefort make a winning pair). Cheddar wants a smooth and sweet whisky, so choose a single grain whisky or a first fill bourbon matured single malt (such as Hedonism or Glenrothes Alba Reserve). Matured cheeses (comté, gouda, gruyère) will like a malty whisky with a honeyed and fruity profile (Balvenie Double Wood, Aberlour 16, Highland Park

Photo: John Paul
Garden Pea and Mint Soup with Aberlour 10 year old

Photo: John Pa
Pears in Ricotta with Scapa 16 year old

12). The more mature the cheese is, the older the whisky will be. I remember an outstanding pairing I made in The Netherlands with a Glenfarclas 40 and a very old gouda – salty and crumbly.

Chocolate

This is probably the favourite ingredient in whisky pairing. Here again, it does not mean that any chocolate pairs with any whisky. The darker, the better with oaky and sherried whiskies. Praline, milk or white chocolate will suit creamy bourbon matured whiskies. A chocolate with a bit of salt will make a good match with a smoky whisky. The textures should also be considered: smooth chocolate fondant, light aerial mousse, crunchy icing... A chocolate fondant (semi-bitter) served with a salty caramel sauce would pair well with Bowmore 12. Try a white chocolate cheesecake with raspberries and Glenlivet Founder's Reserve. A dark chocolate tart (nuts welcome) would marry with Dalmore 15 or Lagavulin Distiller's Edition.

Other desserts

The range is so large that a whole book could be written on the subject! I can only suggest ideas and examples. You will generally find a good connection between Speyside malts and fruit, especially pears and apples or peaches and apricots. Rhubarb is a bit more tricky because of the sour tinge of the fruit which can be challenged by icecream, honey, maple syrup or crystallised ginger. Many malts matured in ex-bourbon casks reveal a custardy creaminess on the nose as well as on the palate. They will love custard. The floating island for instance is a delicious pudding that you can adapt to different whiskies just by flavouring the custard. With coffee, marry it to Glenrothes 1985; with orange zest, prefer Aberlour 18 and with coconut, go for Auchentoshan Classic. I don't like adding whisky to custard. I always find it too rich, almost sickly. Cognac and rum work but not whisky.

Remember that the friends of whisky are... nuts! Choose almond for younger whiskies, toasted hazelnuts for older ones and walnuts for old timers (especially sherried ones).

Now the question. Can whisky be paired with such ingredients as vinegar (in a salad for instance) or onions ? The answer is: yes but... Vinegar can be an excellent enhancer but has to be handled carefully. It is preferable to use balsamic vinegar as it is less sour and offers an interesting aromatic range (from the

Photo: John Paul

Chocolate fondant with Aberlour a'bunadh

Martine Nouet preparing the menu   Photo: Alex Mitchell

youngest to the oldest). I mean here "true" balsamic vinegar, expensive yes but you only need a few drops. This vinegar has to be used raw. I would not recommend any other raw vinegar which will always be too aggressive. When cooked in a sauce though, wine or cider vinegar will lose some of their tanginess, especially when sweetened by honey or dried fruit. Onions follow the same rules. Raw onions are too pungent and will leave a taste in your palate which will clash with whisky. But cooked onions, especially when caramelized (in chutneys, sauces or salads) are a good pairing ingredient.

In fact there is only one ingredient I ban from any pairing or cooking with whisky. It is garlic (whether raw, cooked, pureed). Garlic is said to kill vampires. It does the same for whisky. Full stop.

A question of technique

When adequately chosen as seen above, whisky and food make a delightful pair. One can go even further, using whisky as an ingredient in dishes, on the basis that, if it works between plate and glass, it will naturally work in the plate. Like for pairing, proper cooking with whisky is not just adding any whisky to any dish. It works on the same aromatic

scale as with pairing. As important, if not more, as a good balance of flavours, is the way whisky is incorporated in the preparation. The cook aims for methods which retain the aromas while alcohol evaporates. Single malts work particularly well in cooking as their great range of aromatic profiles allow all sorts of preparations and bring out very interesting flavours.

The first method one will think of is the "flambé technique" which French cooks are famous for. The flame burns out the alcohol fumes but it also takes away most of the aromas. This may be good for show but it is a waste of time and money. The "crêpe Suzette" show has certainly toured the world in Escoffier´s or even Paul Bocuse´s times but that does not mean it is an example to follow. So let's forget about flambé.

Marinades are certainly the best way to combine whisky and food, especially when working with raw ingredients. But they must be quite short. After fifteen minutes, alcohol tends to "cook" the flesh of the fish or meat and gives it a greyish shade. A meat tenderizer, it may break the fibres if used to the excess and reduce the steak to a tattered spongy mass. For a subtle touch, simply brush the surface of a meat or fish tartare with whisky before serving. Some dishes need macerating rather than marinating. Haddock for instance, will better retain the single malt aromas if it is soaked in yogurt and whisky – a floral Lowlands or a malty light Speyside – for 24 hours at least.

Marinades do not only consist of whisky. A proper seasoning includes lemon, olive oils, spices, herbs and special sauces such as Teriyaki or Worcestershire. Fresh ginger, honey, marmalade, balsamic vinegar complement whisky in some preparations. Need I remind you? Forget about garlic! A quick tip: marinate raw langoustines for ten minutes (not more) in lime juice, 2 tablespoons of peated whisky (such as Caol Ila), a teaspoon of grated ginger root and grated lime, adding a pinch of chili. Then cook them in a pan with a touch of olive oil. Deglaze the pan at the very end with the marinade. Marinade speaks for itself when it comes to sweets. Fruit such as pears or apples can be soaked in whisky and honey before being poached. Candied and dried fruit "drink" whisky easily too, whether it be raisins, dates, figs or apricots.

Whisky is also used for the "finishing touch", just like seasoning. According to the above principle of avoiding cooking, the best way is to pour whisky at the last minute, off the stove. It can also be drizzled over roast or steamed fish, just to wake flavours up. The same applies to sweets. The technique of deglazing a very hot pan with a dash of whisky is easy and gives flavours without the alcohol.

As for the proportion of whisky to be used in the recipe, it depends upon its aromatic profile. This is why single malts feature better in cooking than blended whiskies. A light and delicate malt will allow a higher dose whereas a heavily peated or sherried one will argue for a light touch. A matter of chemistry if not alchemy! But that applies to all kinds of ingredients in cooking, doesn't it? Brushing is more a tip than a technique but it works very well, especially for cakes. Just soak the brush in the whisky. Brush the surface of the cooked food. An example, for a barbecued leg of lamb, brush the meat with a mix of honey and a Highland malt (Dalwhinnie 14 or Clynelish 14). Spraying whisky on a dish just before serving adds an intriguing flavour which will surprise your guests. I always have two atomizers by my stove, one filled with a smoky whisky and one filled with a rich Speyside malt. Very important: always do that at the last minute as the fragrances are evanescent. The final touch: spray a peated whisky over a cream of parsnips and toasted almonds. Do the same over a summer fruit salad with a smooth fruity malt.

Feeling hungry? It is time to try your own experiments. If it can help, I will have a cookery book ready by the end of the year under the title "A table", featuring whisky from the glass to the plate. It will include recipes but also an A to Z guide on techniques and ingredients.

Bon appétit!

Guideline for good matches

- smoked food with grassy malty Highlands or Speyside malts or Irish blends (all unpeated)
- light salads, crunchy vegetables with floral Low lands or young Islays
- meat or venison dishes with sherried whiskies, heavily peaty ones, rye whiskeys
- citrus fruit with smoky malts
- shellfish and fish with salty/peaty ones or vanilla tinged whiskies, including bourbons
- all creamy vanilla desserts with malts matured in bourbon casks or bourbons
- chocolate or dried fruit puddings with sherried whiskies, bourbons, rye whiskeys, old pot-still Irish whiskeys

A few principles

- Always cook with the freshest ingredients
- Avoid confusion by combining a whole spectrum of flavours. I always work in a triangular dimension: one main ingredient (whether it be meat, fish, salad, fruit etc..) and two minor ones acting in opposition, fusion or complementing
- Find a bridge between the whisky and the food, a "hyphen" which will facilitate the matching because it enhances a common point. It can be a condiment, a spice, or a bigger ingredient. A common denominator in fact.
- Stick to the season. Forget stews in summer or strawberries in winter. And choose your whisky accordingly as, even within the range of a single distillery, there is a seasonal profile in whiskies.
- To strengthen the marriage between the dish and the whisky, you can (but not systematically) add whisky in the preparation. In this case whisky will act as a seasoning ingredient.

Martine Nouet is the only author and journalist currently working in France and in the UK who writes exclusively about food and spirits. She is a regular contributor to Whisky Magazine UK and she launched Whisky Magazine French edition in 2004. In France she is known as "La reine de l'Alambic" (the Queen of the Still). She is a specialist of matching food and whisky and hosts whisky dinners all over the world. A judge at the annual International Wine & Spirit Competition in London, she has also written a book on malt whisky, "Les Routes du Malt".

Only connect
malts and malt consumers

by Gavin D Smith

Never in the long history of Scotch malt whisky
have consumers enjoyed so much product choice or been armed with so
much knowledge as they are today.

Dedicated websites, blogs, Twitter and Facebook have joined the more traditional tasting
sessions and brand ambassadorial evangelism
to create a remarkably well-informed body of
consumers. They are serious about their malt
whisky and therefore increasingly demanding
of distillers and independent bottlers.

This 'body of consumers' includes bloggers and their followers, whisky journalists,
members of whisky clubs and festival atten-
dees, subscribers to whisky periodicals and, of
course, readers of the Malt Whisky Yearbook!

So just what are these aficionados discussing
and demanding and how are the whisky-makers
responding?

According to Nicola Young, Scotch Malt Whisky
Society Ambassador, internet and social marketer
and contributor to The Whisky Boys blog (www.
whiskyboys.com), "Consumers are enjoying an out-
rageous rise in the choice of whiskies available on
the shelves of specialist drinks shops, supermarkets
and online retailers, is this driven by the consumers'
demands?"

"Today influencers, writers and aficionados' opinions can be heard every day through numerous online
mediums. The top five requests I hear on Twitter,
the blog and Facebook are: un-chill filtration, no
colouring, exclusivity (small batches), distillery-only
bottlings (for investment) and higher ABVs."

Young points out that "Terminology once reserved for the industry professional, now enters the
vocabulary of many whisky lovers, as single malt

whisky drinkers are educating themselves more about whisky – it's a hot topic. The result of this is pressure being added to the distillers to continue to develop, invent and create."

One firm which certainly seems to be responding to that pressure is The Glenmorangie Company, and in particular its Director of Distilling and Whisky Creation Dr Bill Lumsden. Lumsden had conducted Twitter tastings and broke new ground for Scotch whisky by embracing crowd-sourcing to develop Glenmorangie Taghta, released in September 2014. In March the previous year Glenmorangie had launched The Cask Masters programme, which invited aficionados to participate in all elements of developing the new expression, beginning by choosing the liquid from three samples pre-selected by Bill Lumsden, followed by deciding on the whisky's name and packaging design. They were even asked to suggest a launch venue, with Glenmorangie distillery being chosen from the shortlist.

At the time of the launch, Bill Lumsden observed that "No other whisky has ever had consumers involved in all stages of the whisky creation process and we have really enjoyed the experience." He now adds that "We identified crowd-sourcing as another interesting channel through which to liaise with our consumers and allow them to get involved in shaping the product. We are intending to do another as a follow up to Taghta, but get consumers even more involved. Maybe we will start next time by designing it from source – it will be bigger and better."

Such processes, along with the use of platforms like Twitter are clearly valuable, but Lumsden says that "I do PR in the markets and I really look for genuine feedback about products – what do you like, what don't you like, how could we do it better, and so on. For example, I revised the recipe for Glenmorangie Lasanta based on lots of comments from the USA and Asia, in particular. Effectively, I sweetened it up a bit. I've actually changed recipes in quite a few cases for Glenmorangie as the result of feedback."

The same applies to the company's Ardbeg brand, with Lumsden declaring that "In terms of Ardbeg, Galileo was perhaps on the subtle side, and the overall message we were getting was that consumers liked their Ardbegs big and mucky and bold, so now I'm going for expressions that are more classically Ardbeg-like in nature."

Lumsden raises an interesting point, and one shared by a number of fellow senior whisky industry professionals, when he says that "You have to be careful in relation to issues raised on social media, because sometimes they're just an opportunity for people to let off steam. I really like face to face contact – look people in the eye and find out what they really feel. Of course aficionados will want their single malts non-chill-filtered, non-coloured and at varying strengths, but we have to offer a core range of malts for the mainstream. For the aficionados we have the more esoteric Private Edition series, with the sixth and latest being Tùsail, made from floor-malted Maris Otter barley and not chill-filtered. These are developed primarily in response to consumers who want to see that sort of thing."

Remodelling an entire range

If Bill Lumsden and the Glenmorangie team have proved themselves keen to take on board comments from aficionados and even involve them in the creative processes of bringing a new expression to market, one distiller who was actually ahead of the curve in giving them the sort of whiskies they desire is Ian MacMillan. He is Head of Distilleries & Master Blender for Burn Stewart Distillers Ltd and back in 2010 he made the decision not to chill-filter single malts right across the company's portfolio, embracing the Bunnahabhain, Deanston, Ledaig and Tobermory brands, increasing their strengths in the process.

"I realised by comparative sampling before and after chill-filtration that without was better," he says. "That's why I did it, rather than it being in response to any clamour from consumers. It gave enhanced aroma, better texture, and more powerful flavours. The flavour and aroma compounds have taken years to develop, so why would we want to lose some of them just to make the whisky pretty?"

It may well be that MacMillan's transformation of the Burn Stewart malts has added to the social media pressure for higher strength whiskies without chill-filtration, as mentioned by Nicola Young. Effectively, he led consumers in a direction they appreciated. Many aficionados might wonder why other distillers have not followed Ian MacMillan's example, but as he points out, "It worked very well for our single malts, but in certain whiskies it may not have a major difference. Also, you need a process of re-education for consumers if you make the change, and many of the products have been in the market for a long time and consumers know what to expect."

"Also, it's very expensive to do – cartons, other packaging and advertising are all bought in large volumes, so for one of the big distillers the changes to labelling and advertising for the higher strength and stating not chill-filtered and so on would mean major changes and expense. Additionally, the costs rise due to having to pay additional tax due to increased alcohol strength.

Dr Bill Lumsden, Glenmorangie

"But for us the price rises we had to impose when we made the change did not deter consumers, and sales across all our malts have increased. We've had no negative responses, everyone has been very complimentary. It's interesting that when most mainstream distillers do limited editions they are un-chill-filtered."

Like Bill Lumsden, Ian Macmillan does take cognisance of feedback when formulating new expressions, and he says that "Lots of it comes via our websites and from our ambassadors who conduct tastings all over the world.

"Much of the feedback we received concerned Bunnahabhain and how consumers would like a heavily-peated version. We'd been producing peated Bunnahbhain spirit since 2003, and releasing it in a limited way, but last year we were able to launch a 10 year old peated Bunnahabhain. That certainly satisfied a lot of people who wanted a regular peated Bunnahabhain.

Intriguingly, Macmillan declares that "The marketing people work with social media and obviously that gets passed on to me, but I like to listen in person to what people have to say. I don't have a Twitter account and I'm not on Facebook!"

Ian MacMillan, Burn Stewart Distillers

Nicola Young, Scotch Malt Whisky Society Ambassador and Ken Grier, Edrington´s Director of Malts

Catering to collectors

A brand such as The Macallan presents its own set of considerations when it comes to new expressions, as the company has to respond not only to the desires of consumers, with whatever level of knowledge, but also collectors and investors. Edrington's Director of Malts Ken Grier notes that "We are very close to our consumers through our markets, listening to people in bars and stores and carrying out market research. We also keep an ear close to the ground with auction houses and specialist consultants, and we monitor blogs and Twitter and Facebook reactions. We also take into account how quickly a particular product sells!

"The Macallan Royal series – the Coronation Bottling, Royal Marriage and Diamond Jubilee releases – was very much a response to demand. We try to understand the motivation behind what people are looking for as collectors, investors and drinkers. The Macallan is a big auction seller. You are looking at limited editions of a blue chip band which should increase in value."

Grier explains that "With a new product we write down the narrative that goes with it so that everyone is clear about the selling points, with 'sell-sheets' and brand presentations. We communicate the narrative of the product through the specialist press, PR

and social media, and we put lots of information on The Macallan website."

"We always tap into what people are saying on social media and we respond to them. We have listened to consumers talking about the importance to them of strengths, of having differing ABVs, and upcoming products will certainly address that issue."

Ken Grier concludes by echoing Bill Lumsden's cautionary note. "Of course, you've got to take social media in the round, as it were. People's comments are all subjective." However, Nicola Young of The Whisky Boys says that "Social media and whisky seem to be a perfect match. Someone who has a true passion for whisky loves nothing more than to discuss, debate, and compare reviews with the next whisky drinker. Social media is an immediate sounding post for the distillers; this is a ready-made focus group, waiting and willing to share opinions, good and bad, and easy to access."

Inevitably, the largest Scotch whisky companies such as Diageo produce malt whisky for blending purposes first and foremost, and are sometimes accused of not paying too much attention to those aficionados who would like to see innovations among their single malt brands.

Diageo's Head of Whisky Outreach Dr Nick Morgan responds by saying that "Well over 90% of the Scotch we sell is blended Scotch whisky, but we make exceptional malt whiskies in order to make great blended Scotch whiskies. We are very proud to show off our whisky-making skills by bringing these malts to market, in addition to occasional bottlings like our annual Special Releases which are unsurpassed in quality. And of course, malts are a good piece of business to be in."

As far as the issue of chill-filtration is concerned, Nick Morgan contends that "If we were to increase the strength of our single malts like Lagavulin 16 year old they would be different from, but not better than, the current versions. As for chill-filtering, the process prevents the formation of floc which, though harmless, can be troublesome for consumers especially those in very cold climates, and it allows us to deliver a consistent colour and appearance which is what consumers of the most widely available single malt brands expect. Our Special Releases are normally not chill-filtered."

In terms of interacting with single malt aficionados, Morgan explains that "We have Facebook pages for our leading single malt whiskies – Lagavulin, Talisker, Oban, Cardhu and The Singleton – as well as our Friends of the Classic Malts page. We also have active Twitter handles for Lagavulin, Cardhu, Talisker and The Singleton, and for the annual Special Releases. All these platforms offer the opportunity for numerous interactions with our more engaged consumers. Our Brand Ambassadors are also very active on both Facebook and Twitter.

"On top of that, we read and listen very carefully to what consumers and critics say on social media; very little goes unnoticed. It is not our policy normally to engage in this conversation, though in certain circumstances, when we are asked a direct question, we do so."

Far from Diageo in terms of scale are independent bottlers such as Douglas Laing & Co Ltd of Glasgow, but many have seen that the use of social media can be even more significant for them than for the industry's major players.

Managing Director Fred Laing says that "Cara, my daughter, who is our Director of Whisky, is particularly social media-driven

and we have another member of staff doing nothing but Twitter and Facebook work for our key brands such as Big Peat and Scallywag."

However, like so many of his whisky industry colleagues, Laing is also keenly aware that meeting consumers face to face remains an extremely valuable practice, noting that "At shows we try to meet as many consumers as possible and also talk to bloggers, who are important opinion-formers in lots of countries. We take part in festivals, roadshows and tastings, and our staff can be in four countries over a weekend at times."

Participation in such events also allows the independent bottling company to carry out valuable liquid 'market research' among aficionados. Fred Laing says that "At tastings we will often bring along a couple of upcoming releases, maybe a young grain or a 50 year old North British."

"Grains have been very positively received, and there is now sufficient interest that we don't just

A trio of whiskies from independent bottler Douglas Laing

Dr Nick Morgan, Diageo's Head of Whisky Outreach

have the Clan Denny range of grains but also an Old Particular single grain label. That's been led by the enthusiasm we've encountered with single grains. We might also offer a young and feisty six year old Caol Ila, for example, and if it's well received out on the road it gives us the confidence to bottle it."

'The road' is also where whisky industry professionals may encounter aficionados like Oxfordshire-based businessman Peter Arbuthnot, who has been buying and drinking single malts since the 1990s. "I browse company and independent whisky websites now and again," he says, "but I'm not on Twitter or Facebook. I attend tastings and whisky festivals from time to time and really enjoy meeting people from the distillers and having the chance to talk to them, especially if it's someone who's actually involved in making the whisky rather than just a brand person."

Arbuthnot is no whisky obsessive, but he has accumulated a significant amount of knowledge during his two decade-long interest in the subject, and he notes that "I have a copy of Michael Jackson's Malt Whisky Companion and I do buy the Malt Whisky Yearbook each time in order to keep up to date with changes at distilleries and with new expressions."

Arbuthnot raises one point of contention among whisky aficionados which does not figure in Nicola

Recent extension of the Talisker range of single malts

Young's list of most mentioned topics on social media, namely the issue of age statements. "I'm not saying that older is necessarily better," he maintains, but with an age statement you know where you are. More and more non-age-statement whiskies are appearing all the time, and I think it's really because most distillers are running short of mature stocks. They say that the non-age-statement bottlings are all about selecting the whiskies when they're at the perfect age, the equivalent of picking an apple when it's at its best on the tree, but I'm not convinced it isn't just expediency."

Nick Morgan responds to the issue of age statements by declaring that "Like other whisky makers, we are very committed to a policy of innovation, particularly in flavour. This is an area where our innovation teams, working with our coopers, maturation experts and blenders have done us proud. Talisker Storm, Talisker Skye and their siblings in the Talisker family have proved very successful; so have variants like The Singleton Tailfire and Oban Little Bay. More will follow."

"The enthusiasm with which these seductive and interesting new variants have been received gives the lie to those purists who wail that no-age-statement Scotch whiskies herald the end of the world as we know it. And the success of these new expressions perhaps shows that vociferous social media commentators aren't necessarily representative of the majority of regular single malt whisky consumers!"

Just as Glenmorangie formed its Cask Masters 'club' in order to produce what became Taghta, so other brands have created similar organisations such as Diageo's Friends of the Classic Malts, Balvenie's Warehouse 24, The Dalmore Custodians and The Ardbeg Committee.

Bill Lumsden sums up the purpose of all these organisations when he says that "Ardbeg Committee members get exclusive buying opportunities and a heads up on upcoming products. They feel connected and involved and effectively become brand ambassadors for Ardbeg, helping to spread the word."

One brand 'club' which has taken the concept a stage further, and closer to the Glenmorangie Taghta model, is The Guardians of The Glenlivet. Late in 2013 The Glenlivet sent out samples of three single malt expressions selected by Master Distiller Alan Winchester.

The three were Classic – fruity with soft sweet caramel and toffee notes, Exotic – rich, with warm spicy notes, and Revival – fruity with a creamy sweetness. With votes cast by The Guardians, Exotic was pronounced the winner, and was duly bottled

in 2014 as The Glenlivet Guardians Chapter, at a strength of 48.7%, and with no age statement.

Meanwhile distillery-exclusive bottlings continue to proliferate as a means of 'rewarding' consumers for taking the time and trouble to visit the distilleries in question. These have a collectable value as well as offering unique drinking experiences and take the form of ready-bottled offerings such as Glengoyne's Tea Pot Dram and The Dalmore Distillery Exclusive or bottle-your-own single cask expressions, as practised at the likes of Aberfeldy, Aberlour, Benromach and Old Pulteney distilleries.

Nicola Young of The Edinburgh Whisky Boys declares that "I feel it is no longer for the fame or ego boost, or even a novelty, that the highest ranking decision makers of a distiller are put in front of the writers and influencers in place of the local or global ambassador (at a dinner, launch, tasting or on Twitter). It's because they can access unprecedented, independent reviews and opinions about the liquid; they can find out first-hand what we want next and what we thought of the last release."

"The PR and Marketing departments go a long ways to advising and forecasting the next idea, but the men and women on the ground (writers and influencers) can add a different viewpoint, we are the consumer and the influencer."

Young concludes in upbeat style by saying that "It is apparent the single malt whisky distillers are listening, as more and more small-batch or distillery-only releases are being bottled than ever before, and with many of them ticking the boxes of being non-chill filtered, at cask strength and with no colouring. Some of these ticks are even becoming the norm for a few distilleries such as Bowmore, Arran, Glenrothes, Bruichladdich and anCnoc."

With so many literal and virtual channels for communication between distillers and consumers, one thing is certain. The conversations will continue, with ever more voices making their presence heard. The world of single malt Scotch whisky can surely only benefit as a result.

Gavin D Smith is one of Scotland's leading whisky writers and Contributing Editor to www.whisky-pages.com. He regularly undertakes writing commissions for leading drinks companies and produces articles for a wide range of publications, including Whisky Magazine, Whisky Advocate, Whiskeria and Drinks International. He is the author of more than 20 books and collaborated with Dominic Roskrow to produce a new edition of the Michael Jackson's Malt Whisky Companion in 2010. His latest books are The Whisky Opus, co-written with Dominic Roskrow and Let Me Tell You About Whisky, co-written with Neil Ridley.

The Cask is King

by Charles MacLean

Of all the factors which influence
the flavour of malt whisky, the cask in which it matures
is the most influential.
If you think about it, this makes perfect sense.

It takes just four or five days to produce a batch of spirit, from milling the malt to filling the cask; the spirit then lies to the wood for three or five or ten or twenty years. But time is only one factor – as today's Master Blenders are quick to point out when they take age statements off their labels and encourage us to make up our own minds about the whisky's flavour. The species of oak used to construct the cask, how it is seasoned, how many times it has been used for maturing spirit, whether or not it has been rejuvenated, what it was filled with previously and reactions between the spirit and the surrounding atmosphere all play a part. But before we consider these influences upon the flavour of the mature whisky, let me say a little about the history of casks and coopering.

Wooden casks have been around for a very long time. Tubs made from staves, are depicted in Egyptian tomb paintings from 3000BC, and it is reckoned that the Egyptians of the XVIII Dynasty (1570-1544BC) were the first to form a barrel shape – a very strong, double-arch structure which can nevertheless be rolled and pivoted relatively easily, even when full. Signficant advances in barrel-making were achieved with the discovery of how to work iron and steel into both tools and hoops to bind the staves tightly.

These skills arrived in Europe in the late 11th century BC and gradually spread north over the following 500 years. It is generally believed that it was the Northern Celts who first developed water-tight barrels for holding wine, around 350BC, and during the first century AD such containers were being used for transporting wine from Gaul to Rome. The word 'cooper' derives from the wine makers of Illyria and

Cisalpine Gaul, where the wine was strored in wooden vessels called cupals, made by a cuparius.

The Romans themselves favoured earthenware amphorae to store liquids, but soon adopted the barrel. A panel on Trajan's Column in Rome (AD 98) depicts a river-boat being loaded with casks of wine – they look very like today's port pipes – for shipping down the Rhine and across the Channel to Londinium. The oldest cask to have survived in Britain was found at Silchester, a Roman town, and dates from around AD 230. Interestingly, the staves are numbered – an indication that at one time the barrels were re-made: if a cask is broken down and re-assembled, the staves must be raised in the same order as previously if they are to fit together tightly.

"The processes and techniques which were developed at this time…became trade secrets, closely guarded and kept within families, small tribes and later guilds, passing down from father to son and from master to apprentice, right up to the present day". [Kenneth Kilby – The Cooper and His Trade (London, 1971)]

Oakwood

Once upon a time, new Scotch spirit could be filled into any kind of wood, but now only oak may be used: whisky cannot legally be called 'Scotch' unless it has been matured for a minimum of three years in 'oak casks of a capacity not exceeding 700 litres' – to allow optimum contact between the liquid and the wood (the smaller the cask, the greater the liquid to surface ratio) – and this in an excise warehouse in Scotland. Somewhat surprisingly, the requirement was only enshrined in law in 1990: previous definitions demanded only that the spirit be matured in 'wooden casks'. Although beech and chestnut have apparently been used in the past, oak has long been the wood of choice on account of its strength, its malleability – when heated it may be easily bent into shape without cracking – and its porosity, oak's capacity to allow the spirit to 'breathe' and interact with the atmosphere.

There are approximately 600 species of oak tree, the largest number of which grow in North America. Most American oaks are of the genus *Quercus alba* (white oak; the name derives from the colour of the tree's bark, although the vast majority are light grey). They grow mainly in the Eastern States, from Quebec to Florida, and west to Minnesota and Missouri, tolerant of a variety of habitats – dry and moist, in alkaline and acid soils, on low and high ground. They grow well in forests, and when grown in this way the trees' trunks are tall and straight, with few branches (these cause knots in the timber),

which makes them good for coopering. The wood is high in vanillin, (which adds vanilla flavours during maturation), furfural (which adds caramel and marzipan) and oak lactones which add coconut flavours.

Two sub-species of *Q. alba* are *Q. macrocarpa* (burr oak; the Latin name translates as 'large fruit' on account of the size of its acorns, which are penduculate [see below]) and *Q. bicolor* (swamp oak). As with *Q. robur* [below], burr and swamp oaks are lowland trees, growing in open country, where they are able to send forth low branches, which lead to knots and twists in the grain and to leaky casks. As a result these species of oak are not used for barrel-making, although in 2004 Glenmorangie released an expression matured exclusively in virgin burr oak for 11 years.

There are two main species of oak in Europe: *Q. robur* ('robur' means 'strong'; it is also known as 'Pendunculate oak', since the tree's acorns and leaves grow on stalks) and *Q. petraea* ('oak of rocky places', is often referred to as 'Sessile oak' on account of its leaves and acorns being 'stalkless'). The former, is 'the oak of Old England', and also of Spain and the forests of Limousin in France. The trees prefer deep and rich soils at low altitudes, often in open country or is spacious 'forests', as a result of which the timber is prone to having knots and twists. The wood is coarser and looser grained than American white oak and more difficult to cooper, since it must be split, not sawn, and splinters easily. It is also up to ten times more tannic (Spanish oak having the highest levels) – adding colour, astringency and fragrance to the wines and spirits. It is also higher in eugenol (clove-like flavours), guaiacol (spicy, also burnt flavours) and syringealdehyde (spicy, smoky flavours), than *Q. alba*. Limousin oak is especially favoured by cognac distillers and in general Q. robur is preferred for maturing spirits on account of its higher tannin levels.

Q. petraea is the predominant species in France, growing mainly in the Tronçais, Nevers, Allier and Vosges forests, and also in Hungary, Slovenia, Romania, Poland and Russia. A forest species, it reaches upwards during the short summer growing season when the sap rises and the plant fights for light and air, with a long, straight-grain trunk and few lower branches. Ideal for coopering, *Q. petraea* is chosen by French winemakers for their best wines, having a tighter grain than *Q. robur* and lending greater refinement; preferred even over *Q. alba*, which, although it increases the texture of the wine, lends more obviously the oaky flavours.

A sub-species of *Quercus petraea*, known as Gaelic, Cornish, Welsh, Irish and Scottish oak,

Glenmorangie´s Bill Lumsden inspects a tree in Missouri

Kevin O´Gorman, Head of Maturation at Midleton Distillery and Dair Ghaelach, the latest addition to the range

a stalkless species, it grows in upland areas over 300 metres, where the rainfall is higher, the soils shallow and sandy. It is even more difficult to cooper than Q. robur, the trees being smaller and more knotty, but this did not prevent Highland Distillers releasing a 16YO expression of Glengoyne in 2002, finished in Scottish Oak grown on the CEO's estate in Angus.

The final oak species used for maturing whisky is *Quercus mongolica* (Mongolian oak, also known as Mizunara oak) which grows in Eastern Asia and Japan, and has been used by Japanese distillers since the 1930s. The wood is soft and porous and therefore prone to leaking and easily damaged. As a result, casks made from *Q. mongolica* are today usually used for finishing. High in vanillin the wood contributes "vanilla, honey, blossom, fresh fruit (pears, apples), spice (nutmeg, cloves)", and in the words of the well-known Japanese whisky expert, Marcin Miller, it adds elegance: "like the smell of a temple – incense, sandalwood, Oriental spices".

Kevin O'Gorman, Head of Maturation at Midleton Distillery in Cork, sent me a note which perfectly exemplifies how seriously distillers take the oak-wood they use. Irish Distillers have begun to release limited batches of single pot still whiskey matured in virgin Irish oak, harvested from specifc forests, a series they name Midleton Dair Ghaelach (which simply means 'Gaelic Oak'), the first expression

coming from Ballaghtobin Estate, County Kilkenny As part of the planning for this, Kevin and his col-leagues investigated the key differences between Irish oak and Spanish and American oak, in order to understand what effect maturation in native oak bar-rels would have on a Single Pot Still Irish Whiskey. He writes:

" Irish oak grows at a faster rate than Spanish and American oak, due to climate, moisture and the length of growing season. It has lower density and higher porosity, which lead to a more open structure, and this in turn allows more wood compounds to be extracted into the spirit, at a faster rate."

"We identified higher levels of some lignin deriva-tive compounds in Irish oak, such as vanillin and va-nillic acid, which contribute to the enhanced vanilla, caramel and chocolate flavours of Dair Ghaelach. It also contains a greater concentration of furanic com-pounds, such as furfural – formed during the thermal degradation of cellulose and hemicellulose – which tends to give the sweet, caramel and toasted notes that are detectable on the nose of Dair Ghaelach."

Wood Seasoning
Drying

All oak has to be seasoned (i.e. dried) before use. Traditionally this was done in the open air – indeed,

Charring a cask

in the forest manufacture described above "care needed to be taken to avoid the staves drying too quickly in the wind, so it was customary to pile the staves and shield them with branches for some months" (Kenneth Kilby).

The Irish oak mentioned above was staved in Galicia then air dried in Andalucia for just over fifteen months, and it can take longer, even in warm climates of Spain and Kentucky to reduce the moisture content from around 55% to the required 15%. Sometimes this is expedited by kiln drying, but during the late 1980s, Dr. Jim Swan and Dr. Harry Riffkind of Pentlands Scotch Whisky Research proved that kiln drying adversely affects the speed of maturation of Scotch spirit. Their recommendation to the industry was that at least one third of the staves in a cask should be air dried. It would seem that kiln drying is not a problem for maturing bourbon or rye whiskeys, but it is known to affect the flavour of wine. In Understanding Wine Technology, David Bird writes:

"In common with other forms of joinery, kiln drying is sometimes used but this does not give a good result, giving astringency and green characters to the wine from the rough tannins which are cooked into the wood. Natural seasoning adds elegance, richness and complexity to the wine, with a smoother finish."

Heat Treatment

In order to shape a cask, the staves must be heated over an open fire. This is essential for the maturation of whisky: if the staves are bent only with steam, without heat treatment ('toasting'), the spirit will merely acquire a 'green', woody flavour. The greater the degree of toasting, the greater the influence the wood will have on the cask's contents: a 'light' toast takes around twenty minutes, the temperature of the wood reaching between 120°C and 180°C. A further five minutes, with the temperature rising to 200°C, yields a 'medium' toast, and five minutes more a 'heavy' toast, at around 225°C.

Toasting causes chemical changes in the walls of the cask – technically known as 'thermal degradation' – breaking down wood polymers so they will yield flavour extractives and colour. Such treatment also removes undesired oaky, new-wood flavours and astringency, converts hemi-cellulose into, predominantly, furfural (with sweet, caramel, coffee and cocoa flavours; toasted bread, roasted almonds and sweet pepper) and degrades lignin into vanillin and various desirable aromatic aldehydes. However, tannins are accutely sensitive to temperature and are destroyed by even low amounts of heat.

American barrels are then treated further with heat by being set on fire to line the walls of the cask with

Sherry bodega in Jerez, Spain

Photo: Edrington

a layer of char (active carbon). This removes sulphur compounds, increases lignin degradation (in the sub-surface just beneath the char) and also increases the development of extractable colour. The level of charring does not in itself contribute colour, but it does stimulate the release of tannins from the wood. Nor does active carbon make much of a contribution to flavour, but it plays a pivotal role in removing organo-sulphur compounds, which smell of drains, cabbage water and over-cooked vegetables, and is detectable in very low concentrations (from .02 to .1 parts per million).

Wine/Spirit Seasoning

As we have seen, the oak trees grown in Britain are by no means ideal for barrel-making, so from an early date casks were imported from Europe as well as fresh oak staves. Often these had previously been used for transporting wines and spirits, prior to bottling in the U.K., and by the mid-19th century whisky distillers recognised the desirablility of filling casks which had been used to ship sherry from Spain:

"It is well known that Whisky stored in Sherry casks soon acquires a mellow softness which it does not get when put into new casks; in fact, the latter, if not well seasoned, will impart a woodiness, much condemned by the practiced palate. In Sherry casks the Spirit likewise acquires a pleasing tinge of colour which is much sought for; this is frequently imitated by the use of colouring, but it is not creditable to those who adopt such petty deceptions." [Charles Tovey, British and Foreign Spirits (1864)].

Until the 1960s most of these transport casks were made from American White Oak; since then the majority have been European (mainly Spanish) oak casks. The former were preferred: as early as 1948 Manuel Gonzalez Gordon, the head of the great sherry House, Gonzalez Byass, commented: "In recent years some Spanish oak has been used [for shipping sherry], due principally to the difficulties of obtaining American timber". This was confirmed by Alexander Williams of the equally famous sherry House, Williams & Humbert, in conversation with George Urquhart of Gordon & MacPhail, Elgin. Given the chemical differences between the two species, this is significant for those of us interested in the flavour of Scotch whisky in days gone by.

In 1981 Spanish export regulations outlawed the bulk transportation of sherry in casks, and since then whisky distillers who want 'sherry casks' must commission and season them to order from coopers and bodegas in Jerez. The length of time sherry had in contact with the wood in a transport cask was between six and nine months; today, casks made specifically to mature Scotch whisky are first filled with mosto (fermenting grape juice) then with sherry (usually oloroso) for between one and four years before being shipped. Sherry casks are usually either butts (500 litres nominal capacity) or puncheons (460 litres nominal capacity).

The seasoning of casks with sherry was not a new idea. Indeed, W.P. Lowrie – wine and spirits broker in Glasgow (with the Gonzalez Byass agency), pioneer of whisky blending, and supplier of whiskies to James Buchanan of Black & White fame – is reported to have "revolutionised the cooperage trade by arranging to have staves cut, in the forests of America, to standard sizes… The wine seasoning of casks was his patent." [Brian Spiller, The Whisky Barons, xxx p.14]

A variation of this was the use of Paxarette, a very sweet, very dark, vino de color made mainly from Pedro Ximinez grapes, forced into the walls of the cask – 500ml per hogshead; 1 litre per butt – under pressure at around 7 psig for ten minutes. From the 19th Century, Paxarette was also commonly used as a colourant by blenders, in place of the spirit caramel used today. It was very popular, adding sweetness and combining the flavours of the consituent whis-

kies to advantage, but it was banned by the Scotch Whisky Regulations 1990.

American oak ex-bourbon casks only began to be used extensively after the Second World War. The Speyside Cooperage told me that they never saw any bourbon barrels until 1946. Today, 95% of the casks coming into the system arrive from the United States – although, since butts and puncheons last longer, there are many more than 5% being used at any one time.

American Standard Barrels (or ASBs, holding approximately 200 litres of spirit) arrive entire or broken down into 'shooks' and flat packed, to be re-made into hogsheads (five barrels make four hogs-heads), holding around 250 litres. The reason for this is simply that the larger size of barrel is easier to man-handle. Since mechanical handling is now more common, many whisky companies have turned away from 'traditional' hogsheads and embraced ASBs, which, being slightly smaller, mature their contents more rapidly.

Rejuvenation

After a time, depending on how many times the cask has been filled with spirit, but usually around fifty years (three or four fills), the cask activity

Different types of casks - from left to right, bourbon barrel, port pipe and sherry butt   Photo: W Grant & Sons

is close to zero: it is 'exhausted'. Until relatively recently, the cask would then be sold off to make garden furniture or oak chips for smoking salmon. Today, casks are so expensive that it is worthwhile 'rejuvenating' them by scraping the inner walls down to the bare wood, then re-toasting and re-charring – a process called 'de-char/re-char' in American casks and 'de-tartrate/re-toast' in European casks. This re-activates the layer of wood immediately beneath the charred/toasted walls of the cask, but it does not make the cask as good as new.

More than a hundred volatile compounds (oils, tannins, sugars, organic acids and sterols) have been identified and isolated from oak; some of these can be restored but key compounds such as oak lactones and tannins are not replenished, so rejuvenated casks will not mature their contents in the same way as a first-fill cask. Such casks are said to give spicy flavours to the spirit. Ex-sherry casks will sometimes then be seasoned with sherry for a period of weeks or months.

Wood management

Scientific understanding of the chemistry of oak wood and of what is happening during maturation is relatively new – within the last thirty years. Prior to this the general approach was 'if the cask doesn't leak, fill it'. Chemists now talk about 'cask activity', and believe the wood performs three crucial functions (or 'mechanisms'), described as 'subtractive', 'additive' and 'interactive'.

The subtractive machanism is mainly performed by the layer of active carbon which lines most casks, and relates mainly to the removal of sulphur compounds, as we have seen . Since European oak ex-sherry casks are rarely charred, whisky from such casks may sometimes be sulphury.

The most obvious additive mechanism is in relation to colour: European oak, being more tannic, lends its contents a deeper hue than American wood. The degree of colour depends upon how often the cask has been filled, but as a general guide, European wood hue runs from 'old polished oak' to 'young mahogany' (old Oloroso to Amontillado sherry, to put it another way – but beware of thinking the colour comes from the first incumbent), while American wood hue is all gold: 18CT through to 9CT (deep amber to pale straw). The first time the are filled, there will be residues of the previous incumbent – bourbon, sherry, etc. – lurking in the walls of the cask. These leech out into the maturing whisky, adding winey (etc.) notes to the spirit. Both first-fill and refill casks add desirable flavours to the maturing spirit: principally sweetness and colour from caramelised hemicellulose, vanillin and coconut from degraded lignin and astringency, fragrance and delicay from tannins. After three or four fills – depending on the length of time they matured the first and second

fills, and upon the chemistry of the individual staves – they lose their 'activity', their ability to mature their contents, becoming mere vessels, and are termed 'exhausted' or 'spent'.

The third, 'interactive', mechanism is the least understood. Being semi-porous, oakwood allows the contents of the cask to 'breathe' and interact with the air outside. This leads to oxidation, which removes harshness, increases fruitiness and enhances complexity.

It is this aspect of maturation which is most affected by the micro-climate of the warehouse in which the cask rests during maturation. Heat, humidity and atmospheric pressure all play a part. In Scotland, Ireland and Japan most warehouses are cool and damp, even in the summer months, and these conditions are ideal for long maturations, with a liquid loss of 2% per annum allowed against tax (the Angels' Share). In a hot, dry warehouse, such as those in Kentucky, Taiwan and India, the cask will lose water vapour, so the overall volume will go down from between 15% and 25%, while the strength remains high and can even increase. It will be apparent from this that the conditions under which the spirit matures can slow down or speed up the process: in a 'hot' warehouse maturation is achieved more rapidly, but this does not allow for the same degree of interaction, and therefore of complexity.

Time, it is said, is the crucial ingredient in making good whisky, but the time must be passed in a good cask. As the good old boys used to say long before the chemistry of oakwood was understood: 'the wood makes the whisky'.

Charles MacLean has spent the past twenty-five years researching and writing about Scotch whisky and is one of the leading authorities. He spends his time sharing his knowledge around the world, in articles and publications, lectures and tastings, and on TV and radio. His first book (Scotch Whisky) was published in 1993 and since then he has published nine books on the subject. He was elected a Keeper of the Quaich in 1992 and a Master of the Quaich in 2009. In 1997 Malt Whisky won the Glenfiddich Award and in 2003 A Liquid History won 'Best Drinks Book' in the James Beard Awards. In 2012 he also starred in Ken Loach's film The Angel's Share.

Craft and innovation

by Ian Buxton

My email box is crammed with earnest entreaties from small distilleries seeking to persuade me of the 'passion' of their founders; of the 'artisanal' nature of their products and of its 'hand-crafted' character. Generally speaking they've been on a 'journey' as well, and very often their work is 'curated'. Whatever that may mean.

Having read endless variations on these themes I have grown tired and somewhat disillusioned. The craft distilling sector gets a lot of attention on social media; in print and in whisky blogs but increasingly I'm wondering what all the fuss is about. I'm not sure that these words mean very much at all – they're comfortable, reassuring padding at best; lazy, even cynical marketing at worst.

Take 'hand-crafted'. I can't recall visiting any distillery – small, large or enormous – where the work was done by robots. The operation may be on a wholly different scale, and some of laborious work now mechanised, but people still run the process and make the decisions.

As for 'passion' master distillers and blenders such as Richard Paterson, David Stewart, Jim Beveridge, Alan Winchester and countless others in equally large operations exhibit an enthusiasm, commitment and sheer excitement about their work that's the equal of any neophyte toiling over a 40-litre alembic in some light industrial unit or repurposed old farm building. They might make and sell many, many times the volumes of their up-and-coming competitors (and they don't toil late into the night sticking on the labels by hand) but the quality of their products and the joy that they have in making them should not be under-estimated. Are they not passionate craftsmen, albeit on a bigger scale?

And what about scale, because let's face it, the craft distilling category isn't really all that important, at least not in volume terms. Small craft distillers

The young consumers preference for white rum and vodka in the 1970s was one reason that lead to...

might add a little bit of colour and variety to the distilling scene; keep the still-makers busy and they do generate work for the writing and blogging community (for which I suppose one is grateful) but they don't actually amount to very much. A small 'craft' distillery would be generally acknowledged as producing less than 100,000 litres of spirit annually and many produce a great deal less. The very best of luck to them. They provide gainful employment and a few may eventually break through and gain widespread recognition – whereupon they'll likely be snapped up by one of the industry giants who are regularly looking for fresh blood. To be fair, the founders will presumably be happy; after all, a substantial cheque seldom offends.

But let's imagine that in the next year, 100 small craft distilleries get started in the UK (we'll include those producing gin and vodka to help with the numbers) and let's pretend that they all immediately start full production of my hypothetical 100,000 litres, which they won't. Do the maths and you'll quickly see that that would add a reassuringly precise ten million litres of production to the industry's total output. Or, to put it another way, rather less than Diageo are making at Roseisle or Chivas Brothers at their newly-opened Dalmunach. Seen like that the craft industry appears rather insignificant; some of their claims grandiose and the hype attaching

to them, well, just hype. And, just for the sake of clarity, there won't be 100 new craft distillers in the UK in this or any other year.

So why are we so excited? New distilleries certainly appeal to the magpie-like instinct of the collector or hard-core whisky fanatic. It's the lure of the new; something fresh and different to open and impress friends and colleagues. And they certainly represent a refreshing breath of optimism and energy in an industry which has arguably lacked that until very recently.

If I look back around 30 years – not so very long in career terms and no more than a fleeting moment in an industry which claims, somewhat disingenuously, to be more than 500 years old – it's possible to detect some interesting changes and to challenge the view that craft distilling has been the mainspring of innovation in whisky.

Permit me then to take you back to 1985. It was not a happy time. As those of us who lived through it will recall, sales were falling – fast – and the roll-call of distillery closures in the mid-1980s makes dismal reading. Banff, Brora, Coleburn, Convalmore, Dallas Dhu, Glen Albyn, Glen Esk, Glenlochy, Glen Mhor, Glenugie, Glenury Royal, Hillside, Linlithgow, Millburn, Moffat, North Port, Glen Flagler, Garnheath and, of course, the long and loudly lamented Port

...the closure of more than 20 Scotch whisky distilleries in the mid 1980s, like for instance Convalmore

Ellen. All closed, many demolished; none will ever work again.

It was an unholy mess. Whisky – Scotch whisky was, many thought, in terminal decline. As for Irish whiskey… well, basically apart from novelty coffee it was dead and gone. "Good old boys" may have been drinking whiskey and rye – but no one else was. Japanese whisky was confined to its local market and by the end of the 1980s that industry too experienced its own wave of savage closures. White rum and vodka ruled and were expected to take over the whole world. However, strange to relate, before long the whisky market began a slow recovery, which has accelerated faster and faster in the last decade, and which has led us to the present somewhat febrile atmosphere.

When 200 malts was a lot

But let's take another look at the 1980s, when the global market, such as it was, was utterly dominated by blended Scotch. By the end of that decade a pioneering book appeared – Michael Jackson's Malt Whisky Companion. It was first published in 1989 and Jackson was able at that time to include – and I quote – "every malt distillery in Scotland in every current version" as well as throwing in a few from Ireland and Japan. There were approximately 200

entries in the book. "Every malt distillery in Scotland in every current version" – approximately 200! It's inconceivable today. No one person could taste and comment on "every current version". ...even if possessed of a biblical thirst. Indeed, in the 2012 edition of the Malt Whisky Yearbook your Editor began the review of expressions launched during the previous 12 months by stating "It is virtually impossible to list all new bottlings during a year, there are simply too many…" and apologetically offered a selection of 500! A selection. And it's got ever more frenetic since then.

So what's my point? Well, we seem to be making the mistake of looking at some short-term snapshots – today's short-term fad being craft distilleries – and missing the big, underlying picture.

There has, in fact, been massive innovation in the whisky industry over the past 30 years. But it's not apparently fashionable to capture that in a blog, and it's certainly not possible to express it in 140 characters. A lot of the commentary is, bluntly, little more than sound and fury, signifying nothing and it certainly isn't coming from a deep or long appreciation of this industry, its vagaries and historical ebb and flow.

Actually, there has been innovation, and plenty of it. Let's take a look. As the numbers demonstrate, single malt has been innovating at a startling pace, with new expressions, finishes and releases from hitherto unknown distilleries that previously were rendered anonymous in the blending vat.

Remarkably, and almost unbelievably, Ireland's whiskey industry is back from the dead. Indeed, if new distillery start-ups are the yardstick it hasn't been in such good shape since the Easter Rising of 1916. Bourbon is finally reaching world markets in meaningful quantities and rye whiskey is the heart's delight of hipsters from Hoxton to Harlem. And Japanese? Well, one commentator says it's the best whisky in the world! I'm far from certain that there's any such thing but it's certainly very, very good.

We've seen innovation with the arrival of grain whisky. Not exactly a new idea, Cambus grain was being promoted in 1906 under the slogan "Not a headache in a gallon" and Invergordon tried single grain in 1990 but this time round Girvan, Compass Box and Haig Club seem to be getting it right for three very different market segments.

There's been innovation in distillery openings. No longer is the consumer faced with increased consolidation; the plethora of new boutique distillery openings has added choice, colour and variety – perhaps the biggest contribution of the craft distiller. They haven't been afraid to experiment with whisky made from unorthodox grains; whisky made with water from melted snow; whisky with smoke blown through it…you name it, they will have tried it. It hasn't all worked and some of it isn't very nice – but it is innovation. What works will stick and the consumer gains by it.

We've got whisky made by Finns, Australians, Swedes, Belgians, French, Germans, Taiwanese, Indian, Welsh, Dutch, Czech, Spanish and even, perish the thought, English distillers. Again, it hasn't all worked and some of it isn't very nice – but it is innovation. What works will stick and the consumer gains by it.

In the last thirty years we've seen great innovation and imagination in marketing and packaging, allowing whisky to enter hitherto unknown territory. Personally, I still struggle with bottles carrying a £1,000 price tag; my jaw drops at a £10,000 ticket and I gaze in shock and awe, and I'm more than a little depressed if I'm being honest, at the £100,000 behemoths – but I have to acknowledge that, vulgar and ostentatious as I may find them, they're certainly innovative and someone somewhere is buying them. Even if Harrods haven't sold that Richard Paterson £1m collection yet! New whiskies; new styles; new distilleries; new countries; new production techniques; new marketing; new pricing.

And so far I haven't even mentioned "flavoured whiskies"! That's difficult territory for Scotch whisky, with its tight regulation. However, those rules haven't stopped the arrival of 'spirit drinks' based on great whisky names and curiously close in get-up to their parent brand. Think Dewar's Highlander Honey, Ballantine's Brasil and J&B Urban Honey and you can see how the consumer might just be confused, right up until they taste the liquid. Elsewhere, of course, the rules are less strict and Canadian and in particular Bourbon producers have rushed to exploit this particular opportunity. The long-term impact remains to be seen.

Scotch turns to Bourbon

But thinking of Bourbon leads to the curious and very recent phenomenon of Scotch whiskies that seem to be aping that particular style, perhaps in response to Scotch's global slowdown in sales and the burgeoning success of the American offering. Just in the last few months we've seen the arrival of Dewar's Scratched Cask; Chivas Brothers' Barrelhound and – perhaps even more surprising – the Johnnie Walker Select Casks range.

These are, for the moment, primarily exclusively available in the USA so I should explain further. Essentially, these are all Scotch whiskies which have been designed to achieve a sweeter flavour than hitherto has been associated with Scotch whisky.

Speaking at the opening of the Dalmunach distillery Chivas Brothers' CEO, Laurent Lacassagne was asked about Barrelhound, which he said was "a nice little project".

"The idea is to test the appetite, the interest of US consumers," he said. "It is a Scotch whisky that will be much more contemporary, playing a little bit in the territory of the new, edgy and different Bourbon brands, but still a very good Scotch whisky as well."

Lacassagne described Barrelhound as sweeter than other blended Scotches, but stressed that this was achieved solely through the different types of barrels used, allowing it to retain what he called the "quality characteristics" of Scotch.

"What we are seeking is to maybe have a different type of Scotch," he said. "Less serious."

Dewar's employ much the same vocabulary when talking about their Scratched Cask expression which they suggest will allow drinkers "to enjoy the prestige and tradition of Scotch with the accessibility of bourbon" and "push the envelope in Scotch and give Bourbon drinkers a way to bridge one whiskey category to the other" (marketing people actually seem to speak like this). So what is Dewar's Scratched Cask? Basically a four year old blend that has then

Johnnie Walker Select Casks - Rye Cask Finish

been allowed to marry in heavily charred American oak casks, which have been scratched at the char layer to alter the flavour of the whisky. The liquid is then finished for an additional but unspecified period of months.

According to their Master Blender Stephanie Macleod "This is a Scotch produced in a bourbon style. We always age our Blended Scotch Whisky, but in this case, the process of scratching the barrels creates a truly exceptional Scotch whisky with distinctive and accessible notes of sweet vanilla, toasted oak, and a hint of spice with a smooth, satisfying finish."

Finally, the Walker Select Casks. With the first release, the blenders have taken whisky which has been matured for ten years in first-fill American Oak casks that have previously held Bourbon and then finished it for six months in American Rye casks. 'Hallelujah', you might remark 'an age declaration.' There's something going against current trends. But I was surprised to find Walker doing this: on the face of it it's something of a departure from their house style and finishing in rye casks isn't what you would immediately expect from Walker. But that's the point about innovation and, with a 10 year old age declaration; a 46% bottling strength and a relatively modest $45 retail price, I'd expect it to have some considerable appeal.

While currently confined to the USA you can be sure that if these products prove successful they will soon be rolled out to other markets and followed by a raft of imitators. Two things strike me as significant: firstly, while Barrelhound may be unknown, Dewar's and Johnnie Walker are not insignificant, secondary brands so if they make a move like this something significant is happening. More importantly though, these launches could be seen as representing a sudden nervousness; a lack of confidence if you will in Scotch's continued appeal, if only to bearded hipsters.

The definition of craft

Given the long gestation times for new whiskies, especially those from large companies, it would seem that about two years ago a number of different marketing departments came to the same conclusion: Scotch needed to be more like Bourbon. Why else make such a marked move in the direction of previously alien flavours?

These whiskies are very new, of course and it's far from clear if they will appeal and stick around. But let's get back to the 'craft' problem. I'm far from alone in being confused by the paucity of imagination amongst the marketing and PR community.

Alexandre Ricard, CEO of Pernod Ricard

Earlier this year, Alexandre Ricard, CEO of Pernod Ricard (parent of Chivas Brother's, the folk behind Barrelhound) outlined his difficulties thus:

"I'm struggling with the definition of craft. What is craft? Does it have to mean small? Does it have to mean a pot still in a garage? Does it have to come from your close community? Or, is craft what you see when you visit our distilleries?"

"As industry leaders, it is our duty and responsibility to make sure that we engage with consumers and explain to them what real craft is, authentic brands with hundreds of years of history. I mean, look at Jameson, with 235 years of history. Our master cooper is a seventh generation cooper, who uses the same tool as his predecessors. Compare that to a young entrepreneur in an American neighbourhood with a little pot still in his garage. That's our job."

But you'd be forgiven for mistaken Barrelhound as the product of that little pot still in a garage – a case of gamekeeper turned poacher perhaps? Another senior industry executive – from a large distiller naturally – who, sadly, was reluctant to be quoted directly told me "Craft" is a term that has been cynically high jacked by the new wave of marketing-savvy entrepreneurs who have come into the distilling world hungry for profit."

Not that there's anything wrong with a profit (his

And just to confuse matters further, both Diageo (at Stitzel Weller, where a 'barrel a week' plant opened recently) and Pernod Ricard with Our/Vodka are themselves small batch distillers when it suits them and they too can sense a profit.

It's not just me and Pernod Ricard's Alexandre Ricard that's confused. Åsa Caap, who leads the company's Our/Vodka project, has told Just-Drinks.com (a leading on-line news site for the drinks industry) that 'craft' is an abused term.

"I'm tired of the word craft," Caap said. "It's overused. You can't fool consumers."

Well, there's little enough space on a spirits label as it is without confusing consumers with debatable claims. It seems to me that the various mandatory notices and vital information such as strength, age and quantity sit somewhat awkwardly with the more extravagant marketing claims that consumers will find on the other side of the label, embossed into the glass or hung as an enticing leaflet round the neck of the bottle.

But according to Jim Beam's defence in the aforementioned US legal disputes the consumer shouldn't take these claims literally. It's just "commonsense" that the Bourbon isn't "hand crafted".

So let's be clear about the direction of travel: on one side of the bottle the consumer is to be regaled with charming stories that they know (how?) to be little more than pleasing conceits, while on the reverse they can seek out transparent, reliable and scientifically accurate information about the product they are about to consume.

What could possibly go wrong?

Barrelhound - Pernod Ricard's bourbon inspired Scotch

company certainly makes one) – after all no distiller, be they huge, small or even 'craft' is run as a charity and we wouldn't expect that. Nor, I imagine, does the consumer. But that self-same consumer might be confused by 'craft', 'limited edition', 'small batch', 'hand-crafted' and the various other feel-good but ultimately empty verbiage that appears on labels and in advertising. I'm told it's known as "romance copy" by the legal eagles who sign this off.

However, it's coming under attack by an increasingly cynical (that word again) and disenchanted consumer. While our North American cousins may seem overly enthusiastic in their resort to legal action, and ready to sue at even the faintest hint of an ambulance siren, the various US class actions against Tito's Handmade Vodka, Maker's Mark, Templeton Rye, Jim Beam and others should perhaps be considered as an early warning of a gathering storm that could easily blow over the ocean.

While American spirits appear more loosely regulated than here in Europe, producers on this side of the Atlantic need not feel too complacent. As just two random examples, Bowmore have a 'Small Batch' expression and Tanqueray Malacca's (gin) 16,000 case release is apparently a 'limited edition'. Just the sort of limited edition volume a small batch craft distiller might aspire to sell in a year or more…

Keeper of the Quaich and Liveryman of the Worshipful Company of Distillers, Ian Buxton is well-placed to write or talk about whisky, not least because he used to live on the site of a former distillery! Ian began work in the Scotch Whisky industry in 1987 and, since 1991, has run his own strategic marketing consultancy business. In addition, he gives lectures, presentations and tastings and writes regular columns for Whisky Magazine, WhiskyEtc, The Tasting Panel, Malt Advocate and various other titles. He has written the two bestsellers; 101 Whiskies To Try Before You Die and 101 World Whiskies To Try Before You Die and recently published the third book in the series - 101 Legendary Whiskies You're Dying To Try But (Possibly) Never Will

Dalmunach Distillery

Malt distilleries

Including the subsections:
Scottish distilleries | New distilleries | Closed distilleries
Japanese distilleries | Distilleries around the globe

Explanations

Owner: Name of the owning company, sometimes with the parent company within brackets.

Region/district: There are four formal malt whisky regions in Scotland today; the Highlands, the Lowlands, Islay and Campbeltown. Where useful we mention a location within a region e.g. Speyside, Orkney, Northern Highlands etc.

Founded: The year in which the distillery was founded is usually considered as when construction began. The year is rarely the same year in which the distillery was licensed.

Status: The status of the distillery's production. Active, mothballed (temporarily closed), closed (but most of the equipment still present), dismantled (the equipment is gone but part of or all of the buildings remain even if they are used for other purposes) and demolished.

Visitor centre: The letters (vc) after status indicate that the distillery has a visitor centre. Many distilleries accept visitors despite not having a visitor centre. It can be worthwhile making an enquiry.

Address: The distillery's address.

Tel: This is generally to the visitor centre, but can also be to the main office.

Website: The distillery's (or in some cases the owner's) website.

Capacity: The current production capacity expressed in litres of pure alcohol (LPA).

History: The chronology focuses on the official history of the distillery and independent bottlings are only listed in exceptional cases.

Tasting notes: For all the Scottish distilleries that are not permanently closed we present tasting notes of what, in most cases, can be called the core expression (mainly their best selling 10 or 12 year old).

We have tried to provide notes for official bottlings but in those cases where we have not been able to obtain them, we have turned to independent bottlers.

The whiskies have been tasted by Gavin D Smith (GS), a well-known and eperienced whisky profile and author of 20 books on the subject.

There are also tasting notes for Japanese malts and these have been written by Nicholas Coldicott.

All notes have been prepared especially for Malt Whisky Yearbook 2016.

Aberfeldy

[ah•bur•<u>fell</u>•dee]

Owner:
John Dewar & Sons
(Bacardi)

Region/district:
Southern Highlands

Founded: 1896 **Status:** Active (vc) **Capacity:** 3 500 000 litres

Address: Aberfeldy, Perthshire PH15 2EB

Website: aberfeldy.com **Tel:** 01887 822010 (vc)

A major part of a visit to Aberfeldy is about Dewar's blended Scotch. The single malt of the distillery has been in the heart of Dewar's since its inception in 1896.

The magnificent visitor centre, Dewar's World of Whisky, was inaugurated in 2000 but last year saw a major re-development of the facilities and it now boasts a five star attraction with more than 32,000 visitors per year. Dewar's White Label is one of the top ten blends in the world with a strong position in the American market. A total of 32 million bottles were sold in 2014 and apart from White Label, the range consists of a 12 year old, an 18 year old and the exclusive Dewar's Signature. In 2014 a new, limited release was made – Dewar's 15 year old "The Monarch". Initially, this was destined for the Chinese market and Stephanie Macleod, the Dewar's Master Blender since 2006, was anxious to tailor the new bottling for that particular market. Knowing that many Chinese enjoy mixing their whisky with green tea, she had samples of tea sent over from China during the creation process.

The equipment at Aberfeldy consists of a 7.5 ton stainless steel mash tun, eight washbacks made of Siberian larch and three made of stainless steel (the third was installed in 2014) with an average fermentation time of 70 hours and four stills. With the additional washback, production has now escalated to 23 mashes per week and 3.4 million litres of alcohol. The owners have also invested £1.2m in a biomass boiler that will reduce greenhouse gas emissions by up to 90%, thus replacing the heat generated by heavy fuel oil.

Aberfeldy single malt has never been one of the big sellers in the whisky world but this could soon change. In 2014 the owners re-designed the bottle and label for the core range of the **12** and **21 year old** and they also introduced the first Aberfeldy duty free exclusive expression – an **18 year old**. This was followed up by a new **16 year old** core expression in June 2015 and there are also rumours of a limited **30 year old** to be released in the near future.

History:

1896 John and Tommy Dewar embark on the construction of the distillery, a stone's throw from the old Pitilie distillery which was active from 1825 to 1867. Their objective is to produce a single malt for their blended whisky - White Label.

1898 Production starts in November.

1917 The distillery closes.

1919 The distillery re-opens.

1925 Distillers Company Limited (DCL) takes over.

1972 Reconstruction takes place, the floor maltings is closed and the two stills are increased to four.

1991 The first official bottling is a 15 year old in the Flora & Fauna series.

1998 Bacardi buys John Dewar & Sons from Diageo at a price of £1,150 million.

2000 A visitor centre opens and a 25 year old is released.

2005 A 21 year old is launched in October, replacing the 25 year old.

2009 Two 18 year old single casks are released.

2010 A 19 year old single cask, exclusive to France, is released.

2011 A 14 year old single cask is released.

2014 The whole range is revamped and an 18 year old for duty free is released.

2015 A 16 year old is released.

Tasting notes Aberfeldy 12 years old:

GS – Sweet, with honeycombs, breakfast cereal and stewed fruits on the nose. Inviting and warming. Mouth-coating and full-bodied on the palate. Sweet, malty, balanced and elegant. The finish is long and complex, becoming progressively more spicy and drying.

12 years old

Aberlour

[ah•bur•<u>lower</u>]

Owner:
Chivas Brothers Ltd
(Pernod Ricard)

Region/district:
Speyside

Founded: 1826
Status: Active (vo)
Capacity: 3,800,000 litres

Address: Aberlour, Banffshire AB38 9PJ

Website:
aberlour.com

Tel:
01340 881249

Aberlour single malt continues its march towards the very top in terms of sales. Last year it surpassed Glen Grant, and now occupies sixth place with 3.5 million bottles sold in 2014.

The question is if it might not even be in fifth place, because slightly ahead is Singleton. This is the collective name for whiskies from three different distilleries and even if one singles out the big seller Singleton of Glen Ord, it wouldn't be sufficient to match Aberlour. France has been the most important market since way back, and in 2008 Aberlour managed to surpass Glenfiddich as the most sold single malt in the country.

The distillery is equipped with one 12 tonnes semi-lauter mash tun, six stainless steel washbacks and two pairs of stills. There are five warehouses on site (three racked and two dunnage) but only two racked are used for maturation. Aberlour was one of the first distilleries to tailor its distillery tours for the discerning whisky aficionados, instead of for large groups of tourists. The basic tour lasts for two hours and includes a tasting of no less than six expressions. One can also add on an additional tasting (Casks of the Past) of bottlings that are no longer available for purchase.

The core range of Aberlour includes **12, 16** and **18 year olds** – all being matured in a combination of ex-bourbon and ex-sherry casks. Another core expression is **Aberlour a'bunadh** which is based entirely on ex-Oloroso casks. It is always bottled at cask strength and up to 50 different batches have been released by August 2015. With France being the biggest market for the brand, four special expressions have been released for the French consumers (but available in other selected markets as well). These include **10 year old, 12 year old un chill-filtered, 15 year old Select Cask Reserve** and **White Oak Millennium 2004**. The latter was first introduced in 2013 and has been matured solely in American white oak. Two exclusives are available for the duty free market – a **12 year old Sherry Cask** matured and a **15 year old Double Cask** matured.

History:

1826 James Gordon and Peter Weir found the first Aberlour Distillery.

1827 Peter Weir withdraws and James Gordon continues alone.

1879 A fire devastates most of the distillery. The local banker James Fleming constructs a new distillery upstream the Spey river.

1892 The distillery is sold to Robert Thorne & Sons Ltd who expands it.

1898 Another fire rages and almost totally destroys the distillery. The architect Charles Doig is called in to design the new facilities.

1921 Robert Thorne & Sons Ltd sells Aberlour to a brewery, W. H. Holt & Sons.

1945 S. Campbell & Sons Ltd buys the distillery.

1962 Aberlour terminates floor malting.

1973 Number of stills are increased from two to four.

1975 Pernod Ricard buys Campbell Distilleries.

2000 Aberlour a´bunadh is launched.

2001 Pernod Ricard buys Chivas Brothers and merges Chivas Brothers and Campbell Distilleries under the brand Chivas Brothers.

2002 A new, modernized visitor centre is inaugurated in August.

2008 The 18 year old is also introduced outside France.

2013 Aberlour 2001 White Oak is released.

2014 White Oak Millenium 2004 is released.

12 years old

Tasting notes Aberlour 12 year old:

GS – The nose offers brown sugar, honey and sherry, with a hint of grapefruit citrus. The palate is sweet, with buttery caramel, maple syrup and eating apples. Liquorice, peppery oak and mild smoke in the finish.

Allt-a-Bhainne

[alt a•vain]

Owner:	**Region/district:**
Chivas Brothers Ltd	Speyside
(Pernod Ricard)	
Founded: **Status:**	**Capacity:**
1975 Active	4 000 000 litres

Address: Glenrinnes, Dufftown, Banffshire AB55 4DB

Website:	**Tel:**
-	01542 783200

Few places in Scotland are as beautiful as the area in the Cromdale Hills between Tomintoul and Dufftown. A part thereof consists of the Glen of the Livet named after the river which converges into the Avon as it meanders its way towards the coast.

The star distillery in this part of Scotland is Glenlivet but there are a few considerably less famous distilleries in the area, of which Allt-a-Bhainne is one. Built in the mid 1970s this has always been a producer of malt whisky for blends and in particular 100 Pipers, a brand which was introduced in 1966 by the legendary Sam Bronfman of Seagram's. The idea was to challenge the big sellers in the American market at that time and 40 years of constant rising sales-figures soon followed. The last few years have been challenging though and compared to their best year, 2005, when 42 million bottles were sold, sales figures have fallen to 12 million in 2014. Thailand is the biggest market for the brand but the slowing economy, political instability and a big tax hike on spirits during 2013 are the main contributory factors for the decline.

Since the distillery was founded it has been equipped with a traditional mash tun with rakes and ploughs. In summer 2015 however, a new, modern lauter mash gear was fitted into the existing tun. The rest of the equipment consists of eight stainless steel washbacks and two pairs of stills. The distillery is a busy place working 7 days a week with 25 mashes resulting in four million litres of alcohol per year. Chivas Brothers has no distillery on Islay so, to cover their need of peated whisky for their blends, they need to resort to other solutions. During the last few years 50% of the production at Allt-a-Bhainne has therefore been peated spirit with a phenol content in the malted barley of 10ppm. In order to make the distillery more energy efficient, thermal compressors were installed in 2011.

There are no official bottlings of Allt-a-Bhainne single malt but it have been used for bottlings of the Deerstalker brand from time to time. A limited 18 year old, bottled at cask strength was released in summer 2015.

History:

1975 The distillery is founded by Chivas Brothers, a subsidiary of Seagrams, in order to secure malt whisky for its blended whiskies. The total cost amounts to £2.7 million.

1989 Production has doubled.

2001 Pernod Ricard takes over Chivas Brothers from Seagrams.

2002 Mothballed in October.

2005 Production restarts in May.

Tasting notes Deerstalker 18 year old:

GS – Honey, icing sugar, lanolin; becoming buttery. Soft fruits, and finally toffee bonbons. Silky mouth-feel, slightly oily, vanilla, white pepper and tangerines. Relatively long finish and persistently spicy.

Deerstalker 18 years old

Ardbeg

[ard•<u>beg</u>]

Owner:	**Region/district:**
The Glenmorangie Co	Islay
(Moët Hennessy)	
Founded: **Status:**	**Capacity:**
1815 Active (vc)	1 300 000 litres

Address: Port Ellen, Islay, Argyll PA42 7EA

Website:	**Tel:**
ardbeg.com	01496 302244 (vc)

In 2015 Ardbeg celebrated its 200th anniversary. With the reputation the whisky has today, it seems a bit odd to consider that it could just as well have been closed and demolished by now.

In the late 1800s when Alfred Barnard came to Islay on his great tour around the distilleries of the United Kingdom, Ardbeg was the biggest distillery on the island with a yearly production of 1.2 million litres – almost the same as today. It was controlled by the Hay family until 1973 when Hiram Walker took over. They were, in turn, acquired in 1987 by Allied-Lyons (a merger of four breweries in the 1960s) and it was then the problems began. Two years later, Allied acquired yet another business – the Whitbread spirits division. Now Allied had two Islay distilleries under its wings, Ardbeg and Laphroaig. At this point in time, there's no demand for single malt, least of all, the smoky version. All malt whisky is being used for blends, but since the global demand for whisky is decreasing, Allied's management argues that two distilleries producing peated whisky, is one too many! They therefore decided to shut down Ardbeg. During a period of 15 years, up until 1996, minimal quantities were produced before the decision was made to put it up for sale.

The buyer was Glenmorangie Company and here's where the distillery's luck turned. Everyone has agreed upon the fact that Ardberg single malt has always maintained a high quality, but with the new owners, attention to cask quality, as well as marketing skills to build the brand, were needed to transform Ardbeg into a whisky with the huge following and the great reputation that it has today. Sales figures have increased steadily in recent years and 960,000 bottles were sold in 2014, which places Ardbeg fourth amongst the Islay single malts.

The distillery is equipped with a 4.5 ton stainless steel semilauter mash tun, six washbacks made of Oregon pine with a fermentation time of 55 hours and one pair of stills. A purifier is connected to the spirit still to help create the special, fruity character of the spirit. To cope with future demand, the distillery in 2012 went from running 5 days per week to 24/7 production, and in 2015 they will be making 16-17 mashes per week, thereby accounting for 1.25 million litres of pure alcohol.

The core range consists of the **10 year old, Uigeadail** and **Corryvreckan**. Recent, limited releases have included **Galileo**, a 12 year old with a combination of different casks, all filled in 1999, **Auriverdes** matured in bourbon barrels where the cask heads had been replaced by similar ones made from toasted, new American oak and **Ardbeg Kildalton**, a mix of ex-bourbon and new and refill sherry casks. The new expression for Ardbeg Day during Feis Ile 2015 was **Ardbeg Perpetuum**, matured in both bourbon and sherry casks and bottled at 47.4%. Finally, in September, the fifth and final bottling of **Supernova** was released. This time with a phenol specification of 100ppm in the barley and bottled at 54.3%. The first version of Supernova was released in 2009.

History:

1794 First record of a distillery at Ardbeg. It was founded by Alexander Stewart.

1798 The MacDougalls, later to become licensees of Ardbeg, are active on the site through Duncan MacDougall.

1815 The current distillery is founded by John MacDougall, son of Duncan MacDougall.

1853 Alexander MacDougall, John's son, dies and sisters Margaret and Flora MacDougall, assisted by Colin Hay, continue the running of the distillery. Colin Hay takes over the licence when the sisters die.

1888 Colin Elliot Hay and Alexander Wilson Gray Buchanan renew their license.

1900 Colin Hay's son takes over the license.

1959 Ardbeg Distillery Ltd is founded.

1973 Hiram Walker and Distillers Company Ltd jointly purchase the distillery for £300,000 through Ardbeg Distillery Trust.

1974 Widely considered as the last vintage of 'old, peaty' Ardbeg. Malt which has not been produced in the distillery's own maltings is used in increasingly larger shares after this year.

1977 Hiram Walker assumes single control of the distillery. Ardbeg closes its maltings.

1979 Kildalton, a less peated malt, is produced over a number of years.

1981 The distillery closes in March.

1987 Allied Lyons takes over Hiram Walker and thereby Ardbeg.

1989 Production is restored. All malt is taken from Port Ellen.

1996 The distillery closes in July and Allied Distillers decides to put it up for sale.

History continued:

1997 Glenmorangie plc buys the distillery for £7 million. Ardbeg 17 years old and Provenance are launched

1998 A new visitor centre opens.

2000 Ardbeg 10 years is introduced and the Ardbeg Committee is launched.

2001 Lord of the Isles 25 years and Ardbeg 1977 are launched.

2002 Ardbeg Committee Reserve and Ardbeg 1974 are launched.

2003 Uigeadail is launched.

2004 Very Young Ardbeg (6 years) and a limited edition of Ardbeg Kildalton (1300 bottles) are launched. The latter is an un-peated cask strength from 1980.

2005 Serendipity is launched.

2006 Ardbeg 1965 and Still Young are launched. Almost There (9 years old) and Airigh Nam Beist are released.

2007 Ardbeg Mor, a 10 year old in 4.5 litre bottles is released.

2008 The new 10 year old, Corryvreckan, Rennaissance, Blasda and Mor II are released.

2009 Supernova is released, the peatiest expression from Ardbeg ever.

2010 Rollercoaster and Supernova 2010 are released.

2011 Ardbeg Alligator is released.

2012 Ardbeg Day and Galileo are released.

2013 Ardbog is released.

2014 Auriverdes and Kildalton are released.

2015 Ardbeg Perpetuum and Supernova 2015 are released.

Corryvreckan   Supernova 2015   Perpetuum

10 years old   Uigeadail

Tasting notes Ardbeg 10 year old:

GS – Quite sweet on the nose, with soft peat, carbolic soap and Arbroath smokies. Burning peats and dried fruit, followed by sweeter notes of malt and a touch of liquorice in the mouth. Extremely long and smoky in the finish, with a fine balance of cereal sweetness and dry peat notes.

Ardmore

[ard•moor]

Owner:
Beam Suntory

Region/district:
Highland

Founded: 1898
Status: Active
Capacity: 5 550 000 litres

Address: Kennethmont, Aberdeenshire AD54 4NH

Website:
ardmorewhisky.com

Tel:
01464 831213

Founded by the Teacher family in the late 1800s, the distillery has ever since produced malt whisky for the famous Teacher's blend.

It takes more than 30 different single malts to make Teacher's but with Ardmore representing 35% of the total malt content its importance to the character is easy to understand. Teacher's is one of the top selling malts in the UK but the biggest markets are without a doubt India and Brazil. For a long time, Teacher's Highland Cream was the only expression but the range has since been supplemented with Origin (containing a higher malt content), a 25 year old as well as a Teacher's Single Malt (which of course is an Ardmore). A total of 25 million bottles were sold in 2014.

The distillery is equipped with a large (12.5 tonnes) cast iron mash tun with a copper dome, 14 Douglas fir washbacks (4 large and 10 smaller ones) with a fermentation time of 55 hours, as well as four pairs of stills equipped with sub-coolers to give more copper contact. At the moment, Ardmore is doing 25 mashes per week resulting in 5.3 million litres. Traditionally, Ardmore has been the only distillery in the region consistently producing peated whisky with a phenol specification in the barley of 12-14 ppm. For blending purposes they also produce the unpeated Ardlair (around 40% of the yearly output) since 2002. Two years ago trials were initiated with heavily peated spirit during the last couple of weeks in the year.

The only core expression used to be the un chill-filtered **Traditional** with no age statement, but generally made up using a range of ex-bourbon casks from six to thirteen years old and bottled at 46%. This was replaced by **Legacy** in autumn 2014, a mix of 80% peated and 20% unpeated malt, bottled at 40% and chill-filtered. In June 2015 however, Traditional re-appeared as **Tradition** but this time as an exclusive for duty free. Simultaneously **Triple Wood** without age statement and bottled at 46%, was launched with a maturation in bourbon barrels, quarter casks and sherry puncheons. Finally, an expression with 7 years in sherry puncheons previously used to mature Ardmore spirit and a second maturation of 5 years in port pipes, was released later in 2015.

History:

1898 Adam Teacher, son of William Teacher, starts the construction of Ardmore Distillery which eventually becomes William Teacher & Sons' first distillery. Adam Teacher passes away before it is completed.

1955 Stills are increased from two to four.

1974 Another four stills are added, increasing the total to eight.

1976 Allied Breweries takes over William Teacher & Sons and thereby also Ardmore. The own maltings (Saladin box) is terminated.

1999 A 12 year old is released to commemorate the distillery's 100th anniversary. A 21 year old is launched in a limited edition.

2002 Ardmore is one of the last distilleries to abandon direct heating (by coal) of the stills in favour of indirect heating through steam.

2005 Jim Beam Brands becomes new owner when it takes over some 20 spirits and wine brands from Allied Domecq for five billion dollars.

2007 Ardmore Traditional Cask is launched.

2008 A 25 and a 30 year old are launched.

2014 Beam and Suntory merge. Legacy is released.

2015 Traditional is re-launched as Tradition and a Triple Wood is released.

Tasting notes Ardmore Legacy:

GS – Vanilla, caramel and sweet peat smoke on the nose, while on the palate vanilla and honey contrast with quite dry peat notes, plus ginger and dark berries. The finish is medium to long, spicy, with persistently drying smoke.

Legacy

Arran

[ar•ran]

Owner:		Region/district:
Isle of Arran Distillers		Islands (Arran)
Founded:	**Status:**	**Capacity:**
1993	Active (vc)	750 000 litres

Address: Lochranza, Isle of Arran KA27 8HJ

Website:	Tel:
arranwhisky.com	01770 830264

The owners of Arran distillery can proudly look back to a couple of very successful years. A near 60% hike in profits for 2013 was the best-ever annual result. At the same time the development in important markets such as USA and Taiwan were impressive.

The good results and the increased demand have made expansion inevitable. A new blending and vatting store and a fifth warehouse have already been built. In October 2016, two more stills will be installed together with an upgrading of the mash tun equipment.

The distillery is currently equipped with a 2.5 ton semi-lauter mash tun, six Oregon pine washbacks with an average fermentation time of 60 hours and two stills. The total production capacity is 750,000 litres and the plan for 2015 is to do 600,000 litres. Since 2004, the distillery has been producing a share of peated spirit every year. For 2015 it will be 60,000 litres from malt peated to 20ppm and another 30,000 litres of heavily peated spirit (50ppm).

The core range consists of **10 year old, 12 year old cask strength, 14 year old, Robert Burns Malt** and **Lochranza Reserve**. The latter, without age statement, has replaced Arran Original in 2014. Also included in the core range is the peated expression **Machrie Moor** (with a 6th edition being launched in autumn 2015). In spring 2015, an **18 year old**, was also released. The first batch is limited but it will become a core bottling from 2016. In 2014 a number of limited releases were launched; the final chapter of **Devil's Punch Bowl, Machrie Moor cask strength** (with the second edition due in 2015) and the second release of **Bere Barley**. This was followed up in 2015 with an **18 year old single sherry hogshead** which is exclusive to the distillery's White Stag online community. At the same time, the distillery started a new, limited range called The Smuggler's Series reflecting on the island's history of illicit distilling in the late 18th and early 19th century. The first release in the range, **The Illicit Stills**, is a cask strength vatting of port pipes, sherry hogsheads and bourbon barrels spiced up with young peated whisky.

History:

1993 Harold Currie founds the distillery.

1995 Production starts in full on 17th August.

1998 The first release is a 3 year old.

1999 The Arran 4 years old is released.

2002 Single Cask 1995 is launched.

2003 Single Cask 1997, non-chill filtered and Calvados finish is launched.

2004 Cognac finish, Marsala finish, Port finish and Arran First Distillation 1995 are launched.

2005 Arran 1996 and two finishes, Ch. Margaux and Grand Cru Champagne, are launched.

2006 After an unofficial launch in 2005, Arran 10 years old is released as well as a couple of new wood finishes.

2007 Four new wood finishes and Gordon's Dram are released.

2008 The first 12 year old is released as well as four new wood finishes.

2009 Peated single casks, two wood finishes and 1996 Vintage are released.

2010 A 14 year old, Rowan Tree, three cask finishes and Machrie Moor (peated) are released.

2011 The Westie, Sleeping Warrior and a 12 year old cask strength are released.

2012 The Eagle and The Devil's Punch Bowl are released.

2013 A 16 year old and a new edition of Machrie Moor and released.

2014 A 17 year old and Machrie Moor cask strength are released.

2015 A 18 year old and The Illicit Stills are released.

14 years old

Tasting notes Arran 14 year old:

GS – Very fragrant and perfumed on the nose, with peaches, brandy and ginger snaps. Smooth and creamy on the palate, with spicy summer fruits, apricots and nuts. The lingering finish is nutty and slowly drying.

Auchentoshan

[ock•en•tosh•en]

Owner:
Morrison Bowmore (Suntory)

Region/district:
Lowlands

Founded: **Status:**
1823   Active (vc)

Capacity:
2 000 000 litres

Address: Dalmuir, Clydebank, Glasgow G01 4SJ

Website:
auchentoshan.com

Tel:
01389 878561

More and more distilleries are opening up in the Lowlands these days and the number is approaching double digits.

Of these, it is only Auchentoshan that practices what this part of Scotland has become famous for, namely triple distillation. Auchentoshan is also the only distillery in the entire Scotland doing 100% triple distillation. Perhaps this unique selling point is one of the reasons for the distillery´s impressive increase in sales during the last few years, especially in Taiwan, the UK, Scandinavia and USA. In the early years of the new millennium, Auchentoshan was struggling to sell around 300,000 bottles per year. All of a sudden, the Japanese owners decided to do something with their hidden gem outside Glasgow. New expressions were released, bottlings exclusive for duty free saw the light of day and sales figures started to rise in a rather dramatic way. The latest figures for 2014 saw Auchentoshan selling 942,000 bottles – an increase of 23% in the last two years.

Auchentoshan´s triple distillation means, among other things, having a very narrow spirit cut. They start collecting the middle cut at 82% and stop at 80%, long before any other distillery starts collecting. The equipment consists of a semilauter mash tun with a 7 ton mash charge, four Oregon pine washbacks and three made of stainless steel, all with a fermentation time of 50 hours, and three stills. The intermediate still was replaced in January 2015 and currently, the plan is to do 1.35 million litres of alcohol in 2015.

The core range consists of **American Oak**, a first fill bourbon maturation without age statement, **12 years, Three Woods, 18 years** and **21 years**. The duty free range is made up of six expressions; **Springwood, Heartwood, Solera, Cooper´s Reserve, Silveroak** and **Vintage 1975**. The latter three were replaced in August 2015 by two new expressions. Limited bottlings include **Auchentoshan Virgin Oak**, first released in 2013 and fully matured in charred oak casks that have held neither bourbon nor sherry before. The second release of Virgin Oak was launched in June 2014. At the same time a **1988 Bordeaux finish** was also released.

History:

1817 First mention of the distillery Duntocher, which may be identical to Auchentoshan.

1823 An official license is obtained by the owner, Mr. Thorne.

1903 The distillery is purchased by John Maclachlan.

1941 The distillery is severely damaged by a German bomb raid.

1960 Maclachlans Ltd is purchased by the brewery J. & R. Tennant Brewers.

1969 Auchentoshan is bought by Eadie Cairns Ltd who starts major modernizations.

1984 Stanley P. Morrison, eventually becoming Morrison Bowmore, becomes new owner.

1994 Suntory buys Morrison Bowmore.

2002 Auchentoshan Three Wood is launched.

2004 More than a £1 million is spent on a new, refurbished visitor centre. The oldest Auchentoshan ever, 42 years, is released.

2006 Auchentoshan 18 year old is released.

2007 A 40 year old and a 1976 30 year old are released.

2008 New packaging as well as new expressions - Classic, 18 year old and 1988.

2010 Two vintages, 1977 and 1998, are released.

2011 Two vintages, 1975 and 1999, and Valinch are released.

2012 Six new expressions are launched for the Duty Free market.

2013 Virgin Oak is released.

2014 American Oak replaces Classic.

2015 Two new bottlings are released for duty free.

American Oak

Tasting notes Auchentoshan American Oak:

GS – An initial note of rose water, then Madeira, vanilla, developing musky peaches and icing sugar. Spicy fresh fruit on the palate, chilli notes and more Madeira and vanilla. The finish is medium in length, and spicy to the end.

Auchroisk

[ar•thrusk]

Owner:
Diageo

Region/district:
Speyside

Founded: **Status:**   **Capacity:**
1974   Active   5 900 000 litres

Address: Mulben, Banffshire AB55 6XS

Website: **Tel:**
malts.com   01542 885000

If you make a detour off the A95 halfway between Keith and Craigellachie, you will come across Auchroisk, which is situated right at road B9103. This is a big distillery with an unusually large number of warehouses.

It was built towards the end of Scotch whisky's second golden era – the mid 1970s. Only ten years later, the consumer preferences had turned to vodka and white wine and no less than 20 distilleries were forced to close, most of them for good. Auchroisk, being a modern distillery, kept going and has since been functioning as a supplier of malt whisky for blends, most notably J&B.

Auchroisk distillery is equipped with a 12 ton stainless steel semilauter mash tun, eight stainless steel washbacks (with further plans for another two) and four pairs of stills. The character of Auchroisk new-make has in recent years changed between nutty and green/grassy, depending on what style of whisky is needed for Diageo's blends at the time. Currently, the character is nutty which requires a shorter fermentation time of 53 hours. In 2015 the distillery was closed for 5 months due to a substantial upgrade, including an automated control system. This meant that the usual yearly production of 5.8 million litres was reduced to 3.4 million.

Apart from producing malt whisky for Diageo blends, Auchroisk has another important role to play in the company. The site is the location for maturing whiskies from many other Diageo distilleries and also for part of the blending. In order to achieve this there are ten, huge racked warehouses with the capacity of storing 250,000 casks and further planning application for another four has been approved.

The first, widely available release of Auchroisk single malt was in 1986 under the name Singleton, as the Scottish name was deemed unpronounceable by the consumers. In 2001, it was replaced by a **10 year old** in the Flora & Fauna range. Recent, limited bottlings include a **20 year old** from 1990 and a **30 year old**, both launched as a part of the Special Releases.

History:

1972 Building of the distillery commences by Justerini & Brooks (which, together with W. A. Gilbey, make up the group IDV) in order to produce blending whisky. In February the same year IDV is purchased by the brewery Watney Mann which, in July, merges into Grand Metropolitan.

1974 The distillery is completed and, despite the intention of producing malt for blending, the first year's production is sold 12 years later as single malt thanks to the high quality.

1986 The first whisky is marketed under the name Singleton.

1997 Grand Metropolitan and Guinness merge into the conglomerate Diageo. Simultaneously, the subsidiaries United Distillers (to Guinness) and International Distillers & Vintners (to Grand Metropolitan) form the new company United Distillers & Vintners (UDV).

2001 The name Singleton is abandoned and the whisky is now marketed under the name of Auchroisk in the Flora & Fauna series.

2003 Apart from the 10 year old in the Flora & Fauna series, a 28 year old from 1974, the distillery's first year, is launched in the Rare Malt series.

2010 A Manager's Choice single cask and a limited 20 year old are released.

2012 A 30 year old from 1982 is released.

10 years old

Tasting notes Auchroisk 10 year old:

GS – Malt and spice on the light nose, with developing nuts and floral notes. Quite voluptuous on the palate, with fresh fruit and milk chocolate. Raisins in the finish.

Aultmore

[ault•moor]

Owner:	**Region/district:**
John Dewar & Sons (Bacardi)	Speyside
Founded:	**Status:**	**Capacity:**
1896	Active	3 200 000 litres

Address: Keith, Banffshire AB55 6QY

Website:	**Tel:**
aultmore.com	01542 881800

For almost twenty years, Bacardi have been the owners of Dewar´s and their five distilleries but only two of them have been visible as single malts – Aberfeldy and Macduff (Glen Deveron).

They have been well aware of the high quality of the whiskies from Aultmore, Craigellachie and Royal Brackla but because of the success of their blends portfolio, especially Dewar´s, the malt whisky has been used for these instead. Last year the owners presented a range called the Last Great Malts where about twenty malt whiskies from all five distilleries were to be launched. It is an impressive programme where each distillery has its own design when it comes to bottle and label. During the 20th century, locals and fishermen from Buckie, which is situated some 10 kilometres to the north of Aultmore on Moray Firth, asked for "a nip of the Buckie Rd" at inns and bars along the road. That was the secret name for Aultmore single malt and the same words are now embossed at the bottom of the Aultmore bottle.

Aultmore was completely rebuilt at the beginning of the 1970s and nothing is left of the old buildings from 1896. Since 2008 production has been running seven days a week which, for 2015, means 16 mashes per week and 3.1 million litres of alcohol per year. A 10 tonne Steinecker full lauter mash tun, six washbacks made of larch and two pairs of stills are in use.

Most of the output from Aultmore is used in Dewar´s blended whiskies. Until last year, a 12 year old which was launched in 2004, was the only official bottling available. Through the launch of the Last Great Malts in autumn 2014 there is now a new **12 year old** as well a **25 year old** for domestic markets. At the same time a **21 year old** has been reserved for duty free. The range was further expanded with an **18 year old** in spring 2015 and there are plans to release a **30 year old** in the foreseeable future. All the new bottlings are un chill-filtered, without colouring and bottled at 46%.

History:

1896 Alexander Edward, owner of Benrinnes and co-founder of Craigellachie Distillery, builds Aultmore.

1897 Production starts.

1898 Production is doubled; the company Oban & Aultmore Glenlivet Distilleries Ltd manages Aultmore.

1923 Alexander Edward sells Aultmore for £20,000 to John Dewar & Sons.

1925 Dewar's becomes part of Distillers Company Limited (DCL).

1930 The administration is transferred to Scottish Malt Distillers (SMD).

1971 The stills are increased from two to four.

1991 UDV launches a 12-year old Aultmore in the Flora & Fauna series.

1996 A 21 year old cask strength is marketed as a Rare Malt.

1998 Diageo sells Dewar's and Bombay Gin to Bacardi for £1,150 million.

2004 A new official bottling is launched (12 years old).

2014 Three new expression are released – 12, 25 and 21 year old for duty free.

2015 An 18 year old is released.

12 years old

Aultmore 12 years old:

GS – A nose of peaches and lemonade, freshly-mown grass, linseed and milky coffee. Very fruity on the palate, mildly herbal, with toffee and light spices. The finish is medium in length, with lingering spices, fudge, and finally more milky coffee.

Balblair

[bal•blair]

Owner:
Inver House Distillers
(Thai Beverages plc)

Region/district:
Northern Highlands

Founded: 1790
Status: Active (vc)
Capacity: 1 800 000 litres

Address: Edderton, Tain, Ross-shire IV19 1LB

Website: balblair.com
Tel: 01862 821273

Balblair belongs to the exclusive group of distilleries founded in the 1700s that are still operative today. Official records say that production started in 1790, but some facts even suggest 1749 when beer was brewed there at least.

For years Balblair was a single malt that often fell under the radar. The turning point came in 2007 when the entire range of whiskies was revamped and age statements were abandoned in favour of vintages. An added interest was the result in 2012 when the distillery had a prominent role in the movie Angel's Share.

The owners of Balblair and four other distilleries, Inver House, saw profits drop by 20% in 2013 but it was more or less the expectation. A lot of money and energy have been spent over the last few years to build their different single malt brands. The company therefore decided to reduce bulk whisky sales in favour of building stock of their malts for future sales. To cope with the increased volumes, they are also building 12 new warehouses at their headquarters in North Lanarkshire.

The distillery is equipped with a stainless steel, full lauter mash tun, six Oregon pine washbacks and one pair of stills. The production for 2015 is 21 mashes per week which translates to 1.8 million litres of alcohol. In 2011 and 2012, part of the production was heavily peated spirit with a phenol specification of 52ppm in the barley. Since then, however, there has been no peated production and nothing is planned for the near future. A very elegant and contemporary visitor centre/shop was opened in late 2011 and a visit there is well worth the effort.

The current core range consists of four vintages – **1983, 1990, 1999** and **2003**. For the duty free market, three new expressions were released in 2014; a **1999** and two versions of **2004** matured in bourbon and sherry casks respectively. The oldest vintage available from the distillery at the moment is a limited **Vintage 1969**. For visitors to the distillery there is also the opportunity to bottle an exclusive **2000 vintage** second fill bourbon.

History:
1790 The distillery is founded by James McKeddy.
1790 John Ross takes over
1836 John Ross dies and his son Andrew Ross takes over with the help of his sons.
1872 New buildings replace the old.
1873 Andrew Ross dies and his son James takes over.
1894 Balnagowan Estate signs a new lease for 60 years with Alexander Cowan. He builds a new distillery, a few kilometres from the old.
1911 Cowan is forced to cease payments and the distillery closes.
1941 Balnagowan Estate goes bankrupt and the distillery is put up for sale.
1948 Robert Cumming buys Balblair for £48,000.
1949 Production restarts.
1970 Cumming sells Balblair to Hiram Walker.
1988 Allied Distillers becomes the new owner through the merger between Hiram Walker and Allied Vintners.
1996 Allied Domecq sells the distillery to Inver House Distillers.
2000 Balblair Elements and the first version of Balblair 33 years are launched.
2001 Thai company Pacific Spirits (part of the Great Oriole Group) takes over Inver House.
2004 Balblair 38 years is launched.
2005 12 year old Peaty Cask, 1979 (26 years) and 1970 (35 years) are launched.
2006 International Beverage Holdings acquires Pacific Spirits UK.
2007 Three new vintages replace the former range.
2008 Vintage 1975 and 1965 are released.
2009 Vintage 1991 and 1990 are released.
2010 Vintage 1978 and 2000 are released.
2011 Vintage 1995 and 1993 are released.
2012 Vintage 1975, 2001 and 2002 are released. A visitor centre is opened.
2013 Vintage 1983, 1990 and 2003 are released.
2014 Vintage 1999 and 2004 are released for duty free.

Tasting notes Balblair 2003:
GS – The nose offers tinned peaches and apricot jam, with underlying honey and caramel. Early malt, then zesty lemon notes on the palate, which also features freshly-cut grass and hazelnuts. The medium-length finish yields white pepper and cocoa powder.

Vintage 2003

Balmenach

[bal•may•nack]

Owner: **Region/district:**
Inver House Distillers   Speyside
(Thai Beverages plc)

Founded: **Status:**   **Capacity:**
1824   Active   2 800 000 litres

Address: Cromdale, Moray PH26 3PF

Website: **Tel:**
inverhouse.com   01479 872569

Balmenach is one of five distilleries owned by Inver House and definitely the least known one. No official bottlings are released by the owners with the malt going into their many blends instead.

In 1964, Balmenach was one of the first Scottish malt distilleries which started using a technique for malting called Saladin boxes, named after the French inventor Charles Saladin. Although popular on the continent from the early 1900s, it wasn´t until 1948 that the technique was used in Scotland at the North British grain distillery. After steeping, the malt is transferred to large concrete containers. Every 8 hours for 5 days, a crossbar with a large screw attached to it, moved horizontally across the length of the box with the screw raising the barley from the bottom to the top. After that the malted barley was sent to the kiln for drying. Compared to traditional floor malting, much larger quantities could be handled with less labour. In the mid 1980s they stopped using the Saladin boxes at Balmenach.

The distillery´s old cast iron mash tun was replaced in 2014 with a new stainless steel semi-lauter tun but the old copper canopy was fitted to the new tun. There are six washbacks made of Douglas fir with a 52 hour fermentation period and three pairs of stills connected to worm tubs where each worm is 94 metres long. The distillery is doing 20 mashes per week and 2.8 million litres of alcohol annually. Since 2012, a part of the production is heavily peated (50ppm) and for 2015 it will be 400,000 litres. The three dunnage warehouses currently hold 9,500 casks. Since 2009, apart from whisky, gin has been produced at Balmenach distillery and Caorunn is today the third most sold premium gin in the UK.

There is no official bottling of Balemanch single malt. Aberko in Glasgow though, has been working with the distillery for a long time and, over the years, has released Balmenach under the name Deerstalker. The current expression is a 12 year old.

History:

1824 The distillery is licensed to James MacGregor who operated a small farm distillery by the name of Balminoch.

1897 Balmenach Glenlivet Distillery Company is founded.

1922 The MacGregor family sells to a consortium consisting of MacDonald Green, Peter Dawson and James Watson.

1925 The consortium becomes part of Distillers Company Limited (DCL).

1930 Production is transferred to Scottish Malt Distillers (SMD).

1962 The number of stills is increased to six.

1964 Floor maltings replaced with Saladin box.

1992 The first official bottling is a 12 year old.

1993 The distillery is mothballed in May.

1997 Inver House Distillers buys Balmenach from United Distillers.

1998 Production recommences.

2001 Thai company Pacific Spirits takes over Inver House at the price of £56 million. The new owner launches a 27 and a 28 year old.

2002 To commemorate the Queen's Golden Jubilee a 25-year old Balmenach is launched.

2006 International Beverage Holdings acquires Pacific Spirits UK.

2009 Gin production commences.

Deerstalker 12 years old

Tasting notes Deerstalker 12 years old:

GS – The nose is sweet and fruity, with sherry and chilli. Faintly savoury. Fruity and very spicy on the palate, with black pepper and hints of sherry. More chilli in the finish, plus plain chocolate-coated raisins.

Balvenie

[bal•ven•ee]

Owner:
William Grant & Sons

Region/district:
Speyside

Founded: **Status:** **Capacity:**
1892 Active (vc) 6 800 000 litres

Address: Dufftown, Keith, Banffshire AB55 4DH

Website: **Tel:**
thebalvenie.com 01340 820373

Balvenie single malt has made a remarkable journey in recent years with an 85% sales increase during the last decade to almost 3 million bottles. It is now the 8th most sold single malt in the world.

A considerable share of the credits for this success can be attributed to the brands Malt Master, David Stewart. He started his working career at W Grants as an apprentice in 1962 which was one year prior to the launch of Glenfiddich Straight Malt that catapulted the malt whisky boom which we experience today. David has been instrumental in defining the character of the brand and since 2009, when he was succeeded by Brian Kinsman as Master Distiller for W Grants, he has devoted his entire time to Balvenie.

The distillery is equipped with an 11 ton full lauter mash tun, nine wooden and five stainless steel washbacks, five wash stills and six spirit stills. Balvenie is one of few distilleries still doing some of their own maltings (which is around 15%) and there is also a coppersmith and a cooperage. For 2015, the production plan is to do 6.8 million litres of alcohol. The main part is unpeated but each year a smaller part (100,000 litres) is produced from heavily peated barley (45-50 ppm).

The core range consists of **Doublewood 12 years, Doublewood 17 years, Caribbean Cask 14 years, Single Barrel 12 years First Fill, Single Barrel 15 years Sherry Cask, Single Barrel 25 years Traditional Oak, Portwood 21 years, 30 years, 40 years** and the extremely rare **50 years old**. Among the limited releases, Tun 1401, first introduced in 2010, was replaced in 2014 by **Tun 1509**. A variety of casks, some dating back to the 1960s and 1970s, were selected by David Stewart and vatted together in an 8,000 litre marriage vessel. Batch 2 of Tun 1509 was released in 2015. There is also a **Tun 1858** reserved for Asia where batch 3 was also recently released. The Duty Free range was revamped in 2013 when the **Triple Cask** series was launched. All three expressions (**12, 16 and 25 years old**) are vattings of first-fill ex-bourbon barrels, refill American oak casks and first-fill ex-sherry butts.

History:

1892 William Grant rebuilds Balvenie New House to Balvenie Distillery (Glen Gordon was the name originally intended). Part of the equipment is brought in from Lagavulin and Glen Albyn.

1893 The first distillation takes place in May.

1957 The two stills are increased by another two.

1965 Two new stills are installed.

1971 Another two stills are installed and eight stills are now running.

1973 The first official bottling appears.

1982 Founder's Reserve is launched.

1990 A new distillery, Kininvie, is opened on the premises.

1996 Two vintage bottlings and a Port wood finish are launched.

2001 The Balvenie Islay Cask, with 17 years in bourbon casks and six months in Islay casks, is released.

2002 A 50 year old is released.

2004 The Balvenie Thirty is released.

2005 The Balvenie Rum Wood Finish 14 years old is released.

2006 The Balvenie New Wood 17 years old, Roasted Malt 14 years old and Portwood 1993 are released.

2007 Vintage Cask 1974 and Sherry Oak 17 years old are released.

2008 Signature, Vintage 1976, Balvenie Rose and Rum Cask 17 year old are released.

2009 Vintage 1978, 17 year old Madeira finish, 14 year old rum finish and Golden Cask 14 years old are released.

2010 A 40 year old, Peated Cask and Carribean Cask are released.

2011 Second batch of Tun 1401 is released.

2012 A 50 year old and Doublewood 17 years old are released.

2013 Triple Cask 12, 16 and 25 years are launched for duty free.

2014 Single Barrel 15 and 25 years, Tun 1509 and two new 50 year olds are launched.

Tasting notes Balvenie Doublewood 12 years:

GS – Nuts and spicy malt on the nose, full-bodied, with soft fruit, vanilla, sherry and a hint of peat. Dry and spicy in a luxurious, lengthy finish.

Doublewood 12 years old

Ben Nevis

[ben nev•iss]

Owner:	**Region/district:**
Ben Nevis Distillery Ltd	Western Highlands
(Nikka, Asahi Breweries)	

Founded:	**Status:**	**Capacity:**
1825	Active (vc)	2 000 000 litres

Address: Lochy Bridge, Fort William PH33 6TJ

Website:	**Tel:**
bennevisdistillery.com	01397 702476

This year, the last surviving distillery in Fort William celebrates its 25th anniversary of the re-opening of the distillery in 1990.

Production was intermittent during the 1970s and 80s but since Japanese Nikka Whisky bought the distillery in 1989, it has been producing continuously. The anniversary will be highlighted by a bottling of the first cask that was filled in September 1990 by the then owner of Nikka, Takeshi Taketsuru. He was the son of the legendary founder of Nikka, Masataka Taketsuru, often referred to as the father of the Japanese whisky industry. Takeshi Taketsuru died in December 2014, at the age of 90.

Ben Nevis is equipped with one lauter mash tun, six stainless steel washbacks and two made of Oregon pine and two pairs of stills. Fermentation used to be 48 hours in the steel washbacks and 96 hours in the wooden ones. From 2014, however, fermentation is 48 hours in all washbacks. Over the last couple of years production has increased and the plan for 2015 is to do 2 million litres. Part of that (30,000 litres) will be heavily peated. Ben Nevis is an important supplier of whisky for the owner's (Nikka) blends and for 2015, 50% of the newmake will be sent directly to Japan, primarily to be part of the popular blend, Nikka Black.

The core range consists of the **10 year old** and the peated **MacDonald's Traditional Ben Nevis**. The latter, which is an attempt to replicate the style of Ben Nevis single malt from the 1880s, was introduced as a limited expression but has now become a part of the core range. A new range of limited releases, Forgotten Bottlings, was introduced in 2014. The manager since 1989, Colin Ross, discovered stocks of bottles that for some or other reason hadn't been claimed by customers. The latest release in the series is a **40 year old "Blended at Birth" single blend** which was first released 12 years ago. A limited release in 2015 was a **12 year old single white port cask** while last year saw the launch of a **15 year old sherry cask** from 1998, a **21 year old** with 8 years of second maturation in a ruby port bodega butt and a **25 year old first fill sherry cask**.

History:

1825 The distillery is founded by 'Long' John McDonald.

1856 Long John dies and his son Donald P. McDonald takes over.

1878 Demand is so great that another distillery, Nevis Distillery, is built nearby.

1908 Both distilleries merge into one.

1941 D. P. McDonald & Sons sells the distillery to Ben Nevis Distillery Ltd headed by the Canadian millionaire Joseph W. Hobbs.

1955 Hobbs installs a Coffey still which makes it possible to produce both grain and malt whisky.

1964 Joseph Hobbs dies.

1978 Production is stopped.

1981 Joseph Hobbs Jr sells the distillery back to Long John Distillers and Whitbread.

1984 After restoration and reconstruction totalling £2 million, Ben Nevis opens up again.

1986 The distillery closes again.

1989 Whitbread sells the distillery to Nikka Whisky Distilling Company Ltd.

1990 The distillery opens up again.

1991 A visitor centre is inaugurated.

1996 Ben Nevis 10 years old is launched.

2006 A 13 year old port finish is released.

2010 A 25 year old is released.

2011 McDonald's Traditional Ben Nevis is released.

2014 Forgotten Bottlings are introduced.

2015 A 40 year old "Blended at Birth" single blend is released.

10 years old

Tasting notes Ben Nevis 10 years old:

GS – The nose is initially quite green, with developing nutty, orange notes. Coffee, brittle toffee and peat are present on the slightly oily palate, along with chewy oak, which persists to the finish, together with more coffee and a hint of dark chocolate.

Benriach

[ben•ree•ack]

Owner: Benriach Distillery Co	**Region/district:** Speyside

Founded:	**Status:**	**Capacity:**
1897	Active	2 800 000 litres

Address: Longmorn, Elgin, Morayshire IV30 8SJ

Website: benriachdistillery.co.uk	**Tel:** 01343 862888

One can without a doubt say that the best time for BenRiach is the present. After the foundation, it was only working for three years and was then forced to close in the aftermath of the Pattison crash.

For 65 years it lay dormant until it was reopened by Glenlivet Distillers. Under their regime (and later Chivas Bros), the brand was never made known to the public. Instead the whisky was used for blends. It was not until 2004 when the Walker family took over, that BenRiach single malt was launched and not in a modest way either. Some years, more than 20 new bottlings were released and even if the pace has slowed down, the distillery has one of the broadest ranges in the industry.

BenRiach distillery is equipped with a traditional cast iron mash tun with a stainless steel shell, eight washbacks made of stainless steel with a fermentation time between 48 and 66 hours and two pairs of stills. The production for 2015 will be 2.3 million litres of alcohol (which includes 150,000 litres of peated spirit at 35ppm). Seventy percent will be used for single malts while 30% is sold to other customers, most noticeably, Chivas Brothers.

The core range of BenRiach is **Heart of Speyside (no age), 10, 16, 20, 25** and **35 years old** in what the distillery calls Classic Speyside style. The 10 year old, launched in spring 2015, replaces the 12 year old and it is the first core expression where the owners use spirit predominantly produced since the 2004 take-over. Peated varieties include **Birnie Moss, Curiositas 10 year old, Septendecim 17 year old** and **Authenticus 25 year old**. There are four different **wood finishes (12-15 years)** in the Classic Speyside style while the three peated versions (Heredotus, Arumaticus and Maderensis) have now been discontinued. They were replaced in 2015 by three new 18 year old limited expressions; **Dunder** (dark rum), **Albariza** (PX sherry) and **Latada** (madeira). Limited releases from 2013/2014 include **Solstice**, a 17 year old, heavily peated port finish and **Vestige 46 year old**. Every year a number of **single cask** bottlings are released and **batch number 12** was launched in autumn 2015.

History:
1897 John Duff & Co founds the distillery.

1900 The distillery is closed.

1965 The distillery is reopened by the new owner, The Glenlivet Distillers Ltd.

1978 Seagram Distillers takes over.

1983 Production of peated Benriach starts.

1985 The number of stills is increased to four.

1998 The maltings is decommissioned.

2002 The distillery is mothballed in October.

2004 Intra Trading, buys Benriach together with the former Director at Burn Stewart, Billy Walker. The price is £5.4 million.

2004 Standard, Curiositas and 12, 16 and 20 year olds are released.

2005 Four different vintages are released.

2006 Sixteen new releases, i.a. a 25 year old, a 30 year old and 8 different vintages.

2007 A 40 year old and three new heavily peated expressions are released.

2008 New expressions include a peated Madeira finish, a 15 year old Sauternes finish and nine single casks.

2009 Two wood finishes (Moscatel and Gaja Barolo) and nine single casks are released.

2010 Triple distilled Horizons and heavily peated Solstice are released.

2011 A 45 year old and 12 vintages are released.

2012 Septendecim 17 years and ten new vintages are released.

2013 Vestige 46 years is released. The maltings are working again.

2015 Dunder, Albariza, Latada and a 10 year old are released.

10 years old

Tasting notes BenRiach 10 year old:
GS – Earthy and nutty on the early nose, with apples, ginger and vanilla. Smooth and rounded on the palate, with oranges, apricots, mild spice and hazelnuts. The finish is medium in length, nutty and spicy.

Benrinnes

[ben rin•ess]

Owner:　　　　　　　　**Region/district:**
Diageo　　　　　　　　　Speyside

Founded:　**Status:**　**Capacity:**
1826　　　　Active　　　3 500 000 litres

Address: Aberlour, Banffshire AB38 9NN

Website:　　　　　　**Tel:**
malts.com　　　　　　　01340 872600

In Diageo's assortment of blended Scotch there are megabrands such as Johnnie Walker, J&B and Bell's. But there is also room for about 50 additional kinds, often sold locally and some with a rich history behind them.

Crawford's 3 Star is one of them and it has Benrinnes single malt to thank for its character. The brand was launched around 1900 by the blenders A&A Crawford in Leith, It was followed by the deluxe version 5 Stars in 1920 but this has now been discontinued. The brand's glory days were during the 1960s when it sold over 1 million bottles in Scotland alone.

Benrinnes was completely rebuilt in the 1950s and none of the original buildings remains. A major upgrade was made in autumn 2012 which included a full automation of the process and a new control room where one operator can handle all the work. The equipment consists of an 8.5 tonnes semilauter mash tun, eight washbacks made of Oregon pine with a fermentation time of 65 hours and six stills. The composition of stills is rare in the way that there are two wash stills and four spirit stills and, until a few years ago, they were run three and three with a partial triple distillation. This system has been abandoned and one wash still will now serve two spirit stills. The spirit vapours are cooled using six green, cast iron worm tubs which contribute to the character. Since 2012, the distillery is working a 7-day week which translates to 3.5 million litres of alcohol per year. The character of Benrinnes newmake is light sulphury, which can, in part, be attributed to the use of wormtubs. The wide spirit cut (73%-58%) also plays its part in creating a robust and meaty spirit.

Most of the production goes into blended whiskies – J&B, Johnnie Walker and Crawford's 3 Star – and there is only one official single malt, the Flora & Fauna **15 year old**. In May 2010 a **Manager's Choice** from **1996**, drawn from a refill bourbon cask was released and in autumn 2014 it was time for a **21 year old** Special Release bottled at 57%.

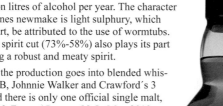

History:

1826 Lyne of Ruthrie distillery is built at Whitehouse Farm by Peter McKenzie.

1829 A flood destroys the distillery and a new distillery is constructed by John Innes a few kilometres from the first one.

1834 John Innes files for bankruptcy and William Smith & Co takes over.

1864 William Smith & Co goes bankrupt and David Edward becomes the new owner.

1896 Benrinnes is ravaged by fire which prompts major refurbishment. Alexander Edward takes over.

1922 John Dewar & Sons takes over ownership.

1925 John Dewar & Sons becomes part of Distillers Company Limited (DCL).

1956 The distillery is completely rebuilt.

1964 Floor maltings is replaced by a Saladin box.

1966 The number of stills doubles to six.

1984 The Saladin box is taken out of service and the malt is purchased centrally.

1991 The first official bottling from Benrinnes is a 15 year old in the Flora & Fauna series.

1996 United Distillers releases a 21 year old cask strength in their Rare Malts series.

2009 A 23 year old is launched as a part of this year's Special Releases.

2010 A Manager's Choice 1996 is released.

2014 A limited 21 year old is released.

15 years old

Tasting notes Benrinnes 15 years old:

GS – A brief flash of caramel shortcake on the initial nose, soon becoming more peppery and leathery, with some sherry. Ultimately savoury and burnt rubber notes. Big-bodied, viscous, with gravy, dark chocolate and more pepper. A medium-length finish features mild smoke and lively spices.

Benromach

[ben•ro•mack]

Owner: Gordon & MacPhail	**Region/district:** Speyside
Founded: 1898	**Status:** Active (vc)	**Capacity:** 500 000 litres

Address: Invererne Road, Forres, Morayshire IV36 3EB

Website: benromach.com	**Tel:** 01309 675968

When top independent bottler, Gordon & MacPhail, bought Benromach distillery in 1983 they were faced with one problem – there was no stock from the past decade as the distillery had been closed.

Apart from that it took a further five years before the production commenced which meant that the new owners either had some very old whisky to sell or newly produced whisky that hadn't even turned 3 years old. Of course this didn't come as a surprise to the experienced company but rhymed well with their business philosophy which had always been characterized by sustainability.

The primary goal with Benromach was to produce a Speyside whisky, just like it had tasted way back in the 1950s and 60s. That meant using barley that had been medium peated which in Benromach's case, meant 12ppm in the barley. This is still the standard level for Benromach but every year a small amount of unpeated and heavily peated newmake is distilled.

Benromach is equipped with a 1.5 tonnes semi-lauter mash tun with a copper dome, four washbacks made of larch and one pair of stills. Almost the entire production is destined to be sold as single malt. New warehouses have been added lately to reach a total of six, all dunnage and holding 18,000 casks.

In 2014, most of the bottles in the range received a new look and new bottlings were added. The core range now consists of a **5 year old** (which replaced the non-aged Traditional in 2014), the **10 year old**, the **15 year old** (launched in spring 2015) and the **100 Proof** (bottled at 57% thus replacing the Cask Strength). Limited releases include a **30 year old, Vintage 1976** and a **Classic 55 year old**. There are also special editions; **Organic**, the first single malt to be fully certified organic by the Soil Association and currently from 2008 and **Peatsmoke**, produced using heavily peated barley and distilled in 2006. Every year wood finishes are released in limited numbers with **Hermitage** and **Sassicaia**, launched in autumn 2015, being the latest. Finally, a duty free bottling called **Traveller's Edition**, was released in 2014. It is bottled at 43% and sold without an age statement.

History:
1898 Benromach Distillery Company starts the distillery.
1911 Harvey McNair & Co buys the distillery.
1919 John Joseph Calder buys Benromach and sells it to recently founded Benromach Distillery Ltd owned by several breweries.
1931 Benromach is mothballed.
1937 The distillery reopens.
1938 Joseph Hobbs buys Benromach and sells it on to National Distillers of America (NDA).
1953 NDA sells Benromach to Distillers Company Limited (DCL).
1966 The distillery is refurbished.
1968 Floor maltings is abolished.
1983 Benromach is mothballed.
1993 Gordon & McPhail buys Benromach from United Distillers.
1998 The distillery is once again in operation.
1999 A visitor centre is opened.
2004 The first bottle distilled by the new owner is released under the name 'Benromach Traditional' in May. Other novelties include a 21 year Tokaji finish and a Vintage 1969.
2005 A Port Wood finish (22 years old) and a Vintage 1968 are released together with the Benromach Classic 55 years.
2006 Benromach Organic is released.
2007 Peat Smoke, the first heavily peated whisky from the distillery, is released.
2008 Benromach Origins Golden Promise is released.
2009 Benromach 10 years old is released.
2010 New batches of Peatsmoke and Origins are released.
2011 New edition of Peatsmoke, a 2001 Hermitage finish and a 30 year old are released.
2013 A Sassicaia Wood Finish is released.
2014 Three new bottlings are launched; a 5 year old, 100 Proof and Traveller's Edition.
2015 A 15 year old and two wood finishes (Hermitage and Sassicaia) are released.

Tasting notes Benromach 10 year old:
GS – A nose that is initially quite smoky, with wet grass, butter, ginger and brittle toffee. Mouth-coating, spicy, malty and nutty on the palate, with developing citrus fruits, raisins and soft wood smoke. The finish is warming, with lingering barbecue notes.

10 years old

Blair Athol

[blair ath•ull]

Owner:
Diageo

Region/district:
Eastern Highlands

Founded: 1798
Status: Active (vc)
Capacity: 2 800 000 litres

Address: Perth Road, Pitlochry, Perthshire PH16 5LY

Website: malts.com
Tel: 01796 482003

Blair Athol distillery, one of few built in the 1700s, is one of the most visited in Scotland with an annual attendance of about 35,000 visitors. It is conveniently situated in the small town of Pitlochry on the busy A9 between Edinburgh and Inverness.

The distillery is also the spiritual home of Bell´s blended whisky, a brand created in the late 1800s by Arthur Bell and which was later taken over by his sons, Robert Duff and Arthur Kinmond. But it was not until 1942, when both brothers died and the company accountant William Govan Farquharson took over, that sales started escalating considerably. Farquharson increased marketing and by 1970 Bell´s had become the best selling whisky in Scotland. The number one position for the entire UK was reached in 1978 and by 1980 the brand had an impressive 35% share of the market in the UK. Today´s leader is Famous Grouse with a 20% market share. Apart from the UK, the best markets for Bell´s are South Africa and Scandinavia with a total of 28 million bottles having been sold in 2014.

The equipment of Blair Athol distillery consists of an 8 tonnes semi-lauter mash tun, six washbacks made of stainless steel and two pairs of stills. The distillery is running seven days a week with 16 mashes giving a production of 2.5 million litres of spirit. The part of the spirit which goes into Bell´s is matured mainly in bourbon casks, while the rest is matured in sherry casks. A cloudy wort, together with a fairly short fermentation time (52 hours) gives Blair Athol newmake a nutty and malty character. In 2015, the distillery will be working a 7-day week with 17 mashes per week and 2.8 million litres of alcohol.

The output today is still used for Bell´s whisky and the only official bottling used to be the **12 year old Flora & Fauna**. In 2010, a **first fill sherry** bottled at cask strength and without age statement was also released as a **distillery exclusive**.

History:

1798 John Stewart and Robert Robertson found Aldour Distillery, the predecessor to Blair Athol. The name is taken from the adjacent river Allt Dour.

1825 The distillery is expanded by John Robertson and takes the name Blair Athol Distillery.

1826 The Duke of Atholl leases the distillery to Alexander Connacher & Co.

1860 Elizabeth Connacher runs the distillery.

1882 Peter Mackenzie & Company Distillers Ltd of Edinburgh (future founder of Dufftown Distillery) buys Blair Athol and expands it.

1932 The distillery is mothballed.

1933 Arthur Bell & Sons takes over by acquiring Peter Mackenzie & Company.

1949 Production restarts.

1973 Stills are expanded from two to four.

1985 Guinness Group buys Arthur Bell & Sons.

1987 A visitor centre is built.

2003 A 27 year old cask strength from 1975 is launched in Diageo's Rare Malts series.

2010 A distillery exclusive with no age statement and a single cask from 1995 are released.

12 years old

Tasting notes Blair Athol 12 years old:

GS – The nose is mellow and sherried, with brittle toffee. Sweet and fragrant. Relatively rich on the palate, with malt, raisins, sultanas and sherry. The finish is lengthy, elegant and slowly drying.

1830 invention of the Coffey still

Until the 1820s, the traditional batch distillation in pot stills was the common practice both when producing malt whisky, as well as grain whisky.

It was a time-consuming and inefficient method though, and different trials to make the distillation process more economical were attempted. One of the first attempts was made by Sir Anthony Perrier, owner of Spring Lane distillery in Cork and mayor of the city. In 1822 he patented one of the first continuous stills in Europe. His invention was not a success, but it inspired others to try and improve the method. Robert Stein was one of them. In 1827 he presented a patent still (as they sometimes were called), which would produce spirit more rapidly and in a cheaper way. It was constructed as a huge column which was divided into separate columns, each of which was separated by plates made of haircloth. The wash was pumped into the still, moving towards the first chamber where it came into contact with steam. The vapour then moved upwards through the chambers until it was condensed at the top as spirit. Water and solids dropped to the bottom where they were removed. The result was a high strength spirit (94-96%) with a light and bland character, because most of the congeners had been lost during the distillation. In accordance with the spirit of the time, this was not a bad feature as the result would be considered "purer and more wholesome" compared to the "oily" spirits produced in a pot still.

John Haig installed a Stein patent still at his Cameron-

bridge distillery in 1827 and Andrew Stein started using one at Kirkliston distillery in 1828. The efficiency of the new still at Cameronbridge became obvious in 1829, when it distilled 150,000 gallons in a year, compared to Macallan distillery which produced 5,000 gallons in their pot stills during the same year. At the same time, and independent of Robert Stein, Aeneas Coffey was working on a similar still in Ireland. Coffey was a former excise officer and owner of Dock distillery in Dublin. His idea started with a huge device, with an analyser at the bottom and a rectifier at the top. Soon this was separated into two, parallel columns with the analyser (corresponding with the wash still) and the rectifier (spirit still) standing next to each other. Like Stein´s still, Coffey´s was also equipped with plates separating the different sections, but his were made of perforated copper. Coffey patented his still in 1830 and the first of its kind in Scotland was installed at Grange distillery in 1834. Aeneas Coffey´s version of the patent still was simpler and more efficient than the Stein still and, although Stein patent stills continued to be installed on a minor scale in Scotland until the 1880s, it was the Coffey still that would revolutionize the whisky industry.

But great changes don´t always happen overnight and the continuous still came at a time when demand for whisky declined because of bad economy due to poor harvests, as well as the great outbreaks of cholera and typhus. In addition to this, the expansion of the railway network in Scotland had also slowed down due to lack of capital. A number of distilleries, mainly in the Lowlands, had made large investments in new patent stills, but many went bankrupt when they couldn't sell their spirit. The only market still functioning was England, where large amounts of the grain whisky was exported as "British Spirits" to be rectified into gin. Over time, the export markets were a good opportunity to sell the mild and pretty tasteless spirit from the column still, but it was because of the new opportunity of blending malt and grain spirit that arouse with the Spirits Act of 1860, that the Coffey still would take its rightful place in the history of whisky.

COFFEY STILL

Bowmore

[bow•moor]

Owner:
Morrison Bowmore (Suntory)

Region/district:
Islay

Founded: 1779

Status: Active (vc)

Capacity: 2 000 000 litres

Address: School Street, Bowmore, Islay, Argyll PA43 7GS

Website:
bowmore.com

Tel:
01496 810441

Bowmore single malt has been selling well, not least after targeting the duty free segment which one started doing a number of years ago.

Lately though, sales have been slipping and the figures are now back to the 2011 level with around 2 million bottles having been sold in 2014. Where the Islay malts are concerned, the brand still holds second place, but Lagavulin is definitely closing the gap. With the merger between Bowmore's owners, Suntory, and Beam last year, it remains to be seen if Bowmore can gain any advantage from Beam's distribution network in the important USA market.

Bowmore is one of only a few Scottish distilleries with its own malting floor, with 30% of the malt requirement produced in-house. The remaining part is bought from Simpson's. Both parts have a phenol specification of 25ppm and are always mixed according to the proportions, 2 ton in-house malt and 6 ton malt from Simpson's before mashing. The distillery has an eight ton stainless steel semi-lauter mash tun, six washbacks of Oregon pine (all named after the six owners predating Suntory) with six short fermentations (48 hours) per week and seven long (100 hours) and two pairs of stills. The 27,000 casks are stored in two dunnage and one racked warehouse. The building closest to the sea, dating back to the 1700s, is probably the oldest whisky warehouse still in use in Scotland. In 2015, they have increased production to 15 mashes per week, which amounts to two million litres of alcohol per year.

The core range for domestic markets includes **Small Batch Reserve** (bourbon matured and with no age statement), **12 years, Darkest 15 years, 18 years** and **25 years**. The duty free line-up has gone through a thorough revamp during the last few years and now consists of **Springtide** (matured in Oloroso casks), **Black Rock** (predominantly matured in first-fill sherry casks), **Gold Reef** (mostly from first-fill bourbon casks), **White Sands** (a 17 year old matured in 100% ex-bourbon casks) and **Vintage 1984**. Limited releases for 2014 included the release of the second batch of **Devil's Cask** and the fifth release of **Tempest** (a 10 year old cask strength from first fill bourbon casks), together with another 50 bottles of the **50 year old** which were first released in 2013. More recent, limited releases (from 2015) were the sixth edition of **Tempest** in July followed by the third release of **Devil's Cask** and more bottles of the **50 year old** in October. A surprise release was the **Mizunara Cask Finish** in September. Having Japanese owners, Bowmore has more easy access to the rare casks made of the Japanese mizunara oak. Parts of this year's release came from three casks made of virgin mizunara oak that had been filled with 12 year old Bowmore malt a couple of years ago, and another three casks that had previously held Yamazaki single malt and had been filled with 20 year old Bowmore malt. There were three special bottlings for Islay Festival 2015; one **unaged** that had been maturing in **virgin oak**, one **12 year old** from an oloroso sherry cask and a **26 year old** sherry maturation from 1988.

History:

1779 Bowmore Distillery is founded by David Simpson and becomes the oldest Islay distillery.

1837 The distillery is sold to James and William Mutter of Glasgow.

1892 After additional construction, the distillery is sold to Bowmore Distillery Company Ltd, a consortium of English businessmen.

1925 J. B. Sheriff and Company takes over.

1929 Distillers Company Limited (DCL) takes over.

1950 William Grigor & Son takes over.

1963 Stanley P. Morrison buys the distillery for £117,000 and forms Morrison Bowmore Distillers Ltd.

1989 Japanese Suntory buys a 35% stake in Morrison Bowmore.

1993 The legendary Black Bowmore is launched. Another two versions are released 1994 and 1995.

1994 Suntory now controls all of Morrison Bowmore.

1996 A Bowmore 1957 (38 years) is bottled at 40.1% but is not released until 2000.

1999 Bowmore Darkest with three years finish on Oloroso barrels is launched.

2000 Bowmore Dusk with two years finish in Bordeaux barrels is launched.

2001 Bowmore Dawn with two years finish on Port pipes is launched.

2002 A 37 year old Bowmore from 1964 and matured in fino casks is launched in a limited edition of 300 bottles (recommended price £1,500).

2003 Another two expressions complete the wood trilogy which started with 1964 Fino - 1964 Bourbon and 1964 Oloroso.

History continued:

2004 Morrison Bowmore buys one of the most out standing collections of Bowmore Single Malt from the private collector Hans Sommer. It totals more than 200 bottles and includes a number of Black Bowmore.

2005 Bowmore 1989 Bourbon (16 years) and 1971 (34 years) are launched.

2006 Bowmore 1990 Oloroso (16 years) and 1968 (37 years) are launched. A new and upgraded visitor centre is opened.

2007 Dusk and Dawn disappear from the range and an 18 year old is introduced. New packaging for the whole range. 1991 (16yo) Port and Black Bowmore are released.

2008 White Bowmore and a 1992 Vintage with Bourdeaux finish are launched.

2009 Gold Bowmore, Maltmen´s Selection, Laimrig and Bowmore Tempest are released.

2010 A 40 year old and Vintage 1981 are released.

2011 Vintage 1982 and new batches of Tempest and Laimrig are released.

2012 100 Degrees Proof, Springtide and Vintage 1983 are released for duty free.

2013 The Devil´s Casks, a 23 year old Port Cask Matured and Vintage 1984 are released.

2014 Black Rock, Gold Reef and White Sands are released for duty free.

2015 New editions of Devil´s Cask, Tempest and the 50 year old are released as well as Mizunara Cask Finish.

Tasting notes Bowmore 12 year old:

GS – An enticing nose of lemon and gentle brine leads into a smoky, citric palate, with notes of cocoa and boiled sweets appearing in the lengthy, complex finish.

The Devil´s Casks 3rd edition   Springtide   Mizunara Cask Finish

12 years old   Small Batch   15 years old

Braeval

[bre•vaal]

Owner:	**Region/district:**
Chivas Brothers (Pernod Ricard)	Speyside
Founded: **Status:**	**Capacity:**
1973 Active	4 000 000 litres

Address: Chapeltown of Glenlivet, Ballindalloch, Banffshire AB37 0JS

Website:	**Tel:**
-	01542 783042

When Chivas Brothers opened Braeval distillery in 1978, the venerable blending company was owned by the Canadian company, Seagram's. Its first entry into the Scottish market, initiated by the legendary owner Sam Bronfman, was the purchase of the Glasgow whisky house Robert Brown Ltd. 1936.

The successful saga continued with further purchases of Chivas Brothers (1949), Strathisla (1950) and Glenlivet, Glen Grant and Longmorn (1977). The good years came to an end and the fall of the once successful Seagram Company, was initiated by Sam's grandson, Edgar Bronfman Jr. who decided to diversify the company into owning film- and record companies as well. The debt grew larger and the company was eventually acquired by the Vivendi Group, who in their turn sold the wine part of the operation to Diageo and the spirits part to Pernod Ricard in 2001. This move together with the acquisition of Allied Domecq four years later, propelled Pernod Ricard to become the second largest producer of alcohol in the world.

Braeval (or Braes of Glenlivet as it is sometimes called) is situated in a very remote, but beautiful part of the Highlands, south of Glenlivet and with Tamnavulin as its closest distillery neighbour. The distillery is both impressive and surprisingly attractive, despite that it is modern and was built only to function as a typical working distillery.

The equipment consists of a 9 ton stainless steel mash tun with traditional rakes and a copper dome, 13 stainless steel washbacks with a fermentation time of 70 hours and six stills. There are two wash stills with aftercoolers and four spirit stills, and with the possibility of doing 26 mashes per week, the distillery can now produce 4 million litres per year.

With its light and floral character, Braeval single malt is used for many of Chivas Brothers' blends. There are no official bottlings but Braeval single malt has, from time to time, been used for the bottling of Deerstalker.

History:

1973 The distillery is founded by Chivas Brothers (Seagram's) and production starts in October.

1975 Three stills are increased to five.

1978 Five stills are further expanded to six.

1994 The distillery changes name to Braeval.

2001 Pernod Ricard takes over Chivas Brothers.

2002 Braeval is mothballed in October.

2008 The distillery starts producing again in July.

Tasting notes Deerstalker 20 year old:

GS – Newsprint, herbal notes, fleeting green apples, sawdust and burgeoning vanilla on the nose. Mouth-coating, initially fruity – apricots and peaches, with an edge of chilli. Spicy, lengthy finish, with a hint of oak.

Deerstalker 20 years old

1860 the Spirits Act of 1860

Historically, whisky had been drunk in Scotland without blending it and in the Highlands that meant malt whisky.

Almost all the production emanated from bere barley - a variety which today is only used for a few experimental batches at a handful of distilleries. In the fertile Lowlands, a wider variety of grains could be cultivated and this led to whisky being distilled from barley, wheat, oat or corn. The whisky was sold cheaply and enjoyed as grain whisky or sold to England for rectification into gin. There were no laws against blending whisky, and merchants and retailers mixed different types of whisky and sometimes even added brandy or rum to the spirit. In 1853, the Forbes-Mackenzie Act was passed which mainly dealt with limiting the consumption in Scotland by demanding that pubs be closed at 10pm weekdays and stayed closed on Sundays. But even more importantly, the new act allowed for blending whisky of different years from the same distillery under bond. This meant that the distillers themselves, for the first time, could blend their own whisky before duty was paid which, in turn, gave them control of their own products and new possibilities on the market.

One of the first producers to make use of this new law was Andrew Usher, a spirits dealer in Edinburgh who, for a long time, had enjoyed close business relations with George Smith, owner of Glenlivet, and who also was the official agent for the distillery. His Old Vatted Glenlivet (OVG) is often referred to as the first branded whisky and was a blend of Glenlivet single malt (and quite possibly grain whisky produced at the distillery) of different ages.

Seven years later a new law was passed – the Spirits Act of 1860. Some of the new law's content proved disadvantageous for the whisky producers. The duty was raised from 8s per proof gallon to 10s, but the bigger problem was that duty on imported wines and spirits from France was reduced. These beverages soon became popular and proved to be a threat for whisky. There was one part of the law, however, that would have a huge impact on the industry in the long run. The vatting of whisky was now expanded to also apply to whiskies from different distilleries. At this time, Irish whisky was popular in Scotland. The taste was not particularly challenging and, above all, it was consistent. You knew what you got every time you bought a bottle. Pot still whisky from Scotland, on the other hand, could vary greatly, something that today is seen as one of Scotch malt whisky's big virtues, but which was not seen as an advantage back in the days.

The Spirits Act of 1860 gave the Scottish producers a tool to customize their whisky to suit the consumers' taste. They could furthermore bottle and label the whisky themselves, hence increasing the power of branding. The first of these blends was probably a vatting of whisky from 57 distilleries and being quite young, perhaps it was similar to the cheapest standard blends that we see today. Andrew Usher had passed away in 1855, but one of his sons, Andrew Jr., quickly remade his father's successful brand OVG to a blend, and other producers soon followed. William P. Lowrie and William Robertson were early pioneers and giants of their time, and they were then followed by even more familiar names whose brands make out the foundation of the entire, modern Scotch whisky industry even today – like Walker, Bell, Buchanan, Dewar, Teacher, Ballantine and Grant. Most of these brands were established during the latter part of the 1800s or early 1900s, but it was actually the Spirits Act of 1860 that laid the foundation.

Bruichladdich

[brook•lad•dee]

Owner: Rémy Cointreau
Region/district: Islay

Founded: 1881
Status: Active (vc)
Capacity: 1 500 000 litres

Address: Bruichladdich, Islay, Argyll PA49 7UN

Website: bruichladdich.com
Tel: 01496 850221

Rémy Cointreau, the French spirit giant, took its first step into the world of whisky in 2012 when they acquired Bruichladdich for £58 million. To judge by the company´s annual results for 2014 this is a step they are not regretting.

Admittedly, whisky is still a very small part of the company´s business, but sales of Bruichladdich single malts and Botanist gin nearly doubled last year to almost £20m. This increase has, of course, been boosted by the opportunity to use the mother company´s global distribution network. Since all whisky from Bruichladdich is maturing on Islay, the rapid growth has created a need for six new warehouses to be built north of Coultorsay. They are also adding a second bottling line and there are further discussions about adding more washbacks and stills in the future.

The distillery is equipped with a cast iron, open mash tun with rakes, dating back to 1881 when the distillery was founded. There are six washbacks of Oregon pine and two pairs of stills. The distillery also has the only functioning Lomond still in the industry which was brought to Bruichladdich from Inverleven distillery. The still is being used for the production of Botanist Gin. The two, traditional pot stills which were also brought in from Inverleven, were bought by Mark Reynier, the former Bruichladdich manager and co-owner, in 2015 to be used at his new distillery in Waterford, Ireland. All whisky produced is based on Scottish barley, 40% of which comes from Islay. During 2015 they will be doing 12 mashes per week, resulting in 1.5 million litres of alcohol per year. The breakdown of the three whisky varieties are 60% Bruichladdich, 20% Port Charlotte and 20% Octomore.

Bruichladdich single malt has been unpeated since the 1960s but, today, there are three main lines in the distillery´s production; unpeated **Bruichladdich**, heavily peated **Port Charlotte** and the ultra-heavily peated **Octomore**. The core expressions for each of the three varieties are **Scottish Barley** and **Islay Barley**. For Bruichladdich there are three duty free exclusives – **Organic Barley**, **Bere Barley** (2009 edition released autumn 2015) and **Black Art 4** (with a maturation in American oak and various wine casks). The duty free expressions for Port Charlotte are the **PC11** and **PC12** (released early in 2015), both bottled at cask strength and for Octomore it is **6.2** with a phenol specification of 167ppm and matured in cognac casks and **7.2**, which is a vatting of whiskies matured in American oak and syrah casks from Rhone. In June 2015, the limited **Octomore 7.1** with a phenol specification of 208ppm was launched as Jim McEwan´s final creation before retiring. Jim has been working in the industry for 52 years, the last 15 as the legendary master distiller at Bruichladdich. The special bottling for Feis Ile 2015 was **High Noon 134** (also called Black Art Valinch) where the number refers to the combined age of whiskies that were married to create it. The aged expressions (**Laddie Ten**, the **16** and the **22 year old**) have been withdrawn from all markets and are only available from the distillery.

History:

1881 Barnett Harvey builds the distillery with money left by his brother William III to his three sons William IV, Robert and John Gourlay.

1886 Bruichladdich Distillery Company Ltd is founded and reconstruction commences.

1889 William Harvey becomes Manager and remains on that post until his death in 1937.

1929 Temporary closure.

1936 The distillery reopens.

1938 Joseph Hobbs, Hatim Attari and Alexander Tolmie purchase the distillery through the company Train & McIntyre.

1938 Operations are moved to Associated Scottish Distillers.

1952 The distillery is sold to Ross & Coulter from Glasgow.

1960 A. B. Grant buys Ross & Coulter.

1961 Own maltings ceases and malt is brought in from Port Ellen.

1968 Invergordon Distillers take over.

1975 The number of stills increases to four.

1983 Temporary closure.

1993 Whyte & Mackay buys Invergordon Distillers.

1995 The distillery is mothballed in January.

1998 In production again for a few months, and then mothballed.

2000 Murray McDavid buys the distillery from JBB Greater Europe for £6.5 million.

2001 The first distillation (Port Charlotte) is on 29th May and the first distillation of Bruichladdich starts in July. In September the owners' first bottlings from the old casks are released, 10, 15 and 20 years old.

2002 The world's most heavily peated whisky is produced on 23rd October when Octomore (80ppm) is distilled.

History continued:

2004 Second edition of the 20 year old (nick-named Flirtation) and 3D, also called The Peat Proposal, are launched.

2005 Several new expressions are launched - the second edition of 3D, Infinity, Rocks, Legacy Series IV, The Yellow Submarine and The Twenty 'Islands'.

2006 Included in a number of new releases is the first official bottling of Port Charlotte; PC5.

2007 New releases include Redder Still, Legacy 6, PC6 and an 18 year old.

2008 More than 20 new expressions including the first Octomore, Bruichladdich 2001, PC7, Golder Still and two sherry matured from 1998.

2009 New releases include Classic, Organic, Black Art, Infinity 3, PC8, Octomore 2 and X4+3 - the first quadruple distilled single malt.

2010 PC Multi Vintage, Organic MV, Octomore/3_152, Bruichladdich 40 year old are released.

2011 The first 10 year old from own production is released as well as PC9 and Octomore 4_167.

2012 Ten year old versions of Port Charlotte and Octomore are released as well as Laddie 16 and 22, Bere Barley 2006, Black Art 3 and DNA4. Rémy Cointreau buys the distillery.

2013 Scottish Barley, Islay Barley Rockside Farm, Bere Barley 2nd edition, Black Art 4, Port Charlotte Scottish Barley, Octomore 06.1 and 06.2 are released.

2014 PC11 and Octomore Scottish Barley are released.

2015 PC12, Octomore 7.1 and High Noon 134 are released.

Tasting notes Bruichladdich Scottish Barley:

GS – Mildly metallic on the early nose, then cooked apple aromas develop, with a touch of linseed. Initially very fruity on the gently oily palate. Ripe peaches and apricots, with vanilla, brittle toffee, lots of spice and sea salt. The finish is drying, with breakfast tea.

Tasting notes Port Charlotte Scottish Barley:

GS – Wood smoke and contrasting bonbons on the nose. Warm Tarmac develops, with white pepper. Finally, fragrant pipe tobacco. Peppery peat and treacle toffee on the palate, with a maritime note. Long in the finish, with black pepper and oak.

Tasting notes Octomore Scottish Barley:

GS – A big hit of sweet peat on the nose; ozone and rock pools, supple leather, damp tweed. Peat on the palate is balanced by allspice, vanilla and fruitiness. Very long in the finish, with chilli, dry roasted nuts and bonfire smoke.

Black Art 4   Bruichladdich Bere Barley 2009   Bruichladdich Islay Barley 2009

The Classic Laddie Scottish Barley

Port Charlotte PC12

Port Charlotte Scottish Barley

Bunnahabhain

[buh•nah•hav•enn]

Owner:
Burn Stewart Distillers
(Distell Group Ltd)

Region/district:
Islay

Founded: 1881

Status:
Active (vc)

Capacity:
2 700 000 litres

Address: Port Askaig, Islay, Argyll PA46 7RP

Website:
bunnahabhain.com

Tel:
01496 840646

In terms of sales, Bunnahabhain is a small player on Islay with only the young Kilchoman selling less. But during the past five years something has happened and volumes have doubled.

In part, this is due to the recreation work of the company´s master blender Ian MacMillan. He decided to stop colouring and chill filtration (for all single malts in the group, not just Bunnahabhain) and also started bottling at a higher strength – 46.3%. At the same time, he released new versions where peated Bunnahabhain defined the character.

The distillery is equipped with a 12.5 tonnes traditional stainless steel mash tun, six washbacks made of Oregon pine and two pairs of stills. The fermentation time varies between 55 and 80 hours. The production for 2015 will be 7-8 mashes per week and around 1.5 million litres, of which 10% will be peated (35ppm). In 2014 the old malting floors were converted into another warehouse, holding 1,500 casks.

The core range consists of **12, 18** and **25 year olds** as well as the 10 year old peated **Toiteach**. A new addition to the range, launched in autumn 2014, was **Ceobanach**, also peated. The difference between the two is that while Toiteach is a blend of younger, peated and some older, unpeated whisky, Ceobanach is 100% peated Bunnahabhain, aged for a minimum of 10 years in bourbon casks. Recent limited releases include a **40 year old** while an even older version (46 years) has been postponed to a possible late 2015 release. There are also two travel retail exclusives – **Cruach-Mhòna** which comprises of young, heavily peated Bunnahabhain matured in ex bourbon casks along with 20-21 years old matured in ex sherry butts and the new **Eirigh Na Greine**. The latter has replaced Darach Ur and is a vatting of whisky from bourbon and sherry casks, as well as red wine casks from France and Italy. For Feis Ile 2015, there was **Rubha A´Mhail**, an 11 year old matured in a manzanilla cask, as well as an **18 year old**, finished for one year in a moscatel cask.

History:

1881 William Robertson of Robertson & Baxter, founds the distillery together with the brothers William and James Greenless, owners of Islay Distillers Company Ltd.

1883 Production starts in earnest in January.

1887 Islay Distillers Company Ltd merges with William Grant & Co. in order to form Highland Distilleries Company Limited.

1963 The two stills are augmented by two more.

1982 The distillery closes.

1984 The distillery reopens. A 21 year old is released to commemorate the 100th anniversary.

1999 Edrington takes over Highland Distillers and mothballs Bunnahabhain but allows for a few weeks of production a year.

2001 A 35 year old from 1965 is released during Islay Whisky Festival.

2002 A 35 year old from 1965 is released during Islay Whisky Festival. Auld Acquaintance 1968 is launched at the Islay Jazz Festival.

2003 Edrington sells Bunnahabhain and Black Bottle to Burn Stewart Distilleries for £10 million. A 40 year old from 1963 is launched.

2004 The first limited edition of the peated version is a 6 year old called Moine.

2005 Three limited editions are released - 34 years old,18 years old and 25 years old.

2006 14 year old Pedro Ximenez and 35 years old are launched.

2008 Darach Ur is released for the travel retail market and Toiteach (a peated 10 year old) is launched on a few selected markets.

2009 Moine Cask Strength is released.

2010 The peated Cruach-Mhòna and a limited 30 year old are released.

2013 A 40 year old is released.

2014 Eirigh Na Greine and Ceobanach are released.

12 years old

Tasting notes Bunnahabhain 12 years old:

GS – The nose is fresh, with light peat and discreet smoke. More overt peat on the nutty and fruity palate, but still restrained for an Islay. The finish is full-bodied and lingering, with a hint of vanilla and some smoke.

Caol Ila

[cull•eel•a]

Owner:
Diageo

Region/district:
Islay

Founded: **Status:**
1846   Active (vc)

Capacity:
6 500 000 litres

Address: Port Askaig, Islay, Argyll PA46 7RL

Website:
malts.com

Tel:
01496 302760

No other distillery on Islay has been so thoroughly rebuilt like Caol Ila. While all the others have most of the original buildings remaining, although refurbished and modernised, the old Caol Ila was completely razed to the ground in 1972 and a new distillery was built.

This is in some way symptomatic of the distillery. When the other distilleries on the island still used worm tubs for condensing, Caol Ila installed shell and tube condensers. They were also the first to heat the stills by steam as opposed to direct firing. Today, it is by far the biggest distillery on Islay, especially after the expansion in 2011. Traditionally, Caol Ila is known for its peated whisky but, in recent years, an increasingly bigger part of unpeated, nutty newmake has been produced. In 2012 this proportion was as large as 30%. Experiments with this style already started in 1999 and, over the years, there have been limited, but regular releases of unpeated Caol Ila.

Caol Ila is equipped with a 13.5 tonnes full lauter mash tun, eight wooden washbacks and two made of stainless steel and three pairs of stills. The fermentation time is 60 hours, except for the unpeated version when it is increased to 80 hours. During 2015 the distillery will be doing 26 mashes per week which amounts to 6.5 million litres of alcohol.

Caol Ila single malt is not one of the biggest sellers from Islay but, considering that it was only just over a decade ago since the owners seriously started to invest in the brand, 600,000 bottles being sold during 2014 is an impressive figure.

The core range consists of **Moch** without age statement, **12, 18** and **25 years old, Distiller's Edition Moscatel finish** and **Cask Strength**. The traditional release for Islay Festival 2015 was a **1998** matured in **American oak, Moscatel casks** and **sherry puncheons** and bottled at 57.3%. Although not confirmed by Diageo at the time of writing, the official registration of a label indicate that a **17 year old** unpeated Caol Ila could be a part of this year's Special Releases.

History:

1846 Hector Henderson founds Caol Ila.

1852 Henderson, Lamont & Co. is subjected to financial difficulties and Henderson is forced to sell Caol Ila to Norman Buchanan.

1863 Norman Buchanan sells to the blending company Bulloch, Lade & Co. from Glasgow.

1879 The distillery is rebuilt and expanded.

1920 Bulloch, Lade & Co. is liquidated and the distillery is taken over by Caol Ila Distillery.

1927 DCL becomes sole owners.

1972 All the buildings, except for the warehouses, are demolished and rebuilt.

1974 The renovation, which totals £1 million, is complete and six new stills are installed.

1999 Experiments with unpeated malt.

2002 The first official bottlings since Flora & Fauna/ Rare Malt appear; 12 years, 18 years and Cask Strength (c. 10 years).

2003 A 25 year old cask strength is released.

2006 Unpeated 8 year old and 1993 Moscatel finish are released.

2007 Second edition of unpeated 8 year old.

2008 Third edition of unpeated 8 year old.

2009 The fourth edition of the unpeated version (10 year old) is released.

2010 A 25 year old, a 1999 Feis Isle bottling and a 1997 Manager's Choice are released.

2011 An unpeated 12 year old and the unaged Moch are released.

2012 An unpeated 14 year old is released.

2013 Unpeated Stitchell Reserve is released.

2014 A 15 year old unpeated and a 30 year old are released.

12 years old

Tasting notes Caol Ila 12 year old:
GS – Iodine, fresh fish and smoked bacon feature on the nose, along with more delicate, floral notes. Smoke, malt, lemon and peat on the slightly oily palate. Peppery peat in the drying finish.

Cardhu

[car•<u>doo</u>]

Owner:	**Region/district:**
Diageo	Speyside
Founded: **Status:**	**Capacity:**
1824 Active (vc)	3 400 000 litres

Address: Knockando, Aberlour, Moray AB38 7RY

Website:	**Tel:**
discovering-distilleries.com	01479 874635

When John Walker & Sons bought Cardhu, their first distillery, in 1893, they were already the biggest blenders in Scotland. Cardhu has ever since been intricately linked to the famous blend and is now the spiritual home of Johnnie Walker.

During the 1800s, the company´s whiskies were sold under the name of Kilmarnock, the town where the company started, or Walker´s Old Highland Whisky. It wasn´t until 1908 that the whisky was named Johnnie Walker and, during that same year, the famous Striding Man was introduced. The elegant Edwardian gentleman with his high hat was created by Tom Browne, a famous cartoonist and artist. Sadly, Browne died the year thereafter, aged 39, but the character he created, although modernized over the years, lives on as the most recognisable figure in the whisky world.

Cardhu distillery is equipped with an 8 tonnes stainless steel full lauter mash tun with a copper dome, ten washbacks (four made of Scottish larch, two of stainless steel and four of Douglas fir), all with a fermentation time of 75 hours and three pairs of stills. As from 2015, Cardhu will be working a 7-day week with a production of 3.4 million litres of alcohol. Located in the old kiln with the two pagoda roofs, is a stylish Johnnie Walker Brand Home used for corporate events.

Even though Cardhu single malt is one of the most important components of Johnnie Walker, at the same time, it is one of the Top Ten single malt brands in the world with a huge following in Spain and France.

The core range from the distillery is **12, 15** and **18 year old** and, released in 2014, two expressions without age statement – **Amber Rock** and **Gold Reserve**, both bottled at 40%. There is also a **Special Cask Reserve** matured in rejuvenated bourbon casks and in 2013, a **21 year old** bottled at 54.2% was launched as part of the Diageo Special Releases.

History:

1824 John Cumming applies for and obtains a licence for Cardhu Distillery.

1846 John Cumming dies and his wife Helen and son Lewis takes over.

1872 Lewis dies and his wife Elizabeth takes over.

1884 A new distillery is built to replace the old.

1893 John Walker & Sons purchases Cardhu for £20,500.

1908 The name reverts to Cardow.

1960 Reconstruction and expansion of stills from four to six.

1981 The name changes to Cardhu.

1998 A visitor centre is constructed.

2002 Diageo changes Cardhu single malt to a vatted malt with contributions from other distilleries in it.

2003 The whisky industry protests sharply against Diageo's plans.

2004 Diageo withdraws Cardhu Pure Malt.

2005 The 12 year old Cardhu Single Malt is relaunched and a 22 year old is released.

2009 Cardhu 1997, a single cask in the new Manager´s Choice range is released.

2011 A 15 year old and an 18 year old are released.

2013 A 21 year old is released.

2014 Amber Rock and Gold Reserve are launched.

Amber Rock

Tasting notes Cardhu 12 years old:

GS – The nose is relatively light and floral, quite sweet, with pears, nuts and a whiff of distant peat. Medium-bodied, malty and sweet in the mouth. Medium-length in the finish, with sweet smoke, malt and a hint of peat.

Websites to watch

whiskyfun.com
Serge Valentin, one of the Malt Maniacs, is almost always first with well written tasting notes on new releases.

nonjatta.com
An excellent blog with a wealth of interesting information on Japanese whisky and Japanese culture.

whiskyreviews.blogspot.com
Ralfy does this video blog with tastings and field reports in an educational yet easy-going and entertaining way.

maltmadness.com
Our all-time favourite with something for everyone. Managed by the founder of Malt Maniacs, Johannes van den Heuvel.

whiskyadvocateblog.com
A first class blog on every aspect of whisky by Whisky Advocate´s John Hansell and others.

edinburghwhiskyblog.com
Lucas, Chris and company review new releases, interview industry people and cover news from the whisky world.

whiskycast.com
The best whisky-related podcast on the internet and one that sets the standard for podcasts in other genres as well.

whiskywhiskywhisky.com
An active forum for whisky friends with lots of daily comments on new whiskies, industry news, whisky events etc.

whiskyintelligence.com
The best site on all kinds of whisky news. The first whisky website you should log into every morning!

whisky-news.com
Apart from daily news, this site contains tasting notes, distillery portraits, lists of retailers, events etc.

dramming.com
Takes a wide-angle view of the whisky world including trip reports, whisky ratings, whisky business, articles etc.

whisky-emporium.com
Keith Wood has created a treasure trove for whisky geeks with more than 1,000 tasting notes.

whiskyforum.se
Swedish whisky forum with more than 1,800 enthusiasts. Excellent debate as well as more than 2,000 tasting notes.

whisky-pages.com
Top class whisky site with features, directories, tasting notes, book reviews, whisky news, glossary and a forum.

whiskynotes.be
This blog is almost entirely about tasting notes (and lots of them, not least independent bottlings) plus some news.

whiskyforeveryone.com
Educational site, perfect for beginners, with a blog where both new releases and affordable standards are reviewed.

blog.thewhiskyexchange.com
A knowledgeable team from The Whisky Exchange write about new bottlings and the whisky industry in general.

recenteats.blogspot.com
Steve Ury serves tasty bits of information (and entertainment) from the world of whisky and other spirits.

whiskymarketplace.com
This is divided into three parts - well written tasting notes, whisky price comparison site and Whisky Marketplace TV.

whisky-distilleries.net
Ernie Scheiner describes more than 130 distilleries in both text and photos and we are talking lots of great images!

connosr.com
This whisky social networking community is a virtual smorgasbord for any whisky lover!

jewmalt.com
An excellent blog by Joshua Hatton who also acts as an independent bottler, check out www.singlecasknation.com.

canadianwhisky.org
Davin de Kergommeaux presents reviews, news and views on all things Canadian whisky. High quality content.

whiskyisrael.co.il
Gal Granov is definitely one of the most active of all bloggers. Well worth checking out daily!

spiritsjournal.klwines.com
Reviews about whiskies and the whisky industry in general by David Driscoll from the US retailer K&L Wines.

thewhiskywire.com
Steve Rush mixes reviews of the latest bottlings with presentations of classics plus news, interviews etc.

bestshotwhiskyreviews.com
Jan van den Ende presents his honest opinions on everything from cheap blends to rare single cask bottlings.

whisky-discovery.blogspot.com
An entertaining blog with tasting notes and event reports from Dave Worthington and his daughter Kat.

whiskiesrus.blogspot.com
Clint Anesbury delivers tasting notes and current reports from the world of Japanese whisky.

scotch-whisky.org.uk
The official site of SWA (Scotch Whisky Association) with i.a. press releases and publications about the industry.

misswhisky.com
Filled with great tasting notes and well-written stories, interviews, event reports and accounts on distillery visits.

whiskysaga.com
Brilliant blog by Norwegian couple Tone and Thomas Öhrbom - not least on every detail relating to Nordic whiskies.

Clynelish

[cline•leash]

Owner:	**Region/district:**
Diageo	Northern Highlands
Founded: **Status:**	**Capacity:**
1967 Active (vc)	4 800 000 litres

Address: Brora, Sutherland KW9 6LR

Website:	**Tel:**
malts.com	01408 623003 (vc)

In January of 2014, Diageo announced that Clynelish distillery would be expanded at a cost of £30m which would, in turn, double its capacity.

Two new mash tuns, ten washbacks and another six stills were to be installed which would increase production capacity to almost 10 million litres. In October 2014, however, the owners said the investment had been put on hold together with a couple of other expansions. The decision was interpreted by industry analysts as a reaction to a sudden slowdown in demand in certain markets for Scotch whisky.

The existing cast iron mash tun with its beautiful copper canopy, is the original dating back to when the distillery was built in 1967. It will be replaced by one made of stainless steel in early 2016. The distillery is also equipped with 8 wooden washbacks and two made of stainless steel. The still room, with its three pairs of stills, has stunning views towards the village of Brora and the North Sea. Currently, Clynelish is working a 7-day week producing 4.8 million litres of alcohol. The distillery will be closed from January to September 2016 due to the replacement of the mashtun and other major upgrades of the equipment. Some 6,000 casks of Clynelish are stored in the two old Brora warehouses next door but most of the production is matured elsewhere. There is also one cask of Brora single malt which was brought back to the distillery for visitors to see during the tour. Most of the old Brora equipment (stills, washbacks and mashtun) is still left in the listed buildings.

Clynelish is the signature malt in Johnnie Walker Gold Label Reserve, a no age statement version of the previous 18 year old expression which has now been phased out. The Clynelish single malt sells around 100,000 bottles per year. Official bottlings include a **14 year old** and a **Distiller's Edition** with an Oloroso Seco finish. There is also an **American oak cask strength** (57.3%) exclusively available at the distillery shop. In 2014, Clynelish **Select Reserve** was launched as part of the annual Special Releases. It carried no age statement but the selected casks were all at least 15 years old.

History:

1819 The 1st Duke of Sutherland founds a distillery called Clynelish Distillery.

1827 The first licensed distiller, James Harper, files for bankruptcy and John Matheson takes over.

1846 George Lawson & Sons become new licensees.

1896 James Ainslie & Heilbron takes over.

1912 James Ainslie & Co. narrowly escapes bankruptcy and Distillers Company Limited (DCL) takes over together with James Risk.

1916 John Walker & Sons buys a stake of James Risk's stocks.

1931 The distillery is mothballed.

1939 Production restarts.

1960 The distillery becomes electrified.

1967 A new distillery, also named Clynelish, is built adjacent to the first one.

1968 'Old' Clynelish is mothballed in August.

1969 'Old' Clynelish is reopened as Brora and starts using a very peaty malt.

1983 Brora is closed in March.

2002 A 14 year old is released.

2006 A Distiller's Edition 1991 finished in Oloroso casks is released.

2009 A 12 year old is released for Friends of the Classic Malts.

2010 A 1997 Manager's Choice single cask is released.

2014 Clynelish Select Reserve is released.

14 years old

Tasting notes Clynelish 14 year old:

GS – A nose that is fragrant, spicy and complex, with candle wax, malt and a whiff of smoke. Notably smooth in the mouth, with honey and contrasting citric notes, plus spicy peat, before a brine and tropical fruit finish.

Cragganmore

[crag•an•moor]

Owner:	**Region/district:**
Diageo	Speyside
Founded: **Status:**	**Capacity:**
1869 Active (vc)	2 200 000 litres

Address: Ballindalloch, Moray AB37 9AB

Website:	**Tel:**
malts.com	01479 874700

Cragganmore single malt has always been known as one of the more complex malts coming from Speyside. The reason behind this is to be found in the distillation process.

The style of the newmake is characterised as heavy and sulphury. The large size of the wash stills could indicate a delicate spirit but sharply descending lyne arms allow for little reflux. The spirit stills are considerably smaller with boil balls and long, slightly descending lyne arms, giving more copper contact and a lighter character. The spirit finally ends up in worm tubs which, by definition, gives a robust flavour and even more so at Cragganmore, where the worms are only 15 metres but very thick.

The interesting flavour profile of the single malt has given character to the Old Parr blend for many years. This brand, established by the Greenlees brothers in 1909, is what the owners, Diageo, call a local star with a strong presence in a few markets as opposed to global giants such as Johnnie Walker and J&B. The main markets are to be found in Mexico, Colombia and Venezuela and the brand has been a successful climber on the Top 20 list with a sales increase of 250% during the last decade.

The distillery is equipped with a 6.8 tonnes stainless steel full lauter mash tun with a copper canopy. There are also six washbacks made of Oregon pine with a 60 hour fermentation time and two pairs of stills. The two spirit stills are most peculiar with flat tops, which had already been introduced during the times of the founder, John Smith. The distillery is doing 16 mashes per week which means a production of 2.2 million litres during 2015.

Cragganmore is one of the original Classic Malts and the core range of single malts is made up of a **12 year old** and a **Distiller's Edition** with a finish in Port pipes. In autumn 2014, a **25 year old** appeared as a Special Release and a **triple matured expression** without age statement was launched for sale at the visitor centre, as well as for Friends of the Classic Malts.

History:

1869 John Smith, who already runs Glenfarclas distillery, founds Cragganmore.

1886 John Smith dies and his brother George takes over operations.

1893 John's son Gordon, at 21, is old enough to assume responsibility for operations.

1901 The distillery is refurbished and modernized with help of the famous architect Charles Doig.

1912 Gordon Smith dies and his widow Mary Jane supervises operations.

1917 The distillery closes.

1918 The distillery reopens and Mary Jane installs electric lighting.

1923 The distillery is sold to the newly formed Cragganmore Distillery Co. where Mackie & Co. and Sir George Macpherson-Grant of Ballindalloch Estate share ownership.

1927 White Horse Distillers is bought by DCL which thus obtains 50% of Cragganmore.

1964 The number of stills is increased from two to four.

1965 DCL buys the remainder of Cragganmore.

1988 Cragganmore 12 years becomes one of six selected for United Distillers' Classic Malts.

1998 Cragganmore Distillers Edition Double Matured (port) is launched for the first time.

2002 A visitor centre opens in May.

2006 A 17 year old from 1988 is released.

2010 Manager's Choice single cask 1997 and a limited 21 year old are released.

2014 A 25 year old is released.

12 years old

Tasting notes Cragganmore 12 years old:

GS – A nose of sherry, brittle toffee, nuts, mild wood smoke, angelica and mixed peel. Elegant on the malty palate, with herbal and fruit notes, notably orange. Medium in length, with a drying, slightly smoky finish.

Craigellachie

[craig•<u>ell</u>•ack•ee]

Owner:	**Region/district:**
John Dewar & Sons	Speyside
(Bacardi)	

Founded:	**Status:**	**Capacity:**
1891	Active	4 100 000 litres

Address: Aberlour, Banffshire AB38 9ST

Website:	**Tel:**
craigellachie.com	01340 872971

The whisky boom in the late 19th century put pressure on the big, blending companies. Stocks were running low due to the rapidly increased demand and building or buying a distillery of their own, became necessary.

Peter Mackie, admittedly, already owned Lagavulin, but the sales of his White Horse blend increased so rapidly that he needed another distillery and, together with Alexander Edward, he built Craigellachie. He established a sales office in London but became annoyed with the competition he got from the large producers of grain whisky which, according to him, sold "young, cheap, fiery whisky". During his entire life he was an advocate for quality products and his opinion was crystal clear. "What the public wants is age and plenty of it". In 1909 he fought for a minimum age of three years, but it took until 1969 before it became law.

Craigellachie distillery is equipped with a modern Steinecker full lauter mash tun, installed in 2001, which replaced the old, open cast iron mash tun. There are also eight washbacks made of larch with a fermentation time of 56-60 hours and two pairs of stills. Both stills are attached to worm tubs. The old cast iron tubs were exchanged for stainless steel in 2014 and the existing copper worms were moved to the new tubs. Production during 2015 will be 21 mashes per week and 4.1 million litres of alcohol.

Most of the production goes into Dewar's blends and a 14 year old has, until now, been the only official bottling available. Hence it was even more gratifying that Craigellachie single malt was re-launched in a major way in autumn 2014.

No less than three new expressions (**13, 17 and 23 year old**) were released for selected domestic markets and a **19 year old** was launched for duty free. There is also a **31 year old** which will probably be available in autumn 2015. All the new single malts are un chill-filtered, without colouring and bottled at either 43% or 46%.

History:

1891 The distillery is built by Craigellachie–Glenlivet Distillery Company which has Alexander Edward and Peter Mackie as part-owners.

1898 Production does not start until this year.

1916 Mackie & Company Distillers Ltd takes over.

1924 Peter Mackie dies and Mackie & Company changes name to White Horse Distillers.

1927 White Horse Distillers are bought by Distillers Company Limited (DCL).

1930 Administration is transferred to Scottish Malt Distillers (SMD), a subsidiary of DCL.

1964 Refurbishing takes place and two new stills are bought, increasing the number to four.

1998 United Distillers & Vintners (UDV) sells Craigellachie together with Aberfeldy, Brackla and Aultmore and the blending company John Dewar & Sons to Bacardi Martini.

2004 The first bottlings from the new owners are a new 14 year old which replaces UDV's Flora & Fauna and a 21 year old cask strength from 1982 produced for Craigellachie Hotel.

2014 Three new bottlings for domestic markets (13, 17 and 23 years) and one for duty free (19 years) are released.

13 years old

Tasting notes Craigellachie 13 years old:

GS – Savoury on the early nose, with spent matches, green apples and mixed nuts. Malt join the nuts and apples on the palate, with sawdust and very faint smoke. Drying, with cranberries, spice and more subtle smoke.

Dailuaine

[dall•yoo•an]

Owner: Diageo	**Region/district:** Speyside
Founded: 1852	**Status:** Active	**Capacity:** 5 200 000 litres

Address: Carron, Banffshire AB38 7RE

Website: malts.com	**Tel:** 01340 872500

Only one out of ten bottles of Scotch is sold as single malt. The rest consists of blended whisky. It is therefore not difficult to understand the important role the single malt plays in every bottle of blended Scotch.

It is rarely as obvious as when taking a look at Dailuaine. Almost the entire production is used for blends, but unlike many other distilleries, the distillate from Dailuaine differs from year to year. The character is determined by what the blending team expects to need within a few years. For a long while, the newmake from the distillery was sulphury due to the rare use of condensers made of stainless steel. Later on, the style has shifted between green/grassy and nutty. Since 2014, the distillery has diversified into a fourth character – that of being waxy. Clynelish is the only distillery so far that has accounted for this style which is so important for some blends. A delayed expansion of Clynelish is now solved by extending the fermentation time at Dailuaine to 85 hours and extending the run times in the stills. At the same time they are allowing a waxy residue to build up in the foreshots and feints receiver.

Dailuaine distillery is equipped with a stainless steel full lauter mash tun, eight washbacks made of larch, plus two stainless steel ones placed outside and three pairs of stills. Around 3.9 million litres of alcohol will be produced in 2015 but depending on the chosen style of the spirit, the distillery could do as much as 5.2 million litres. On the site also lies a dark grains plant processing draff and pot ale into cattle feed. To help power the plant and the distillery, Diageo recently invested £6m in a ground breaking bio-energy plant. Through anaerobic digestion, spent lees and waste water from the distillation are converted into biogas, which provides 40% of the electrical demand for the entire site.

The only core bottling is the **16 year old** Flora & Fauna release. Although not confirmed by Diageo at the time of writing, the official registration of a label indicate that a **34 years old** from 1980 could be a part of this year's Special Releases.

History:

1852 The distillery is founded by William Mackenzie.

1865 William Mackenzie dies and his widow leases the distillery to James Fleming, a banker from Aberlour.

1879 William Mackenzie's son forms Mackenzie and Company with Fleming.

1891 Dailuaine-Glenlivet Distillery Ltd is founded.

1898 Dailuaine-Glenlivet Distillery Ltd merges with Talisker Distillery Ltd and forms Dailuaine-Talisker Distilleries Ltd.

1915 Thomas Mackenzie dies without heirs.

1916 Dailuaine-Talisker Company Ltd is bought by the previous customers John Dewar & Sons, John Walker & Sons and James Buchanan & Co.

1917 A fire rages and the pagoda roof collapses. The distillery is forced to close.

1920 The distillery reopens.

1925 Distillers Company Limited (DCL) takes over.

1960 Refurbishing. The stills increase from four to six and a Saladin box replaces the floor maltings.

1965 Indirect still heating through steam is installed.

1983 On site maltings is closed down and malt is purchased centrally.

1991 The first official bottling, a 16 year old, is launched in the Flora & Fauna series.

1996 A 22 year old cask strength from 1973 is released as a Rare Malt.

1997 A cask strength version of the 16 year old is launched.

2000 A 17 year old Manager's Dram matured in sherry casks is launched.

2010 A single cask from 1997 is released.

2012 The production capacity is increased by 25%.

16 years old

Tasting notes Dailuaine 16 years old:

GS – Barley, sherry and nuts on the substan-tial nose, developing into maple syrup. Medium-bodied, rich and malty in the mouth, with more sherry and nuts, plus ripe oranges, fruitcake, spice and a little smoke. The finish is lengthy and slightly oily, with almonds, cedar and slightly smoky oak.

Dalmore

[dal•moor]

Owner:
Whyte & Mackay Ltd
(Emperador Inc)

Region/district:
Northern Highlands

Founded: 1830

Status: Active (vc)

Capacity: 4 000 000 litres

Address: Alness, Ross-shire IV17 0UT

Website: thedalmore.com

Tel: 01349 882362

Seven years ago, sales volumes for Dalmore started falling as a direct consequence of the decision to invest more on a premium range of whiskies in a small number of key markets.

In 2008 figures were as low as 285,000 bottles. The sales have increased ever since and in 2014, more than 1 million bottles of Dalmore single malt were sold. This, however, only seems to be the start of a bright, new future. The new owners since 2014, the Philippine brandy producer Emperador, have decided that the sales of Dalmore and Jura should increase properly. The majority owner and chairman of Emperador, Andrew Tan, has put up a target for Dalmore of 1 million cases in ten years from now. This means an incredible10-dubbling by today's standards!

A completely upgraded visitor centre and shop was opened in summer 2011. It is very elegant and contemporary, with excellent exhibitions showing, for instance, the importance of the wood. One of the best features is the opportunity for the visitor to nose the different steps in the distillation process from the first drops from the wash still down to the feints.

The distillery is equipped with a 9.9 ton stainless steel, semi-lauter mash tun, eight washbacks made of Oregon pine with a fermentation time of 50 hours and four pairs of stills. The spirit stills have water jackets, which allow cold water to circulate between the reflux bowl and the neck of the stills, thus increasing the reflux. The owners expect to do 23 mashes per week during 2015, thus producing 4 million litres. The whisky is normally unpeated, but until 2010, a heavily peated spirit was also produced for use in the blend, Black Watch.

The core range consists of **12, 15, 18, 25 year old** (limited to 3,000 bottles per year), **1263 King Alexander III** and **Cigar Malt**. In 2013 **Dalmore Valour** was released as a duty free exclusive and another three expressions without age statement have been announced for release later in 2015. Dalmore has also become known for some very exclusive bottlings, including the 30 year old **Dalmore Ceti** and the **Constellation Collection**.

History:
1839 Alexander Matheson founds the distillery.
1867 Three Mackenzie brothers run the distillery.
1891 Sir Kenneth Matheson sells the distillery for £14,500 to the Mackenzie brothers.
1917 The Royal Navy moves in to start manufacturing American mines.
1920 The Royal Navy moves out and leaves behind a distillery damaged by an explosion.
1922 The distillery is in production again.
1956 Floor malting replaced by Saladin box.
1960 Mackenzie Brothers (Dalmore) Ltd merges with Whyte & Mackay.
1966 Number of stills is increased to eight.
1982 The Saladin box is abandoned.
1990 American Brands buys Whyte & Mackay.
1996 Whyte & Mackay changes name to JBB (Greater Europe).
2001 Through management buy-out, JBB (Greater Europe) is bought from Fortune Brands and changes name to Kyndal Spirits.
2002 Kyndal Spirits changes name to Whyte & Mackay.
2007 United Spirits buys Whyte & Mackay. A 15 year old, and a 40 year old are released.
2008 1263 King Alexander III and Vintage 1974 are released.
2009 New releases include an 18 year old, a 58 year old and a Vintage 1951.
2010 The Dalmore Mackenzie 1992 Vintage is released.
2011 More expressions in the River Collection and 1995 Castle Leod are released.
2012 The visitor centre is upgraded and Constellaton Collection is launched.
2013 Valour is released for duty free.
2014 Emperador Inc buys Whyte & Mackay.

12 years old

Tasting notes Dalmore 12 years old:
GS – The nose offers sweet malt, orange marmalade, sherry and a hint of leather. Full-bodied, with a dry sherry taste though sweeter sherry develops in the mouth along with spice and citrus notes. Lengthy finish with more spices, ginger, Seville oranges and vanilla.

Dalwhinnie

[dal•whin•nay]

Owner: Diageo   **Region/district:** Northern Highlands

Founded: 1897   **Status:** Active (vc)   **Capacity:** 2 200 000 litres

Address: Dalwhinnie, Inverness-shire PH19 1AB

Website: malts.com   **Tel:** 01540 672219 (vc)

It is impossible to miss the Dalwhinnie distillery when you travel the A9 between Perth and Inverness. One of the most striking features are the worm tubs which were exchanged for new ones in 2015.

Dalwhinnie is one of 18 distilleries in Scotland still using this old technique of cooling the spirit vapours. Worm tubs were first introduced in Scotland in the mid 1500s. Before that, the distillers relied on inefficient air cooling and, later, leading the tube from the still straight through a tub of water. When the worm, a copper spiral emerged in water through which the vapours were lead, was introduced it resulted in a much more efficient cooling and a larger yield. In the second half on the 19th century, distilleries began to use shell and tube condensers which are the most common ones used today. Of the distilleries that have opened up in recent years, only two (Abhainn Dearg and Ballindalloch) have installed worm tubs.

The distillery is equipped with a 7.3 ton full lauter mash tun, six wooden washbacks with a fermentation time of 60 hours and just the one pair of stills attached to worm tubs. The production for 2014 will be 10 mashes per week which gives 1.4 million litres of alcohol in the year. Dalwhinnie is the signature malt in one of the fastest increasing Scotch blends in the world right now – Buchanan's. The brand already made its debut in the 1890s but during the last ten years, volumes have rocketed by 120% to 20 million bottles sold in 2014. The main markets are USA, Mexico and Venezuela and the range is represented by a 12 year old, an 18 year old, Red Seal 21 year old and the most recent expression, Buchanan's Master.

The core range is made up of a **15 year old** and a **Distiller's Edition** with a finish in oloroso casks. A new addition, **Dalwhinnie Winter's Gold**, was released in July 2015. A vatting of both American and European oak, it is designed to be enjoyed straight from the freezer. Although not confirmed by Diageo at the time of writing, the official registration of a label indicate that a **25 years old** from 1989 could be a part of this year's Special Releases.

History:

1897 John Grant, George Sellar and Alexander Mackenzie commence building the facilities. The first name is Strathspey.

1898 The owner encounters financial troubles and John Somerville & Co and A P Blyth & Sons take over and change the name to Dalwhinnie.

1905 Cook & Bernheimer in New York, buys Dalwhinnie for £1,250 at an auction. The administration of Dalwhinnie is placed in the newly formed company James Munro & Sons.

1919 Macdonald Greenlees & Willliams Ltd headed by Sir James Calder buys Dalwhinnie.

1926 Macdonald Greenlees & Williams Ltd is bought by Distillers Company Ltd (DCL) which licences Dalwhinnie to James Buchanan & Co.

1930 Operations are transferred to Scottish Malt Distilleries (SMD).

1934 The distillery is closed after a fire in February.

1938 The distillery opens again.

1968 The maltings is decommissioned.

1987 Dalwhinnie 15 years becomes one of the selected six in United Distillers' Classic Malts.

1991 A visitor centre is constructed.

1992 The distillery closes and goes through a major refurbishment costing £3.2 million.

1995 The distillery opens in March.

1998 Dalwhinnie Distillers Edition 1980 (oloroso) is introduced.

2002 A 36 year old is released.

2006 A 20 year old is released.

2010 A Manager's Choice 1992 is released.

2012 A 25 year old is released.

2014 A triple matured bottling without age statement is released for The Friends of the Classic Malts.

2015 Dalwhinnie Winter's Gold is released.

15 years old

Tasting notes Dalwhinnie 15 years old:

GS – The nose is fresh, with pine needles, heather and vanilla. Sweet and balanced on the fruity palate, with honey, malt and a very subtle note of peat. The medium length finish dries elegantly.

Deanston

[deen•stun]

Owner:	**Region/district:**
Burn Stewart Distillers	Southern Highlands
(Distell Group Ltd)	

Founded:	**Status:**	**Capacity:**
1965	Active (vc)	3 000 000 litres

Address: Deanston, Perthshire FK16 6AG

Website:	**Tel:**
deanstonmalt.com	01786 843010

When South African drinks giant Distell took over Burn Stewart Distillers in 2013, the main reason wasn't to get their hands on the three single malt brands in the portfolio – Deanston, Bunnahabhain and Tobermory.

What really enticed them was the blended Scotch, Scottish Leader. The brand is the most sold Scotch in the important Taiwanese market and also very successful in South Africa. Towards the end of 2014, the owners announced a new packaging for the brand and also a new expression called Signature to complement the existing Original. The new bottling is richer and smokier which means more peated Bunanhabhain and also Ledaig has gone into the recipe. Scottish Leader, created in the 1800s and re-launched in 1970, is selling around 6 million bottles every year.

Deanston distillery is equipped with an 11 ton traditional, open top, cast iron mash tun, eight stainless steel washbacks with an average fermentation time of 80 hours and two pairs of stills with ascending lyne arms. During 2015, the distillery will be operating 24/6, doing 10 mashes per week and producing 2 million litres of alcohol. Starting in 2000, a small part of organic spirit has been produced yearly. The single malt from Deanston has no artificial colouring and it is also un chill-filtered.

The core range is a **12 year old** and **Virgin Oak**. The latter is a non-age statement malt with a finish in virgin oak casks. A third addition to the range appeared in May 2015 when an **18 year old**, matured in hogsheads and finished in first fill bourbon barrels, was released. Exclusive to the American market, an **18 year old** finished in cognac for more than 6 years, was released in 2014. The first version of Organic Deanston is planned for release in late 2015 or 2016 for select markets only. For those travelling to the distillery, there is a reward in the form of some exclusive bottlings – **Spanish Oak, Vintage 1974** and the new **11 year old Marsala finish**. You also have the unique opportunity of filling your own bottle of a **10 year old sherry maturation**.

History:

1965 A weavery from 1785 is transformed into Deanston Distillery by James Finlay & Co. and Brodie Hepburn Ltd Brodie Hepburn also runs Tullibardine Distillery.

1966 Production commences in October.

1971 The first single malt is named Old Bannockburn.

1972 Invergordon Distillers takes over.

1974 The first single malt bearing the name Deanston is produced.

1982 The distillery closes.

1990 Burn Stewart Distillers from Glasgow buys the distillery for £2.1 million.

1991 The distillery resumes production.

1999 C L Financial buys an 18% stake of Burn Stewart.

2002 C L Financial acquires the remaining stake.

2006 Deanston 30 years old is released.

2009 A new version of the 12 year old is released.

2010 Virgin Oak is released.

2012 A visitor centre is opened.

2013 Burn Stewart Distillers is bought by South African Distell Group for £160m

2014 An 18 year old cognac finish is released in the USA.

2015 An 18 year old is released.

12 years old

Tasting notes Deanston 12 years old:

GS – A fresh, fruity nose with malt and honey. The palate displays cloves, ginger, honey and malt, while the finish is long, quite dry and pleasantly herbal.

Dufftown

[duff•town]

Owner:		Region/district:
Diageo		Speyside

Founded:	Status:	Capacity:
1896	Active	6 000 000 litres

Address: Dufftown, Keith, Banffshire AB55 4BR

Website:	Tel:
malts.com	01340 822100
thesingleton.com	

For every major Scotch whisky producer, the master blender is a key factor for its success. Their work is just as essential for single malts as for blends.

Diageo, with a wide range of brands, has numerous blenders. Jim Beveridge is responsible for Johnnie Walker. Caroline Martin looks after Bell´s and J&B, while Maureen Robinson has taken Singleton of Glen Ord to become one of the top five malts in the world. One of the younger talents is Matthew Crow, who has worked alongside Jim Beveridge as an apprentice for many years. Recently, Crow has been responsible for the re-launch of Mortlach and he is now also the caretaker of Singleton of Dufftown.

Dufftown distillery is equipped with a 13 tonnes full lauter mash tun, 12 stainless steel washbacks and three pairs of stills. All stills furthermore have sub coolers. The style of Dufftown single malt is green and grassy which is achieved by a clear wort from the mash tun and long fermentation (75 hours minimum) in the washbacks. Add to that a slow distillation and the fact that the stills are filled with small volumes to allow as much copper contact as possible, you have what gives it its real character. Dufftown has been working 24/7 since 2007 and during 2015 they will be doing 6 million litres of alcohol.

The core range consists of **Singleton of Dufftown 12, 15** and **18 year old**. In 2013 a new sub-range, The Singleton Reserve Collection, exclusive to duty free was launched. The first two releases were **Singleton of Dufftown Trinité** and **Liberté** and yet a third release, **Artisan**, was added in spring of 2014. But new releases didn´t stop there. March 2014 saw the launch of two new Singleton of Dufftown for domestic markets in Western Europe; **Tailfire** with a maturation predominantly in European oak, and **Sunray** which gets its character from toasted ex bourbon casks. The two were followed in autumn 2014 by a new entry level bottling called **Spey Cascade**. In 2013, the first official cask strength bottling from the distillery was launched as part of the Special Releases, the **Singleton of Dufftown 28 year old**.

History:

1895 Peter Mackenzie, Richard Stackpole, John Symon and Charles MacPherson build the distillery Dufftown-Glenlivet in an old mill.

1896 Production starts in November.

1897 The distillery is owned by P. Mackenzie & Co., who also owns Blair Athol in Pitlochry.

1933 P. Mackenzie & Co. is bought by Arthur Bell & Sons for £56,000.

1968 The floor maltings is discontinued and malt is bought from outside suppliers. The number of stills is increased from two to four.

1974 The number of stills is increased from four to six.

1979 The stills are increased by a further two to eight.

1985 Guinness buys Arthur Bell & Sons.

1997 Guinness and Grand Metropolitan merge to form Diageo.

2006 The Singleton of Dufftown 12 year old is launched as a special duty free bottling.

2008 The Singleton of Dufftown is made available also in the UK.

2010 A Manager´s Choice 1997 is released.

2013 A 28 year old cask strength and two expressions for duty free - Unité and Trinité - are released.

2014 Tailfire, Sunray and Spey Cascade are released.

Singleton of Dufftown
Sunray

Tasting notes Dufftown 12 years old:

GS – The nose is sweet, almost violet-like, with underlying malt. Big and bold on the palate, this is an upfront yet very drinkable whisky. The finish is medium to long, warming, spicy, with slowly fading notes of sherry and fudge.

Edradour

[ed•ra•**dow**•er]

Owner:
Signatory Vintage
Scotch Whisky Co. Ltd

Region/district:
Southern Highland

Founded: 1825
Status: Active (vc)
Capacity: 130 000 litroc

Address: Pitlochry, Perthshire PH16 5JP

Website:
edradour.com

Tel:
01796 472095

For many years Edradour distillery was marketed as Scotland's smallest, but lately, at least seven distilleries have started with lower capacities than Edradour.

This doesn't seem to concern the owners. The success of the last few years has created a need for a new and bigger distillery. A planning application was approved in May 2015 for a new distillery which will work in tandem with the existing Edradour. The plan is to replicate the current equipment but also to provide space for two, additional washbacks. The time-scale for the expansion is 2-5 years and the long term goal is to have a capacity of 400,000 litres of alcohol. In the shorter term, a new warehouse complex will be built and is due to be completed sometime during 2015.

The distillery is equipped with an open, traditional cast iron mash tun with a mash size of 1.15 tonnes. The two washbacks are made of Oregon pine and two stills are connected to a more than 100 year old wormtub. In 2015 they will be doing 6 mashes per week and 130,000 litres of alcohol in the ensuing year, of which 26,000 litres will be heavily peated.

The core range consists of the **10 year old**, the **12 year old Caledonia Selection** (oloroso finish), **Cask Strength Sherry 14 year old**, **Cask Strength Bourbon 12 year old** and the recently launched **Fairy Flag**, a 15 year old Oloroso finish. A series of wood finishes under the name **Straight From The Cask** (SFTC), are all bottled at cask strength in various types of casks. The current range is **Burgundy, Marsala, Sherry** (full maturation) and **Barolo**. In addition to this, there is a range of wood maturations, aged 8-12 years; **2003 Ruby Port**, **2006 Super Tuscan** and **2006 Barolo**. The first release of the heavily peated Ballechin was in 2006 and the final bottling in that series was a Sauternes maturation released in 2013. **Ballechin** is now represented by a **10 year old** which was launched in 2014. Limited releases in 2015 were a **1993 Sauternes cask finish** and a **2006 sherry cask**, bottled for the Nepal earthquake disaster relief.

History:
1825 Probably the year when a distillery called Glenforres is founded by farmers in Perthshire.
1837 The first year Edradour is mentioned.
1841 The farmers form a proprietary company, John MacGlashan & Co.
1886 John McIntosh & Co. acquires Edradour.
1933 William Whiteley & Co. buys the distillery.
1982 Campbell Distilleries (Pernod Ricard) buys Edradour and builds a visitor centre.
1986 The first single malt is released.
2002 Edradour is bought by Andrew Symington from Signatory for £5.4 million. The product range is expanded with a 10 year old and a 13 year old cask strength.
2003 A 30 year old and a 10 year old are released.
2004 A number of wood finishes are launched as cask strength.
2006 The first bottling of peated Ballechin is released.
2007 A Madeira matured Ballechin is released.
2008 A Ballechin matured in Port pipes and a 10 year old Edradour with a Sauternes finish are released.
2009 Fourth edition of Ballechin (Oloroso) is released.
2010 Ballechin #5 Marsala is released.
2011 Ballechin #6 Bourbon and a 26 year old PX sherry finish are relased.
2012 A 1993 Oloroso and a 1993 Sauternes finish as well as the 7th edition of Ballechin (Bordeaux) are released.
2013 Ballechin Sauternes is released.
2014 The first release of a 10 year old Ballechin.
2015 Fairy Flag is released.

12 years old

Tasting notes Edradour 10 years old:
GS – Cider apples, malt, almonds, vanilla and honey ar present on the nose, along with a hint of smoke and sherry. The palate is rich, creamy and malty, with a persistent nuttiness and quite a pronounced kick of slightly leathery sherry. Spices and sherry dominate the medium to long finish.

Fettercairn

[fett•er•cairn]

Owner:	**Region/district:**
Whyte & Mackay (Emperador)	Eastern Highlands
Founded:	**Status:**	**Capacity:**
1824	Active (vc)	3 200 000 litres

Address: Fettercairn, Laurencekirk, Kincardineshire AB30 1YB

Website:	**Tel:**
fettercairndistillery.co.uk	01561 340205

It seems that Fettercairn and the rest of Whyte & Mackay acquired a prosperous company as new owners in summer 2014, when Emperador Inc., the world's largest producer of brandy, took over.

The Philippine company announced in spring 2015 that, in spite of spending $800m on acquisitions the previous year, they would pre-pay their loans to become debt-free. At the same time, they have made an offer to buy Louis Royer cognac from Beam Suntory. Albeit on a smaller scale, Emperador will also be investing in their Scottish distilleries. In summer 2015, a new mill and another three washbacks will be installed at Fettercairn, which means that its capacity could increase to 3.2 million litres.

Fettercairn distillery is equipped with a traditional cast iron mash tun with a copper canopy, eight washbacks made of Douglas fir and two pairs of stills. One feature makes it unique among Scottish distilleries – when collecting the middle cut, cooling water is allowed to trickle along the outside of the spirit still necks and is collected at the base for circulation towards the top again. This is done in order to increase reflux and thereby produce a lighter and cleaner spirit. For some time, the owners have produced around 200,000 litres of heavily peated spirit (55ppm), but there has been no peated production in the last couple of years. There are 14 dunnage warehouses on site and the oldest whisky dates back to 1965. Two quarter casks and three hogsheads from 1965 were re-racked into a sherry butt in 2003.

The core range consists of **Fettercairn Fior** without age statement and three older whiskies – **24, 30** and **40 year old**. The greatest part of Fior comprises of whiskies aged 14 and 15 years with an addition of 5 year old peated whisky. In 2011 another whisky without age statement was launched as an exclusive for Tesco. It is called **Fettercairn Fasque** and contains slightly younger whiskies than Fior of which only 5% is peated whisky. There is also a distillery exclusive bottling, a **9 year old**, heavily peated single cask.

History:

1824 Sir Alexander Ramsay founds the distillery.

1830 Sir John Gladstone buys the distillery.

1887 A fire erupts and the distillery is forced to close for repairs.

1890 Thomas Gladstone dies and his son John Robert takes over. The distillery reopens.

1912 The company is close to liquidation and John Gladstone buys out the other investors.

1926 The distillery is mothballed.

1939 The distillery is bought by Associated Scottish Distillers Ltd. Production restarts.

1960 The maltings discontinues.

1966 The stills are increased from two to four.

1971 The distillery is bought by Tomintoul-Glenlivet Distillery Co. Ltd.

1973 Tomintoul-Glenlivet Distillery Co. Ltd is bought by Whyte & Mackay Distillers Ltd.

1974 The mega group of companies Lonrho buys Whyte & Mackay.

1988 Lonrho sells to Brent Walker Group plc.

1989 A visitor centre opens.

1990 American Brands Inc. buys Whyte & Mackay for £160 million.

1996 Whyte & Mackay and Jim Beam Brands merge to become JBB Worldwide.

2001 Kyndal Spirits buys Whyte & Mackay from JBB Worldwide.

2002 The whisky changes name to Fettercairn 1824.

2003 Kyndal Spirits changes name to Whyte & Mackay.

2007 United Spirits buys Whyte & Mackay. A 23 year old single cask is released.

2009 24, 30 and 40 year olds are released.

2010 Fettercairn Fior is launched.

2012 Fettercairn Fasque is released.

Fior

Tasting notes Fettercairn Fior:

GS – A complex, weighty nose of toffee, sherry, ginger, orange and smoke. More orange and smoke on the palate, with a sherried nuttiness and hints of treacle toffee. Mild, spicy oak and a touch of liquorice in the lengthy finish.

Glencadam

[glen•ka•dam]

Owner:
Angus Dundee Distillers

Region/district:
Eastern Highlands

Founded: **Status:**
1826   Active

Capacity:
1 300 000 litres

Address: Brechin, Angus DD9 7PA

Website:
glencadamdistillery.co.uk

Tel:
01356 622217

The first distillery that family-owned Angus Dundee bought was Tomintoul in the year 2000, but it could just as well have been Glencadam.

The owner, Terence Hillman, has had his eyes on Glencadam since 1999, but decided that the capacity was too small for the expansion plans the company had in mind. Hillman, however, couldn't keep his mind off the small distillery based in Brechin and so in 2003 he had put in an offer for that one too. The two distilleries were in good hands. Angus Dundee has a huge range of blended Scotch brands in its portfolio and is represented in more than 70 countries, not least of all China, where The Angus blend has recently been launched. But the single malts from Glencadam have also received a real boost in recent years. The range has expanded and an agreement was concluded last year with Camus Wine & Spirits to distribute the brand in the USA.

At one time there were eight distilleries on the east coast between Aberdeen and Dundee. Two of them, Auchinblae and Glencoull, having already closed in the 1920s and with the big whisky crisis in the mid 1980s, another four were forced to cease production - Glenesk, Glenury Royal, North Port and Lochside. The only two surviving distilleries are Fettercairn and, further to the south in the town of Brechin, Glencadam.

Glencadam distillery is equipped with a traditional cast iron mash tun, six stainless steel washbacks with a fermentation time of 52 hours and one pair of stills. The distillery is currently working seven days a week, which enables 16 mashes per week and 1.3 million litres of alcohol per year. Glencadam is not only a busy distillery, but also hosts a huge filling and bottling plant with 16 large tanks for blending malt and grain whisky.

The core range consists of a **10 year old**, a **14 year old oloroso finish**, a **15 year old** and a **21 year old**. Recent limited editions included a **17 year old port finish** and a **32 year old single cask**. Finally, a **25 year old** has been announced for release by the end of 2015.

History:

1825   George Cooper founds the distillery.

1827   David Scott takes over.

1837   The distillery is sold by David Scott.

1852   Alexander Miln Thompson becomes the owner.

1857   Glencadam Distillery Company is formed.

1891   Gilmour, Thompson & Co Ltd takes over.

1954   Hiram Walker takes over.

1959   Refurbishing of the distillery.

1987   Allied Lyons buys Hiram Walker Gooderham & Worts.

1994   Allied Lyons changes name to Allied Domecq.

2000   The distillery is mothballed.

2003   Allied Domecq sells the distillery to Angus Dundee Distillers.

2005   The new owner releases a 15 year old.

2008   A re-designed 15 year old and a new 10 year old are introduced.

2009   A 25 and a 30 year old are released in limited numbers.

2010   A 12 year old port finish, a 14 year old sherry finish, a 21 year old and a 32 year old are released.

2012   A 30 year old is released.

2015   A 25 year old is launched.

10 years old

Tasting notes Glencadam 10 years old:

GS – A light and delicate, floral nose, with tinned pears and fondant cream. Medium-bodied, smooth, with citrus fruits and gently-spiced oak on the palate. The finish is quite long and fruity, with a hint of barley.

GlenDronach

[glen•dro•nack]

Owner:	**Region/district:**
Benriach Distillery Co	Highlands

Founded:	**Status:**	**Capacity:**
1826	Active (vc)	1 400 000 litres

Address: Forgue, Aberdeenshire AB54 6DB

Website:	**Tel:**
glendronachdistillery.com	01466 730202

Today, GlenDronach single malt may not be amongst the top 30 malts of the world in terms of sales, but there was a short period some decades ago when it was actually in sixth place.

This was important to Billy Walker when he bought the distillery in 2008. The brand already had loyal followers who enjoyed the sherried character and there was a huge potential to expand on this. One drawback which is now beginning to show, is that the distillery was closed between 1996 and 2002. Stocks of the popular 15 year old are almost depleted and it will take a couple of years before the supply can meet the demand.

The distillery equipment consists of a small (3.7 tonnes) cast iron mash tun with rakes, nine washbacks made of larch with a fermentation time of 60 to 90 hours, two wash stills with heat exchangers and two spirit stills. The expectation is to produce 1.2 million litres of alcohol during 2015. Traditionally focused on unpeated malt, the distillery has lately made small volumes of peated spirit and for 2015 it will be around 40,000 litres with a phenol specification in the barley of 38ppm. GlenDronach was the last Scottish distillery to fire the stills with coal, a process which continued until 2005.

The core range is the **12 year Original** (a combination of PX and Oloroso sherry casks), **15 year Revival** (Oloroso-matured), **18 year Allardice** (Oloroso-matured), **21 year Parliament** (a combination of PX and Oloroso sherry casks) and **24 year old Grandeur** (Oloroso-matured). A new addition to the core range was made in June 2015 with the **8 year old The Hielan** (replacing the now discontinued Octarine). The Hielan has been matured predominantly in bourbon casks with an addition of sherry-matured whisky. There are three wood finishes – **Sauternes 12 year old, Marsala 18 year old** and **Tawny Port 18 year old**. The owners released their first **cask strength** expression in 2012 and this was followed by batch number 4 in late 2014. Batch 12 of the **single casks** was released in summer 2015, while the **44 year old Recherché** released in 2013, represents the oldest GlenDronach still left in the warehouses.

History:

1826 The distillery is founded by a consortium with James Allardes as one of the owners.

1837 Parts of the distillery is destroyed in a fire.

1852 Walter Scott (from Teaninich) takes over.

1887 Walter Scott dies and Glendronach is taken over by a consortium from Leith.

1920 Charles Grant buys Glendronach for £9,000.

1960 William Teacher & Sons buys the distillery.

1966 The number of stills is increased to four.

1976 A visitor centre is opened.

1976 Allied Breweries takes over William Teacher & Sons.

1996 The distillery is mothballed.

2002 Production is resumed on 14th May.

2005 The distillery closes to rebuild from coal to indirect firing by steam. Reopens in September. Chivas Brothers (Pernod Ricard) becomes new owner through the acquisition of Allied Domecq.

2008 Pernod Ricard sells the distillery to the owners of BenRiach distillery.

2009 Relaunch of the whole range - 12, 15 and 18 year old including limited editions of a 33 year old and five single casks.

2010 A 31 year old, a 1996 single cask and a total of 11 vintages and four wood finishes are released. A visitor centre is opened.

2011 The 21 year old Parliament and 11 vintages are released.

2012 A number of vintages are released.

2013 Recherché 44 years and a number of new vintages are released.

2014 Nine different single casks are released.

2015 The Hielan, 8 years old, is released.

12 years old Original

Tasting notes GlenDronach 12 years old:

GS – A sweet nose of Christmas cake fresh from the oven. Smooth on the palate, with sherry, soft oak, fruit, almonds and spices. The finish is comparatively dry and nutty, ending with bitter chocolate.

Glendullan

[glen•dull•an]

Owner: Diageo

Region/district: Speyside

Founded: 1897

Status: Active

Capacity: 5 000 000 litres

Address: Dufftown, Keith, Banffshire AB55 4DJ

Website: www.malts.com

Tel: 01340 822100

Glendullan distillery is one of the seven, original distilleries of Dufftown and the last of them to open. The distillery that we see today, however, is not the original.

The current plant was built in 1972 and the two distilleries were working parallel to each other for 13 years until 1985, when the old one was closed. The whisky from the two distilleries was vatted together before bottling. The old distillery, which was equipped with one pair of stills and had a capacity of one million litres, is now used as a workshop for Diageo´s distillery engineering team. Glendullan distillery is situated just a one minute`s drive east of Glenfiddich near a river which, despite of the distillery`s name isn´t Dullan, but Fiddich. The confluence of the two rivers lies just a mile to the south of Glendullan.

The distillery is equipped with a 12 ton full lauter stainless steel mash tun, 8 wooden washbacks made of larch and two made of stainless steel with a fermentation time of 75 hours to promote a green/grassy character of the whisky as well as three pairs of stills. In 2014 the distillery will be doing 21 mashes per week, producing 5 million litres of alcohol.

Until 2007, Glendullan single malt was just sold as a 12 year old in the Flora & Fauna range. That year it was re-launched, together with Dufftown and Glen Ord, under the name Singleton with The Singleton of Glendullan being targeted for the American market. The core bottling is **Singleton of Glendullan 12 year old**. In summer 2013 a new sub-range, The Singleton Reserve Collection, exclusive to duty free was launched. All three Singleton distilleries (Glendullan, Dufftown and Glen Ord) were blessed with new expressions and for Glendullan it was the **Liberty** and **Trinity** – both bottled at 40%. Finally, in autumn 2014, a Glendullan single malt was, for the first time, launched as part of the Special Releases – a **38 year old** distilled in 1975, matured in re-fill European oak and bottled at 59,8%.

History:

1896 William Williams & Sons, a blending company with Three Stars and Strahdon among its brands, founds the distillery.

1902 Glendullan is delivered to the Royal Court and becomes the favourite whisky of Edward VII.

1919 Macdonald Greenlees buys a share of the company and Macdonald Greenlees & Williams Distillers is formed.

1926 Distillers Company Limited (DCL) buys Glendullan.

1930 Glendullan is transferred to Scottish Malt Distillers (SMD).

1962 Major refurbishing and reconstruction.

1972 A brand new distillery is constructed next to the old one and both operate simultaneously during a few years.

1985 The oldest of the two distilleries is mothballed.

1995 The first launch of Glendullan in the Rare Malts series is a 22 year old from 1972.

2005 A 26 year old from 1978 is launched in the Rare Malts series.

2007 Singleton of Glendullan is launched in the USA.

2013 Singleton of Glendullan Liberty and Trinity are released for duty free.

2014 A 38 year old is released.

12 years old

Tasting notes Singleton of Glendullan 12 years:

GS – The nose is spicy, with brittle toffee, vanilla, new leather and hazelnuts. Spicy and sweet on the smooth palate, with citrus fruits, more vanilla and fresh oak. Drying and pleasingly peppery in the finish.

Glen Elgin

[glen el•gin]

Owner:	**Region/district:**
Diageo	Speyside
Founded: **Status:**	**Capacity:**
1898 Active	2 700 000 litres

Address: Longmorn, Morayshire IV30 3SL

Website:	**Tel:**
malts.com	01343 862100

Founded in 1898, Glen Elgin was one of the last distilleries to be built in the 19th century and it was the first to start production in the 20th century.

The two partners, William Simpson and James Crale, did not get off to a good start. The financing was insufficient and the effects of the great whisky crash caused by the Pattison brothers, left the project reeling with a myriad of problems. The founders were unable to pay the contractors and the opening was therefore delayed until 1st of May 1900. The whisky business had still not recovered and the source of water also started to dry up because the Coleburn distillery a few kilometres upstream had used the same water source. The distillery was sold, eventually closed and did not reopen again until 1906. It has, however, gone significantly better since then. The whisky that was produced quickly got a good reputation among blenders and Glen Elgin has, for a long time, been a signature malt in the blend White Horse, a major brand especially in Japan. The strong position in that segment of the market inspired Diageo to make a special bottling by producing a single malt named White Horse Glen Elgin.

The distillery is equipped with an 8.2 tonnes Steinecker full lauter mash tun from 2001, nine washbacks made of larch and six small stills. Three of the washbacks were installed as late as 2012 in the extended tun room, which meant that the production capacity had increased by 50%. Since the site had progressed from a 5 day operation to 7 days, the fermentation time went from a combination of short and longs to an even 90 hours for all. The stills are connected to six wooden worm tubs where the spirit vapours are condensed. A new boiler was installed in December 2014, replacing the two, old existing ones. During 2015, they will be doing 16 mashes per week for the first six months and from July, 12 mashes which translates to a production of 1.9 million litres.

The only official bottling is a **12 year old**, but older expressions (up to 32 years) have been released in limited numbers during the last decade.

History:

1898 The former manager of Glenfarclas, William Simpson and banker James Carle found Glen Elgin.

1900 Production starts in May but the distillery closes just five months later.

1901 The distillery is auctioned for £4,000 to the Glen Elgin-Glenlivet Distillery Co. and is mothballed.

1906 The wine producer J. J. Blanche & Co. buys the distillery for £7,000 and production resumes.

1929 J. J. Blanche dies and the distillery is put up for sale again.

1930 Scottish Malt Distillers (SMD) buys it and the license goes to White Horse Distillers.

1964 Expansion from two to six stills plus other refurbishing takes place.

1992 The distillery closes for refurbishing and installation of new stills.

1995 Production resumes in September.

2001 A 12 year old is launched in the Flora & Fauna series.

2002 The Flora & Fauna series malt is replaced by Hidden Malt 12 years.

2003 A 32 year old cask strength from 1971 is released.

2008 A 16 year old is launched as a Special Release.

2009 Glen Elgin 1998, a single cask in the new Manager´s Choice range is released.

12 years old

Tasting notes Glen Elgin 12 years old:

GS – A nose of rich, fruity sherry, figs and fragrant spice. Full-bodied, soft, malty and honeyed in the mouth. The finish is lengthy, slightly perfumed, with spicy oak.

Glenfarclas

[glen•fark•lass]

Owner: J. & G. Grant	**Region/district:** Speyside
Founded: 1836	**Status:** Active (vc)	**Capacity:** 3 500 000 litres

Address: Ballindalloch, Banffshire AB37 9BD

Website: glenfarclas.co.uk

Tel: 01807 500257

Since 1865, when John Grant took over the lease of Rechlerich farm and bought Glenfarclas, this distillery has been in the hands of the same family.

At the moment it is run by the 5th generation through the Chairman, John Grant and the 6th generation, George Grant, who is working as Director of Sales. Many years ago, the stills in all of Scotland's distilleries would have been directly fired, i. e. with a flame beneath the bottom of the still, using either peat, coal, oil or gas as fuel. Over time this regime was replaced by steam coils and pans being built into the stills and only Macallan, Glenfiddich, Glenfarclas and Springbank (one of the stills) continued in the traditional manner. For some time now, Macallan has heated all their stills with steam and Glenfiddich has moved to indirect firing in one of their still houses. The chances of Glenfarclas following are highly unlikely. They tried doing it in 1981 but stopped when it changed the character of the whisky. This says a lot about the family-owned distillery where traditions mean a lot simply because it works.

The distillery is equipped with a 16.5 tonnes semi-lauter mash tun and twelve stainless steel washbacks with a fermentation time of 48 hours. There are three pairs of stills and the wash stills are equipped with rummagers. This is a copper chain rotating at the bottom of the still to prevent solids from sticking to the copper, something that otherwise easily happens with a direct fired still. There are 34 dunnage warehouses on-site with the capacity of holding 65,000 casks. In 2015, the distillery will produce 7 days a week for 43 weeks, resulting in 3.5 million litres – a record for Glenfarclas. Glenfarclas single malt has always been highly ranked among whisky aficionados and close to 800,000 bottles are sold annually.

The Glenfarclas core range consists of the **10, 12, 15, 21, 25, 30** and **40 year old**, as well as the **105 Cask Strength**. There is also a **17 year old** destined for the USA, Japan and Sweden. Also in the core range, is the lightly sherried **Glenfarclas Heritage** without age statement. It was originally launched for the French hypermarket trade but is now also available in other countries. An **18 year old** exclusive to travel retail was launched in 2014. Limited releases in 2014 included the oldest whisky ever released by the owners. It had been maturing for **60 years** (and thereby eclipsed the 58 year old single cask which was released in 2012) and only 360 bottles became available. Slightly younger was the single cask **1966** which was launched in September 2014. The fact that it came from a fino sherry cask and not oloroso, made it unusual. This was followed up in 2015 with the **1956 Sherry Cask** and there will be other editions in this series in the future. Another new release this year, was a bottling to celebrate the 150th anniversary of the Grant family taking over Glenfarclas. It was named **£511.19s.0d Family Reserve**, being the amount paid for the distillery by John Grant on 8 June 1865. The owners also continue to release bottlings in their **Family Casks** series with vintages from 1954 to 1999.

History:

1836 Robert Hay founds the distillery on the original site since 1797.

1865 Robert Hay passes away and John Grant and his son George buy the distillery. They lease it to John Smith at The Glenlivet Distillery.

1870 John Smith resigns in order to start Cragganmore and J. & G. Grant Ltd takes over.

1889 John Grant dies and George Grant takes over.

1890 George Grant dies and his widow Barbara takes over the license while sons John and George control operations.

1895 John and George Grant take over and form The Glenfarclas-Glenlivet Distillery Co. Ltd with the infamous Pattison, Elder & Co.

1898 Pattison becomes bankrupt. Glenfarclas encounters financial problems after a major overhaul of the distillery but survives by mortgaging and selling stored whisky to R. I. Cameron, a whisky broker from Elgin.

1914 John Grant leaves due to ill health and George continues alone.

1948 The Grant family celebrates the distillery's 100th anniversary, a century of active licensing. It is 9 years late, as the actual anniversary coincided with WW2.

1949 George Grant senior dies and sons George Scott and John Peter inherit the distillery.

1960 Stills are increased from two to four.

1968 Glenfarclas is first to launch a cask-strength single malt. It is later named Glenfarclas 105.

1972 Floor maltings is abandoned and malt is purchased centrally.

1973 A visitor centre is opened.

1976 Enlargement from four stills to six.

History continued:

2001 Glenfarclas launches its first Flower of Scotland gift tin which becomes a great success and increases sales by 30%.

2002 George S Grant dies and is succeeded as company chairman by his son John L S Grant

2003 Two new gift tins are released (10 years old and 105 cask strength).

2005 A 50 year old is released to commemorate the bi-centenary of John Grant´s birth

2006 Ten new vintages are released.

2007 Family Casks, a series of single cask bottlings from 43 consecutive years, is released.

2008 New releases in the Family Cask range. Glenfarclas 105 40 years old is released.

2009 A third release in the Family Casks series.

2010 A 40 year old and new vintages from Family Casks are released.

2011 Chairman´s Reserve and 175th Anniversary are released.

2012 A 58 year old and a 43 year old are released.

2013 An 18 year old for duty free is released as well as a 25 year old quarter cask.

2014 A 60 year old and a 1966 single fino sherry cask are released.

2015 A 1956 Sherry Cask and Family Reserve are released.

105 Cask Strength

Family Reserve

Tasting notes Glenfarclas 10 year old:

GS – Full and richly sherried on the nose, with nuts, fruit cake and a hint of citrus fruit. The palate is big, with ripe fruit, brittle toffee, some peat and oak. Medium length and gingery in the finish.

18 years old   Family Cask 1959

10 years old   Heritage   40 years old

Glenfiddich

[glen•fidd•ick]

Owner: **Region/district:**
William Grant & Sons   Speyside

Founded: **Status:**   **Capacity:**
1886   Active (vc)   14 000 000 litres

Address: Dufftown, Keith, Banffshire AB55 4DH

Website: **Tel:**
glenfiddich.com   01340 820373 (vc)

What everyone deemed impossible just a few years ago became reality in 2014 – Glenfiddich lost its spot as the world´s most sold single malt, a position the brand has held at least since 1963.

It was in 1963 that Glenfiddich Straight Malt was introduced, a launch that would not only lead to Glenfiddich's leading position, but would also off-set the global interest in single malts. After Glenfiddich in 2011 became the first single malt to sell more than 1 million cases (12 million bottles) in a year, volumes have increased negligibly. Official figures from IWSR now show that Glenlivet sold 12,760,000 bottles in 2014 which is 200,000 more than Glenfiddich.

In autumn 2014, Glenfiddich introduced a unique opportunity for dedicated fans to select and purchase limited single cask expressions. Malt Master, Brian Kinsman, has selected 36 different casks from 1958 to 1996 and bottled some of the content. They are all presented in detail on their website in the Glenfiddich Gallery and prices range from £300 to £100,000. There is also the possibility to get suggestions of bottles that will suit you best by answering four questions determining your preferred flavour profile. To top it all, after you have selected a bottle, you can design your own personal label and box. Included in the range, apart from some very old bottlings, are a couple of unusual expressions such as virgin oak 1982, rum cask 1992 and madeira finish 1993.

Glenfiddich distillery is equipped with two, stainless steel, full lauter mash tuns – both with a 10 ton mash. There are 24 Douglas fir washbacks and eight stainless steel washbacks which were installed in late 2012, all with a fermentation time of 68 hours. There are two still rooms with a total of 11 wash stills and 20 spirit stills. Three of them were added in autumn 2014 but did not add to the capacity, instead it had the desired effect of making the production smoother. The capacity is limited by the mash tuns. The production for 2015 will be 68 mashes per week and 13.5 million litres of pure alcohol.

The core range consists of **12, 15, 18, 21** (also called Gran Reserva and finished in rum casks), **Rich Oak 14 year old** and **15 year old Distillery Edition** (bottled at 51%). Included in the core range, although limited, are **Malt Master´s Edition** and the **26 year old Glenfiddich Excellence**. Some older expressions have also been released over the years – **30, 40** and **50 years old**, as well as the **38 year old Glenfiddich Ultimate**. A new range for Duty Free called **Age of Discovery** was introduced in 2011. So far, three expressions have been released; **Madeira cask, Bourbon cask** and **Red Wine cask finish**. The focus on duty free continued in 2013 when the range Cask Collection was launched with **Select Cask**, a vatting from bourbon, European oak and California red wine casks and **Reserve Cask**, matured in sherry casks were introduced. They were followed in 2014 by **Vintage Cask** made using a proportion of peated barley. Later in 2014 **Glenfiddich Rare Oak 25 years** was launched for duty free.

History:

1886 The distillery is founded by William Grant, 47 years old, who had learned the trade at Mortlach Distillery. The equipment is bought from Mrs. Cummings of Cardow Distillery. The construction totals £800.

1887 The first distilling takes place on Christmas Day.

1892 William Grant builds Balvenie.

1898 The blending company Pattisons, largest customer of Glenfiddich, files for bankruptcy and Grant decides to blend their own whisky. Standfast becomes one of their major brands.

1903 William Grant & Sons is formed.

1957 The famous, three-cornered bottle is introduced.

1958 The floor maltings is closed.

1963 Glennfiddich becomes the first whisky to be marketed as single malt in the UK and the rest of the world.

1964 A version of Standfast's three-cornered bottle is launched for Glenfiddich in green glass.

1969 Glenfiddich becomes the first distillery in Scotland to open a visitor centre.

1974 16 new stills are installed.

2001 1965 Vintage Reserve is launched in a limited edition of 480 bottles. Glenfiddich 1937 is bottled (61 bottles).

2002 Glenfiddich Gran Reserva 21 years old, finished in Cuban rum casks is launched. Caoran Reserve 12 years is released. Glenfiddich Rare Collection 1937 (61 bottles) is launched and becomes the oldest Scotch whisky on the market.

2003 1973 Vintage Reserve (440 bottles) is launched.

2004 1991 Vintage Reserve (13 years) and 1972 Vintage Reserve (519 bottles) are launched.

History continued:

2005 Circa £1.7 million is invested in a new visitor centre.

2006 1973 Vintage Reserve, 33 years (861 bottles) and 12 year old Toasted Oak are released.

2007 1976 Vintage Reserve, 31 years is released in September.

2008 1977 Vintage Reserve is released.

2009 A 50 year old and 1975 Vintage Reserve are released.

2010 Rich Oak, 1978 Vintage Reserve, the 6th edition of 40 year old and Snow Phoenix are released.

2011 1974 Vintage Reserve and a 19 year old Madeira finish are released.

2012 Cask of Dreams and Millenium Vintage are released.

2013 A 19 year old red wine finish and 1987 Anniversary Vintage are released. Cask Collection with three different expressions is released for duty free.

2014 The 26 year old Glenfiddich Excellence and Rare Oak 25 years are released.

Select Cask   21 years old   Reserve Cask

Tasting notes Glenfiddich 12 year old:

GS – Delicate, floral and slightly fruity on the nose. Well mannered in the mouth, malty, elegant and soft. Rich, fruit flavours dominate the palate, with a developing nuttiness and an elusive whiff of peat smoke in the fragrant finish.

12 years old   Rich Oak   Distillery Edition

Glen Garioch

[glen gee•ree]

Owner:	**Region/district:**
Morrison Bowmore (Suntory)	Eastern Highlands
Founded: **Status:**	**Capacity:**
1797 Active (vc)	1 370 000 litres

Address: Oldmeldrum, Inverurie, Aberdeenshire
AB51 0ES

Website:	**Tel:**
glengarioch.com	01651 873450

Glen Garioch, situated in the small town of Old-meldrum west of Aberdeen, is the easternmost distil-lery in Scotland and one of the oldest.

Generally it is said to have been founded in 1797, but some evidence suggests that it might already have been operational by 1785. It is also one of the few distilleries in Scotland where you can drive by in the street and look right onto the stills just a couple of metres away.

The owners, Morrison Bowmore, have a range of five single malts called McClelland´s, which represents different whisky re-gions in Scotland. Over the last four years, sales of McClelland´s have risen by more than 30% and one of them, McClelland´s Highland, is a young Glen Garioch. The single malt is typically unpeated, but smoky notes can easily be detected in expressions distilled before 1994 when their own floor maltings closed. Until then, the malt was peated with a specification of 8-10ppm.

The distillery is equipped with a 4,4 tonne full lauter mash tun, eight stainless steel washbacks with a fermentation time of 48 hours and one pair of stills. There is also a third still, which has not been used for a long time. The spirit is tankered to Glasgow, filled into casks and returned to the distillery´s four warehouses. During 2015 the production will be 15 mashes per week and almost 1,2 million litres in the year.

The core range is the **1797 Founder´s Reserve** (without age statement) and a **12 year old**, both of them bottled at 48% and un chill-filtered. There have also been a number of limited cask strength vintages which have been released over the years – the latest, a **Vintage 1999**, in 2013. Other limited releases include **Virgin Oak** – the first Glen Garioch fully matured in virgin Ameri-can white oak – and a **15 year old Renaissance**. The latter was the first in a new range called Glen Garioch Renaissance Collection, with the second release due in October 2015. A sherry-matured **15 year old** was also laun-ched in 2015 for the duty free market.

History:

1797 Thomas Simpson founds Glen Garioch.

1837 The distillery is bought by John Manson & Co., owner of Strathmeldrum Distillery.

1908 Glengarioch Distillery Company, owned by William Sanderson, buys the distillery.

1933 Sanderson & Son merges with the gin maker Booth's Distilleries Ltd.

1937 Booth´s Distilleries Ltd is acquired by Distillers Company Limited (DCL).

1968 Glen Garioch is decommissioned.

1970 It is sold to Stanley P. Morrison Ltd.

1973 Production starts again.

1978 Stills are increased from two to three.

1994 Suntory controls all of Morrison Bowmore Distillers Ltd.

1995 The distillery is mothballed in October.

1997 The distillery reopens in August.

2004 Glen Garioch 46 year old is released.

2005 15 year old Bordeaux Cask Finish is launched. A visitor centre opens in October.

2006 An 8 year old is released.

2009 Complete revamp of the range - 1979 Founders Reserve (unaged), 12 year old, Vintage 1978 and 1990 are released.

2010 1991 vintage is released.

2011 Vintage 1986 and 1994 are released.

2012 Vintage 1995 and 1997 are released.

2013 Virgin Oak, Vintage 1999 and 11 single casks are released.

2014 Glen Garioch Renaissance Collection 15 years is released.

2015 The second release of the Rennaisance Collection as well as a 15 year old duty free bottling are launched.

Tasting notes Glen Garioch 12 years old:

GS – Luscious and sweet on the nose, peaches and pineapple, vanilla, malt and a hint of sherry. Full-bodied and nicely textured, with more fresh fruit on the palate, along with spice, brittle toffee and finally dry oak notes.

12 years old

1863 Phylloxera arrives in France

In 1863, the wine-makers in the village of Pujaut in Languedoc, discovered a blight that was causing damage to their vines.

The disease soon spread to vineyards all over France, but it was not until 1868 that botanist, Jule Émile Planchon, discovered what caused the blight. Examining the roots of the affected vines, he found what he first thought were lice, sucking sap from the plant. After examination, it was determined that it was an aphid which had come from American vine brought to France. American vine was resistant to the attacks of the aphid which was named grape phylloxera. It spread quickly across Europe and it took until 1888 before a remedy was found. By grafting the resistant American vine to the susceptible European Vitis vinifera, a new varfiety was created that could withstand the attacks of phylloxera. By that time, it was estimated that 40% of the French vineyards had been destroyed and the effect on the French economy was catastrophic.

Now, what has this to do with Scotch whisky? Cognac or brandy had already been exported to England in the 1600s and during the 18th century, cognac houses were established in France by the English and the Irish, many of whom are still dominant players, like Hennessy and Hine. Then in 1860, a commercial treaty between England and France was signed, which for instance reduced the duty on imported wines and spirits from France and the import of cognac exploded. A few years after that, sales in England had increased to 65 million bottles. Phylloxera did not strike in the Cognac region until 1871 but, when it did, it was a tremendous blow to the Cognac business. What had become the favourite tipple of the English middle classes, soon became expensive and sometimes unobtainable. In search of a replacement, the consumers then discovered Scotch whisky. With the customers having been used to cognac being slightly sweeter, it was the Speyside whiskies that were the ultimate winners at the expense of the more pungent style produced in Campbeltown and on Islay. The blenders saw that a high percentage of Speyside malt in their blends, created a flavour that was appropriate for the period of time and just small amounts of peated whisky were needed. The consumption of Scotch whisky on home grounds had dropped during the 1880s due to a bad economy, but come the 90s with the shortage of cognac, the demand rose again quickly and no less than 33 new distilleries were built during that last decade of the 19th century. Exports of whisky from Scotland to England had increased by more than 200% from 1888 to 1900 and several of the producers expanded their business to other markets across the world – as far afield as Australia, South America and the USA. This was indeed the first golden era for Scotch whisky.

The decline of the cognac markets had a further debilitating effect on the Scottish whisky industry. Apart from turning to Scotch, many customers discovered sherry and the import of sherry in casks from Jerez in Spain to England grew quickly. This, in turn, led to the fact that England was overloaded with sherry casks which the whisky producers soon began to use for maturation. Sherry wood continued to dominate the market until the late 1930s when bourbon barrels began to be used. Sherry consumption started to decrease, as did the volume of empty casks and prices started to increase. Today, nine out of ten casks come from the American bourbon industry.

The cognac industry eventually recovered from the phylloxera attack and, today, it is a thriving business producing one of the most recognised spirits in the world. It is interesting to note, however, that France, being the origin of cognac, has been the largest market in the world for Scotch whisky for many years.

Glenglassaugh

[glen•glass•ock]

Owner: **Region/district:**
Glenglassaugh Distillery Co   Highlands
(BenRiach Distillery Co.)

Founded: **Status:**   **Capacity:**
1875   Active (vc)   1 100 000 litres

Address: Portsoy, Banffshire AB45 2SQ

Website: **Tel:**
glenglassaugh.com   01261 842367

The history of Glenglassaugh distillery clearly follows the ebb and flow of Scotch whisky. It opened during whisky's first golden era, around the end of the 1800s.

Blending of whisky had become a fashion and the malt was used for famous brands such as Teachers. It was then closed for more or less 50 years when two World Wars and the prohibition in the USA slowed down sales of Scotch. When it was re-opened in 1960, the industry was in the midst of its second golden era. Blended Scotch ruled the world and Glenglassaugh could be found in Famous Grouse and Cutty Sark. The market deflated in the mid 1980s and Glenglassaugh, together with around 20 other distilleries, was forced to close again. After 22 years, and well into the third golden era, new owners have taken over and the emphasis is now on single malt.

The equipment of the distillery consists of a Porteus cast iron mash tun with rakes, four wooden washbacks and two stainless steel ones and one pair of stills. After BenRiach took over the production, the pace has increased considerably. During the first year the production increased from 200,000 to 600,000 litres and for 2015 they have planned for 800,000 litres of pure alcohol of which 40,000 litres will be peated (30ppm). The main part (85%) is filled to be used as single malt while the rest is sold externally.

The core range is **Revival**, a 3 year old with a 6 month Oloroso finish, followed by **Evolution**, slightly older, matured in American oak and bottled at 50% and then there is **Torfa** which is peated (20ppm), matured in bourbon casks, bottled at 50% and without age statement. Limited releases include **30, 40** and **51 year old**, as well as two bottlings with a finish in wine casks from the famous Crimean Massandra winery – a **35 year old madeira** and a **41 year old sherry**. True to their bottling regime for BenRiach and GlenDronach, the owners released the first batch of **single cask Glenglassaugh** in 2014 and followed it up with a second batch in 2015.

History:

1873 The distillery is founded by James Moir.

1887 Alexander Morrison embarks on renovation work.

1892 Morrison sells the distillery to Robertson & Baxter. They in turn sell it on to Highland Distilleries Company for £15,000.

1908 The distillery closes.

1931 The distillery reopens.

1936 The distillery closes.

1957 Reconstruction takes place.

1960 The distillery reopens.

1986 Glenglassaugh is mothballed.

2005 A 22 year old is released.

2006 Three limited editions are released - 19 years old, 38 years old and 44 years old.

2008 The distillery is bought by the Scaent Group for £5m. Three bottlings are released - 21, 30 and 40 year old.

2009 New make spirit and 6 months old are released.

2010 A 26 year old replaces the 21 year old.

2011 A 35 year old and the first bottling from the new owners production, a 3 year old, are released.

2012 A visitor centre is inaugurated and Glenglassaugh Revival is released.

2013 BenRiach Distillery Co buys the distillery and Glenglassaugh Evolution and a 30 year old are released.

2014 The peated Torfa is released as well as eight different single casks and Massandra Connection (35 and 41 years old).

2015 The second batch of single casks is released.

Torfa

Tasting notes Glenglassaugh Evolution:

GS – Peaches and gingerbread on the nose, with brittle toffee, icing sugar, and vanilla. Luscious soft fruits dipped in caramel figure on the palate, with coconut and background stem ginger. The finish is medium in length, with spicy toffee.

Glengoyne

[glen•goyn]

Owner:
Ian Macleod Distillers

Region/district:
Southern Highlands

Founded: 1833
Status: Active (vc)
Capacity: 1 100 000 litres

Address: Dumgoyne by Killearn, Glasgow G63 9LB

Website:
glengoyne.com

Tel:
01360 550254 (vc)

According to the Scotch Whisky Association, a record number of visitors of 1.5 million, travelled to Scotland´s distilleries during 2014.

The key to the rising interest is the ability of distilleries to offer a variety of tours at different levels. One of the forerunners is Glengoyne which yearly hosts over 60,000 visitors (with only two distilleries that have more). There are no fewer than seven different tours to choose from, including a whisky and chocolate pairing, as well as a master class with the possibility of creating your own personal Glengoyne single malt. The latest addition is a Maturation Experience where you can learn about different casks and how time affects the flavour of the whisky.

Glengoyne has a magnificent location in the scenic region of Trossarchs, which is situated at the base of Dumgoyne Hill and right on the border between the Lowlands and the Highlands. As a matter of fact, the A81 road with the distillery on the one side and the warehouses on the other, is the dividing line.

The distillery is equipped with a 3.8 ton semi lauter mash tun. There are also six Oregon pine washbacks, as well as the rather unusual combination of one wash still and two spirit stills. Both short (56 hours) and long (110 hours) fermentations are practised. In 2015, the production will be split between 12 and 16 mashes per week which constitutes 920,000 litres of alcohol. Since the owners took over the distillery in 2003, sales have increased by 150% to 480,000 bottles in 2014.

The core range of Glengoyne single malts consists of **10, 12, 15, 18, 21** and **25 year old**. The latter was launched in spring 2014 and is bottled at a higher strength (48%) than the rest (40-43%). There is also a **cask strength** (currently at 58.7%) without age statement. In 2013 a very limited quantity of a **35 year old** was released. The line-up for duty free is a **15 year old Distiller´s Gold** and a **First Fill 25 year old**. A distillery exclusive bottling is also available, a **26 year old single cask**.

History:

1833 The distillery is licensed under the name Burnfoot Distilleries by the Edmonstone family.

1876 Lang Brothers buys the distillery and changes the name to Glenguin.

1905 The name changes to Glengoyne.

1965 Robertson & Baxter takes over Lang Brothers and the distillery is refurbished. The stills are increased from two to three.

2001 Glengoyne Scottish Oak Finish (16 years old) is launched.

2003 Ian MacLeod Distillers Ltd buys the distillery plus the brand Langs from the Edrington Group for £7.2 million.

2005 A 19 year old, a 32 year old and a 37 year old cask strength are launched.

2006 Nine "choices" from Stillmen, Mashmen and Manager are released.

2007 A new version of the 21 year old, two Warehousemen´s Choice, Vintage 1972 and two single casks are released.

2008 A 16 year old Shiraz cask finish, three single casks and Heritage Gold are released.

2009 A 40 year old, two single casks and a new 12 year old are launched.

2010 Two single casks, 1987 and 1997, released.

2011 A 24 year old single cask is released.

2012 A 15 and an 18 year old are released as well as a Cask Strength with no age statement.

2013 A limited 35 year old is launched.

2014 A 25 year old is released.

15 years old

Tasting notes Glengoyne 15 years old:

GS – A nose of vanilla, ginger, toffee, vintage cars leather seats, and sweet fruit notes. The somewhat oily palate features quite lively spices, raisins, hazelnuts, and oak. The finish is medium in length and spicy to the end, with cocoa powder.

Glen Grant

[glen grant]

Owner:
Campari Group Speyside

Region/district:

Founded: **Status:**   **Capacity:**
1840   Active (vc)   6 200 000 litres

Address: Elgin Road, Rothes, Banffshire AB38 7BS

Website: **Tel:**
glengrant.com   01340 832118

The Scotch whisky history is inundated with strong individuals whose character and entrepreneurship helped create an industry that would later take the world by storm.

The two brothers, James and John Grant, who founded Glen Grant distillery, are definitely amongst them, but the younger James Grant, who inherited the distillery from his uncle John, was probably the most colourful of them all. He took over the distillery in 1872 and ran it for almost 60 years until his death in 1931. James was an officer in the Rifle Volunteers, advanced to the rank of major and was always referred to, even after his retirement from the army, as The Major. He had a great interest in technology and in 1883 Glen Grant became the first distillery in Scotland to install electricity for the purpose of lighting. He also actively took part in the building of the Strahtspey railway which ran through Rothes. As a true Victorian gentleman, he was also a keen hunter and often went on expeditions to India and Africa. During one of his travels in Africa, he found an abandoned boy who he brought home to Rothes to become his page boy. Biawa Makalaga continued to live in Glen Grant house long after the Major had passed away, until he himself died in 1972.

The international breakthrough for Glen Grant single malt came in 1961 when the Italian, Armando Giovinetti, visited the distillery. He was so impressed by the product that he persuaded Douglas Mackessak, the Major's grandson, to appoint him agent for Glen Grant in Italy. Glen Grant soon became the best selling malt whisky in Italy and still remains so today. Global sales have declined in the last couple of decades, but as many as 3.3 million bottles were nevertheless sold during 2014.

The distillery is equipped with a 12.3 tonnes semi-lauter mash tun, 10 Oregon pine washbacks with a minimum fermentation time of 48 hours and four pairs of stills. The wash stills are peculiar in that they have vertical sides at the base of the neck and all eight stills are fitted with purifiers. This gives an increased reflux and creates a light and delicate whisky. A new, extremely efficient £5m bottling hall was inaugurated in 2013. It has a capacity of 12,000 bottles an hour and Glen Grant is the only one of the larger distillers bottling the entire production on site. During 2015, whisky will be produced for a total of two thirds of the year, while the same operators will be working in the bottling hall for the remainder of the year. In litres of alcohol, this translates to 2.5 million in 2015.

The Glen Grant core range of single malts consists of **Major's Reserve** with no age statement, a **5 year old** sold in Italy only, a **10 year old**, as well as a **16 year old**. Recent limited editions include a **25 year old** from sherry butts in 2011 and **Five Decades** which was launched in 2013 to celebrate distillery manager, Dennis Malcolm's 53 years in the industry. The oldest official bottling of Glen Grant (at a cost of £8,000) was a **50 year old** which was released in Hong Kong in March of 2014.

History:

1840 The brothers James and John Grant, managers of Dandelaith Distillery, found the distillery.

1861 The distillery becomes the first to install electric lighting.

1864 John Grant dies.

1872 James Grant passes away and the distillery is inherited by his son, James junior (Major James Grant).

1897 James Grant decides to build another distillery across the road; it is named Glen Grant No. 2.

1902 Glen Grant No. 2 is mothballed.

1931 Major Grant dies and is succeeded by his grandson Major Douglas Mackessack.

1953 J. & J. Grant merges with George & J. G. Smith who runs Glenlivet distillery, forming The Glenlivet & Glen Grant Distillers Ltd.

1961 Armando Giovinetti and Douglas Mackessak found a friendship that eventually leads to Glen Grant becoming the most sold malt whisky in Italy.

1965 Glen Grant No. 2 is back in production, but renamed Caperdonich.

1972 The Glenlivet & Glen Grant Distillers merges with Hill Thompson & Co. and Longmorn-Glenlivet Ltd to form The Glenlivet Distillers. The drum maltings ceases.

History continued:

1973 Stills are increased from four to six.

1977 The Chivas & Glenlivet Group (Seagrams) buys Glen Grant Distillery. Stills are increased from six to ten.

2001 Pernod Ricard and Diageo buy Seagrams Spirits and Wine, with Pernod acquiring the Chivas Group.

2006 Campari buys Glen Grant for €115m.

2007 The entire range is re-packaged and re-launched and a 15 year old single cask is released. Reconstruction of the visitor centre.

2008 Two limited cask strengths - a 16 year old and a 27 year old - are released.

2009 Cellar Reserve 1992 is released.

2010 A 170th Anniversary bottling is released.

2011 A 25 year old is released.

2012 A 19 year old Distillery Edition is released.

2013 Five Decades is released and a bottling hall is built.

2014 The Rothes Edition 10 years old is released.

16 years old   The Rothes Edition   Five Decades
                                    10 years old

Tasting notes Glen Grant 10 year old:

GS – Relatively dry on the nose, with cooking apples. Fresh and fruity on the palate, with a comparatively lengthy, malty finish, which features almonds and hazelnuts.

10 years old   The Major´s Reserve

Glengyle

[glen•gajl]

Owner:
Mitchell's Glengyle Ltd

Region/district:
Campbeltown

Founded: **Status:**
2004   Active

Capacity:
750 000 litres

Address: Glengyle Road, Campbeltown,
Argyll PA28 6LR

Website:
kilkerran.com

Tel:
01586 551710

The most prosperous time for Campbeltown was in the Victorian days when shipbuilding, fishing and coal mining made most of the inhabitants quite prosperous.

This was also the time when at least 25 whisky distilleries were operating in this little town with, today, only 5,000 people. The downturn came during the first world war and by 1934, when Rieclachan closed, only two distilleries were left – Glen Scotia and Springbank. Seventy years later the number grew to three when Glengyle (also closed in 1925) was re-opened by the owners of Springbank.

The distillery is equipped with a 4 tonnes semilauter mash tun, four washbacks made of larch with a fermentation time of 80-100 hours and one pair of stills. Malt is obtained from the neighbouring Springbank whose staff also runs operations. Its capacity is 750,000 litres, but considerably smaller amounts have been produced during the years. The plan for 2015 is only to produce 30,000 litres.

Although the bottlings so far haven't revealed it, there is a lot of experimentation going on at Glengyle where different wood maturations, peated barley and even quadruple distillation are being used. The idea is to make limited releases from these experiments once the first 12 year old has been released in summer 2016. The name Kilkerran is used for the whisky, as Glengyle was already in use for a vatted malt produced by Loch Lomond Distillers.

In anticipation of the first core 12 year old, the owners have since 2009 released a limited edition called **Work in Progress**, which will show-case what the distillery is capable of. In June 2015 it was time for the 11 year old Kilkerran **Work in Progress 7** and, just like the last couple of years, two versions were released – one matured in **bourbon** casks and the other in **sherry** casks. For the sherry version there were 12,000 bottles and the bourbon matured 6,000. Usually bottled at 46%, this year's bourbon expression was bottled at cask strength (54.1%).

History:

1872 The original Glengyle Distillery is built by William Mitchell.

1919 The distillery is bought by West Highland Malt Distilleries Ltd.

1925 The distillery is closed.

1929 The warehouses (but no stock) are purchased by the Craig Brothers and rebuilt into a petrol station and garage.

1941 The distillery is acquired by the Bloch Brothers.

1957 Campbell Henderson applies for planning permission with the intention of reopening the distillery.

2000 Hedley Wright, owner of Springbank Distillery and related to founder William Mitchell, acquires the distillery.

2004 The first distillation after reconstruction takes place in March.

2007 The first limited release - a 3 year old.

2009 Kilkerran "Work in progress" is released.

2010 "Work in progress 2" is released.

2011 "Work in progress 3" is released.

2012 "Work in progress 4" is released.

2013 "Work in progress 5" is released and this time in two versions - bourbon and sherry.

2014 "Work in progress 6" is released in two versions - bourbon and sherry

2015 "Work in progress 7" is released in two versions - bourbon and sherry

Tasting notes Kilkerran WiP VII (Bourbon):

GS – Creamy malt, marzipan and vanilla on the nose, with pears, linseed and a hint of brine. Soft and slightly oily on the warm, spicy palate, with ripe apples, caramel and slight smokiness. Long and slowly drying, with a touch of aniseed and oak.

Tasting notes Kilkerran WiP VII (Sherry):

GS – The nose offers fragrant wood fires, old leather, damp tweed and autumn berries. The palate is rich and rounded, with zesty spice, then peaches in syrup and sweet smoke. The spices persist. The finish is long and warming with chilli and worn leather.

Work in Progress VII Bourbon

1905 the "What is Whisky" inquiry

The definition of Scotch whisky today is governed by the Scotch Whisky Regulation which was adopted in 2009.

This regulation, in turn, is based on other laws which have emerged during the 1900s, but is also based on industry practice over the years. The Regulation stipulates a number of strict terms that have to be met regarding raw material, production, maturation and labeling. Prior to 1909, however, there were no such rules. Instead, the Sale of Food and Drugs Act of 1875 was used in those cases where it was suspected that a certain product didn't maintain sufficient quality. Such a case appeared in London in 1905 when guests had been complaining about bad whisky that had been served at local pubs. Two whisky retailers were heard by Snow Fordham, the local magistrate in Islington Borough, and it turned out that the Irish whiskey and the Scotch whisky that the guests had been served, had been made of 90% grain spirit which was made less than a year before. Fordham gave judgment in favour of the complainants and fined the two pub owners £100 each, but it was his second ruling in the case that would stir up the whole whisky industry. He stated that whisky, whether malt, grain or blended, must be made in a pot still, otherwise it could not be called whisky.

At this time, the industry had become divided into two groups, with the grain distillers in the Lowlands on one side and the distillers in the Highlands who commonly produced malt whisky in pot stills on the other. For the latter, the ruling was encouraging since they considered their business threatened by the big companies in the south

who were able to produce huge quantities of grain whisky in their column stills. The grain distillers and blending houses, on the other hand, claimed that most English consumers preferred either the softer blend of malt and grain whiskies or indeed a single grain with its pure taste. They even launched advertising campaigns where they marketed grain spirit under the slogan "not a headache in a gallon", implicating that pure malt whisky was too powerful to be enjoyed on its own. Distillers Company Limited (DCL), an association of grain distillers, appealed the ruling made by Fordham, but the appeal court couldn't reach an agreement, so no decision was taken. In 1908 a Royal Commission was appointed by the Parliament instead, and a year later they reached a decision. The outcome was a victory for the blenders and grain distillers – in that all spirits made from either malted or unmalted grain and not just barley and distilled in either pot stills or column stills, were to be considered whisky. The demand by the malt whisky producers that there should be at least be a law regulating the amount of malt whisky in a blend and that the whisky should be matured for at least two years, was left without action.

Both sides of the whisky industry quickly buried the hatchet since a greater threat had appeared that needed cohesion. The PrimeMinister, Lloyd George, introduced his "People´s Budget" of 1909 where he, by way of a severe increase of the duty on spirit, wanted to decrease the consumption of whisky which, he felt, hurt the British population and the economy.

The decisions made after what became known as the "What is Whisky" inquiry in Islington really saved the day for blended whisky, which would later become the cornerstone on which the Scotch whisky industry was to be built. Even if the pot distillers of the Highlands temporarily felt hard done in by, they would eventually realise the perks of an exploding interest for blended Scotch which demanded increased production of their own malt whisky.

Glen Keith

[glen <u>keeth</u>]

Owner:
Chivas Brothers
(Pernod Ricard)

Region/district:
Speyside

Founded:
1957

Status:
Active

Capacity:
6 000 000 litres

Address: Station Road, Keith, Banffshire AB55 3BU

Website:
-

Tel:
01542 783042

When Seagram's (owners at the time of Chivas Brothers) founded Glen Keith in 1957, it was in order to resolve a number of challenges that the company was facing.

They had already taken ownership of Chivas Regal, a well-known deluxe blend, as well as Strathisla, a distillery dating back to the 1700s. Sam Bronfman, the owner of Seagram's, now wanted to add a cheaper, but easy-drinking whisky to the repertoire. He had seen the success that J&B and Cutty Sark had in the USA with its light flavour, but Strathisla single malt had a stronger character and was further needed for Chivas Regal. At first he tried to acquire Cutty Sark from Berry Brothers & Rudd but the chairman, Hugh Rudd, rejected the offer with the announcement: "I have something you want, but you have nothing I need." At the beginning of the 1960s, he instead created the 100 Pipers blend. In order to make it really light, he used triple distillation at Glen Keith for a period of time. During certain times in the 1970s, in the absence of its own distillery on Islay, they also distilled peated whisky for use in other whiskies. This smoky variety has on a few occasions been released by independent bottlers as single malts under the names of Craigduff and Glenisla.

Following 13 years, where production at Glen Keith had ground to a standstill, Chivas Brothers started to reignite the work at the distillery in spring 2012. The old Saladin maltings were demolished and part of that area now holds a new building with a Briggs 8 tonnes full lauter mash tun and six stainless steel washbacks. In the old building there are nine new washbacks made of Oregon pine and six, old but refurbished stills. The stills have extremely long lyne arms and the desired character of the new make spirit is fruity. The distillery was re-opened in April 2013 and now has the capacity to do an impressive 6 million litres with the possibility of producing 40 mashes per week.

The only current, official bottling is a **cask strength** distilled in **1995** and bottled in 2014 which is available at Chivas' visitor centres.

History:

1957 The Distillery is founded by Chivas Brothers (Seagrams).

1958 Production starts.

1970 The first gas-fuelled still in Scotland is installed, the number of stills increases from three to five.

1976 Own maltings (Saladin box) ceases.

1983 A sixth still is installed.

1994 The first official bottling, a 10 year old, is released as part of Seagram's Heritage Selection.

1999 The distillery is mothballed.

2001 Pernod Ricard takes over Chivas Brothers from Seagrams.

2012 The reconstruction and refurbishing of the distillery begins.

2013 Production starts again.

1995 19 years old

Tasting notes Glen Keith 19 years old:

GS – Malt, cereal, figs and gingery banana on the nose. Smooth and viscous on the palate, white pepper, sweet sherry, fudge, and Madeira cake. Slighlty drying in the lengthy finish.

Glenkinchie

[glen•kin•chee]

Owner: Diageo		**Region/district:** Lowlands
Founded: 1837	**Status:** Active (vc)	**Capacity:** 2 500 000 litres

Address: Pencaitland, Tranent, East Lothian EH34 5ET

Website: malts.com	**Tel:** 01875 342004

Unlike many of the distilleries in the Highlands, there's nothing wild and rugged about Glenkinchie's surroundings.

Just 15 miles from Edinburgh, this is a well-maintained agricultural landscape and has so been for almost 2000 years. Some of the best barley in Britain is cultivated here. This part of Scotland is called Lothian after Lot, the brother-in-law of King Arthur, who reigned here.

Glenkinchie is one of the original six Classic Malts and also the one that sells the least (around 250,000 bottles per year). The single malt isn't associated with one particular blend like Cardhu (Johnnie Walker), Strathmill (J&B) or Blair Athol (Bell's). Instead, it is part of several of Diageo's blended brands and the company has many. Apart from the three already mentioned (which are also the best-selling) there are a further seven blends, each selling more than 5 million bottles every year (Buchanan's, Old Parr, White Horse, Windsor, Black & White, VAT 69 and Haig).

Glenkinchie is equipped with a full lauter mash tun (9 tonnes) and six wooden washbacks. There is only one pair of stills but they are, on the other hand, very big – in fact, the wash still (30,963 litres) is the biggest in Scotland. Steeply descending lyne arms give very little reflux and condensation of the spirit vapours take place in a cast iron worm tub. Since 2008, the distillery has been working 7 days and 14 mashes per week which amounts to 2,5 million litres of alcohol per year. Three dunnage warehouses on site have 10,000 casks maturing, the oldest dating from 1952. The proximity to Edinburgh is one reason why more than 40,000 visitors find their way to the distillery and its excellent visitor centre each year.

The core range consists of a **12 year old** and a **Distiller's Edition** with a finish in amontillado sherry casks. There is also a cask strength without age statement which is sold exclusively at the visitor centre. Recent limited editions include a **Manager's Choice 1992** single cask and a **20 year old cask strength**, both released in 2010.

History:

1825 A distillery known as Milton is founded by John and George Rate.

1837 The Rate brothers are registered as licensees of a distillery named Glenkinchie.

1853 John Rate sells the distillery to a farmer by the name of Christie who converts it to a sawmill.

1881 The buildings are bought by a consortium from Edinburgh.

1890 Glenkinchie Distillery Company is founded. Reconstruction and refurbishment is on-going for the next few years.

1914 Glenkinchie forms Scottish Malt Distillers (SMD) with four other Lowland distilleries.

1939-
1945 Glenkinchie is one of few distilleries allowed to maintain production during the war.

1968 Floor maltings is decommissioned.

1969 The maltings is converted into a museum.

1988 Glenkinchie 10 years becomes one of selected six in the Classic Malt series.

1998 A Distiller's Edition with Amontillado finish is launched.

2007 A 12 year old and a 20 year old cask strength are released.

2010 A cask strength exclusive for the visitor centre, a 1992 single cask and a 20 year old are released.

Tasting notes Glenkinchie 12 years old:

GS – The nose is fresh and floral, with spices and citrus fruits, plus a hint of marshmallow. Notably elegant. Water releases cut grass and lemon notes. Medium-bodied, smooth, sweet and fruity, with malt, butter and cheesecake. The finish is comparatively long and drying, initially rather herbal.

12 years old

Glenlivet

[glen•liv•it]

Owner: **Region/district:**
Chivas Brothers   Speyside
(Pernod Ricard)

Founded: **Status:**   **Capacity:**
1824   Active (vc)   10 500 000 litres

Address: Ballindalloch, Banffshire AB37 9DB

Website: **Tel:**
theglenlivet.com   01340 821720 (vc)

In 2014 there was an accession of a new monarch in the whisky world which would have been hard to anticipate only five years ago. Glenlivet is now the best selling single malt in the world – a spot reserved for Glenfiddich from 1963 at least.

According to figures from the IWSR, Glenlivet sold 12,768,000 bottles in 2014, which are almost 200,000 more than Glenfiddich. The growth shown by Glenlivet over the past five years is nothing less than astonishing with an increase of 72%! Glenlivet is also the number one single malt in the important USA market.

Following a substantial expansion in 2010 the distillery is equipped with a Briggs mash tun with six arms (13.5 ton capacity) and 16 wooden washbacks. There are seven pairs of stills, three of which are lined up in a beautiful, new stillhouse with a stunning view of Glen of the Livet. The plan for 2015 is to do 43 mashes per week which equates to 10.5 million litres of alcohol. The bright prospects for Glenlivet single malt, has strengthened the owners´ belief in that a further expansion of the capacity is inevitable. In 2013, a planning application was submitted to the council and in 2014 it was approved. The size of the expansion is astounding. With no time frame announced, the plan is to construct two new distillery units on the back of the existing warehouses. Each unit will consist of one mash tun, 16 washbacks and seven pairs of stills. When the entire expansion is ultimately completed, it would mean that Glenlivet´s capacity would have tripled to well over 30 million litres!

Until spring 2015, Glenlivet's core range used to be **12 year old, 15 year old French Oak Reserve, 18 year old, 21 year old Archive** and **Glenlivet XXV.** Now it has been complemented by **Founder´s Reserve** without age statement, which will be replacing the 12 year old, at least in "mature" markets. A special range of un chill-filtered whiskies called Nadurra, comes in four expressions; **Nadurra Oloroso Cask Strength** and **Nadurra First Fill Selection Cask Strength** for domestic markets and the same two, bottled at **48%** and available in duty free. The owners have decided to expand the Nadurra range into a series of what they call different cask experiences and more versions will follow.

From the current duty free range of First Fill Sherry Cask, a 12 year old, a 15 year old, an 18 year old Batch Reserve and a **Master Distiller´s Reserve,** only the latter will remain. The other four were replaced in July 2015 by **Master Distiller´s Reserve, Solera Vatted** and **Master Distiller´s Reserve Small Batch,** both without age statement. In autumn 2014, the inaugural bottling in a new range, **The Winchester Collection,** was launched. It was a very limited (only 100 bottles) **50 year old** matured in a hogshead. Alan Winchester, with 40 years in the industry, is Glenlivet´s Master Distiller, as well as Distilling Manager for Chivas Brothers, responsible for all the distilleries in the company. Finally, **Glenlivet Carn Mor,** one of Glenlivet´s yearly single cask editions was released in spring 2015 in Hong Kong.

History:
1817 George Smith inherits the farm distillery Upper Drummin from his father Andrew Smith who has been distilling on the site since 1774.
1840 George Smith buys Delnabo farm near Tomintoul and leases Cairngorm Distillery.
1845 George Smith leases three other farms, one of which is situated on the river Livet and is called Minmore.
1846 William Smith develops tuberculosis and his brother John Gordon moves back home to assist his father.
1858 George Smith buys Minmore farm and obtains permission to build a distillery.
1859 Upper Drummin and Cairngorm close and all equipment is brought to Minmore which is renamed The Glenlivet Distillery.
1864 George Smith cooperates with the whisky agent Andrew P. Usher and exports the whisky with great success.
1871 George Smith dies and his son John Gordon takes over.
1880 John Gordon Smith applies for and is granted sole rights to the name The Glenlivet.
1890 A fire breaks out and some of the buildings are replaced.
1896 Another two stills are installed.
1901 John Gordon Smith dies.
1904 John Gordon's nephew George Smith Grant takes over.
1921 Captain Bill Smith Grant, son of George Smith Grant, takes over.
1953 George & J. G. Smith Ltd merges with J. & J. Grant of Glen Grant Distillery and forms the company Glenlivet & Glen Grant Distillers.
1966 Floor maltings closes.
1970 Glenlivet & Glen Grant Distillers Ltd merges with Longmorn-Glenlivet Distilleries Ltd and Hill Thomson & Co. Ltd to form The Glenlivet Distillers Ltd.

History continued:

1978 Seagrams buys The Glenlivet Distillers Ltd. A visitor centre opens.

1996 The visitor centre is expanded, and a multimedia facility installed.

2000 French Oak 12 years and American Oak 12 years are launched

2001 Pernod Ricard and Diageo buy Seagram Spirits & Wine. Pernod Ricard thereby gains control of the Chivas group.

2004 This year sees a lavish relaunch of Glenlivet. French Oak 15 years replaces the previous 12 year old.

2005 Two new duty-free versions are introduced – The Glenlivet 12 year old First Fill and Nadurra. The 1972 Cellar Collection (2,015 bottles) is launched.

2006 Nadurra 16 year old cask strength and 1969 Cellar Collection are released. Glenlivet sells more than 500,000 cases for the first time in one year.

2007 Glenlivet XXV is released.

2000 Four more stills are installed and the capacity increases to 8.5 million litres. Nadurra Triumph 1991 is released.

2010 Another two stills are commissioned and capacity increases to 10.5 million litres. Glenlivet Founder's Reserve is released.

2011 Glenlivet Master Distiller's Reserve is released for the duty free market.

2012 1980 Cellar Collection is released.

2013 The 18 year old Batch Reserve and Glenlivet Alpha are released.

2014 Nadurra Oloroso, Nadurra First Fill Selection and The Glenlivet Guardian's Chapter are released.

2015 Founder's Reserve is released as well as two new expressions for duty free; Solera Vatted and Small Batch.

Tasting notes Glenlivet 12 year old:

GS – A lovely, honeyed, floral, fragrant nose. Medium-bodied, smooth and malty on the palate, with vanilla sweetness. Not as sweet, however, as the nose might suggest. The finish is pleasantly lengthy and sophisticated.

Tasting notes Glenlivet Founder's Reserve:

GS – The nose is fresh and floral, with ripe pears, pineapple, tangerines, honey and vanilla. Medium-bodied, with ginger nuts, soft toffee and tropical fruit on the smooth palate. Soft spices and lingering fruitiness in the finish.

Master Distiller's Reserve Solera Vatted

Master Distiller's Reserve

Master Distiller's Reserve Small Batch

21 years old Archive

Nadurra Oloroso

Founder's Reserve

15 years old

12 years old

Glenlossie

[glen•loss•ay]

Owner:	**Region/district:**
Diageo	Speyside
Founded: **Status:**	**Capacity:**
1876 Active	3 700 000 litres

Address: Birnie, Elgin, Morayshire IV30 8SS

Website:	**Tel:**
malts.com	01343 862000

The strong ties between Glenlossie and Haig´s blended whisky dates back to 1919 when John Haig & Co, themselves being a part of DCL, became the new owners.

Glenlossie has ever since provided a key malt for Haig Gold Label, which in the 1970s was the first spirits brand to sell one million cases (12 million bottles) in the UK alone. Haig´s is a renowned brand and the Haig family stands out among the real pioneers within the Scotch whisky fraternity. In 1667 Robert Haig was already distilling and over the years, the family has been involved in more than 20 different distilleries. Their bonds to another whisky dynasty, the Stein family, were strong, not least by marriage but, at the same time, they were fierce competitors. John Haig founded the Cameronbridge grain distillery and used this as basis to establish the Haig blended whisky. His four sons, in turn, founded Haig & Haig with the intention of expanding the business in America. They also created their own deluxe blend, Dimple, and this, as well as Haig Gold Label, is still available today.

The distillery is equipped with one stainless steel full lauter mash tun (8 tonnes) installed in 1992 and eight washbacks made of larch. There are also plans to install a further two stainless steel, exterior washbacks sometime in the future. There are three pairs of stills with the spirit stills equipped with purifiers between the lyne arms and the condensers, thus increasing the reflux which, together with the 75-80 hour fermentation time, gives Glenlossie newmake its light and green/grassy character. In 2013, the distillery moved from a 5-day to a 7-day week production, enabling them to do 17 mashes per week and 3 million litres of alcohol per year. For the year of 2015, they have gone back to a 5-day production with 12 mashes and 2 million litres for the year. Next to Glenlossie lies the much younger Mannochmore distillery and except for the two distilleries, a dark grains plant and a newly constructed bio-plant, the site also holds fourteen warehouses that can store 250,000 casks of maturing whisky.

The only official bottling of Glenlossie available today is a **10 year old**. During 2010 a first fill bourbon cask distilled in **1999** was released as a part of the **Manager´s Choice** range.

History:

1876 John Duff, former manager at Glendronach Distillery, founds the distillery. Alexander Grigor Allan (to become part-owner of Talisker Distillery), the whisky trader George Thomson and Charles Shirres (both will co-found Longmorn Distillery some 20 years later with John Duff) and H. Mackay are also involved in the company.

1895 The company Glenlossie-Glenlivet Distillery Co. is formed. Alexander Grigor Allan passes away.

1896 John Duff becomes more involved in Longmorn and Mackay takes over management of Glenlossie.

1919 Distillers Company Limited (DCL) takes over the company.

1929 A fire breaks out and causes considerable damage.

1930 DCL transfers operations to Scottish Malt Distillers (SMD).

1962 Stills are increased from four to six.

1971 Another distillery, Mannochmore, is constructed by SMD on the premises. A dark grains plant is installed.

1990 A 10 year old is launched in the Flora & Fauna series.

2010 A Manager´s Choice single cask from 1999 is released.

10 years old

Tasting notes Glenlossie 10 years old:

GS – Cereal, silage and vanilla notes on the relatively light nose, with a voluptuous, sweet palate, offering plums, ginger and barley sugar, plus a hint of oak. The finish is medium in length, with grist and slightly peppery oak.

Books and Magazines

This page doesn´t have the intention of reporting on a comprehensive list of whisky books that are available. That can be found anywhere on the internet. Instead, I would like to put forward two titles that have been published recently and that I have found particularly interesting.

Laphroaig, one of the most iconic distilleries in Scotland celebrated its 200th anniversary in 2015 and this will be acknowledged in October by way of a book written by Hans Offringa and complemented by the work of photographer, Marcel van Gils. The two gentlemen from The Netherlands are not unfamiliar with the distillery. They already published "The Legend of Laphroaig" in 2007, but with their new book "200 Years of Laphroaig" they probe the subject even deeper. Offringa has apparently done his homework when it comes to the previous history of the distillery. Archives have obviously been an important source but the research also includes interviews with descendants of the first owners, the Johnston's, some of whom are today living in Australia and USA. Laphroaig`s path to glory over the past two centuries is described in detail with special reference to Ian Hunter and Bessie Williamson - two of the most prominent owners over the years. One aspect which especially pleased me while I was reading the book, was the portraits of many current employees at the distillery. Offringa is not satisfied with just interviewing the distillery manager, but also includes interviews with maltmen, brewers, warehousemen, technicians and tour guides. It gives the reader a rare chance to meet with many of the people who have worked to create the legendary spirit, but are rarely given recognition in books or magazines. Even retired managers get their share of attention. The photos by van Gils are stunning! Not only does he have an eye for all the beautiful details in a working distillery, but is also a master of capturing the breathtaking scenery and culture of Islay. To conclude, "200 Years of Laphroaig" is one of the most charming and comprehensive distillery portraits I´ve had the pleasure to read.

"Scotch Missed" by Brian Townsend was one of my favourite whisky books back in the nineties. I was fascinated by the stories of distilleries that have closed long ago, and some even forgotten. It is often said that we need to know about our history in order to understand the present. I couldn´t agree more. Reading Townsend´s book gave me clues as to why an industry like Scotch whisky had its ups and downs and to what extent, traditions have helped to shape the whisky business we see today. Relying on the first edition of the book, I skipped the following two (in 1997 and 2000), assured that nothing new could have been added. After all, we were talking about distilleries that were closed for good. In 2015, a fourth edition was published and looking at my well used copy of Scotch Missed, I thought I might as well invest in a new that was in mint condition. I´m glad that I did. First of all, the 76 pictures in black and white that are spread throughout my old book, have now been expanded to 355 - many of them in colour. Distillery pictures from various archives are now intermingled with beautiful old ads and pictures of both old and new bottles. A wonderful (and geeky) addition to this edition are the Victorian maps which show the location of almost all the distilleries. The bulk of the text remains unchanged, for obvious reasons, but some of the distillery entries have been revised, such as Glengyle, Glenglassaugh and Annandale. The latter distillery was closed in 1918 - for good it seemed. But as it turned out, the final chapter of Annandale distillery had not been written. Almost a century after the last distillation, it was re-opened, thanks to a time-consuming and costly initiative by Professor David Thomson. This is his own explanation for taking such a bold step - "Scotch Missed helped me to realise that much of Scotland´s architectural heritage can be saved. That is why I bought and restored Annandale Distillery." I admit, reading "Scotch Missed" doesn't inspire me to take the same grandiose decision but that´s me - not the book. It does however provide me with a good insight into an industry that is constantly evolving.

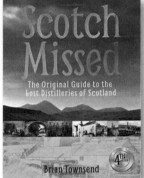

200 Years of Laphroaig
ISBN 978-9078668312

Scotch Missed
ISBN 978-1906000820

Whisky Advocate
www.whiskyadvocate.com

Whisky Passion
www.whiskypassion.nl

Whisky Magazine
www.whiskymag.com

Allt om Whisky
www.alltomwhisky.se

Der Whisky-Botschafter
www.whiskybotschafter.com

Whisky Time
www.whiskytime-magazin.

Whisky Magazine
www.whiskymag.fr

Whisky & Bourbon
www.livetsgoda.se

Glenmorangie

[glen•mor•run•jee]

Owner:
The Glenmorangie Co
(Moët Hennessy)

Region/district:
Northern Highlands

Founded: 1843
Status: Active (vc)
Capacity: 6 000 000 litres

Address: Tain, Ross-shire IV19 1PZ

Website:
glenmorangie.com

Tel:
01862 892477 (vc)

To specify with any degree of accuracy what it is that gives a certain malt whisky its flavour is impossible. Generally speaking though, it is often claimed that for a core bottling that has not been extra matured, 60% of the character comes from the cask.

Bill Lumsden, head of whisky creation at Glenmorangie, has always placed great importance on the selection of casks, not least for various finishes. The past 10 years though, he has also been experimenting with different types of yeast, since it is assumed that they contribute up to 20% of the flavour profile. This is something that has always been high on the agenda for Japanese whisky makers, but not in Scotland, where just a few types of yeast are used by the distilleries. Lumsden´s research has brought him to the point where Glenmorangie, probably in the near future, will be using its own, unique yeast for all of the production.

Glenmorangie single malt continues to show strength. Of the top 4 malts, with Glenmorangie in fourth place, it is the fastest growing brand with an increase since 2009 by 82%. A total of 5.9 million bottles were sold in 2014 and the USA is a market where the brand is increasing exceptionally fast.

The distillery is equipped with a full lauter mash tun with a charge of 10 tonnes, 12 stainless steel washbacks with a fermentation time of 52 hours and six pairs of stills. They are the tallest in Scotland and the still room is one of the most magnificent to be seen. Production for 2015 will be 32 mashes per week which equates to 6 million litres in the year.

The core range consists of **Original** (10 year old), **18** and **25 year old**. There are three 12 year old wood finishes: **Quinta Ruban** (port), **Nectar D´Or** (Sauternes) and **Lasanta** (sherry). Added to the core range is **Signet**, an unusual piece of work with 20% of the whisky having been made using chocolate malt. A series of bottlings, called Private Edition, started in 2009 with the release of the sherried **Sonnalta PX**. This has been followed up once a year with lightly peated **Finealta**, **Artein** extra matured in Sassicaia casks, **Ealanta** which was fully matured in virgin American oak and **Companta** which was a selection of whiskies matured for 14-18 years in a variety of casks. The sixth edition, released early in 2015, was **Tùsail** which was made from floor malted Maris Otter, a barley variety which was common decades ago, but has since fallen out of use. In 2013 Glenmorangie launched the Cask Masters Programme where consumers around the world were given the opportunity to select the next Glenmorangie bottling from three samplings. The winner, **Taghta**, had received an extra maturation in Manzanilla casks and was released in September 2014. One month later, **Dornoch**, extra matured in Amontillado sherry casks, was released as a duty free exclusive and this was followed up in early 2015 by **Duthac**. This is the first in a new duty free range called Glenmorangie´s Legends. Duthac, bottled at 43%, is a vatting of whiskies matured in bourbon casks, PX sherry casks and charred, virgin oak casks.

History:

1843 William Mathesen applies for a license for a farm distillery called Morangie, which is rebuilt by them. Production took place here in 1738, and possibly since 1703.

1849 Production starts in November.

1880 Exports to foreign destinations such as Rome and San Francisco commence.

1887 The distillery is rebuilt and Glenmorangie Distillery Company Ltd is formed.

1918 40% of the distillery is sold to Macdonald & Muir Ltd and 60 % to the whisky dealer Durham. Macdonald & Muir takes over Durham's share by the late thirties.

1931 The distillery closes.

1936 Production restarts in November.

1980 Number of stills increases from two to four and own maltings ceases.

1990 The number of stills is doubled to eight.

1994 A visitor centre opens. September sees the launch of Glenmorangie Port Wood Finish which marks the start of a number of different wood finishes.

1995 Glenmorangie´s Tain l´Hermitage (Rhone wine) is launched.

1996 Two different wood finishes are launched, Madeira and Sherry. Glenmorangie plc is formed.

1997 A museum opens.

2001 A limited edition of a cask strength port wood finish is released in July, Cote de Beaune Wood Finish is launched in September and Three Cask (ex-Bourbon, charred oak and ex-Rioja) is launched in October for Sainsbury's.

History continued:

2002 A Sauternes finish, a 20 year Glenmorangie with two and a half years in Sauternes casks, is launched.

2003 Burgundy Wood Finish is launched in July and a limited edition of cask strength Madeira-matured (i. e. not just finished) in August.

2004 Glenmorangie buys the Scotch Malt Whisky Society. The Macdonald family decides to sell Glenmorangie plc (including the distilleries Glenmorangie, Glen Moray and Ardbeg) to Moët Hennessy at £300 million. A new version of Glenmorangie Tain l'Hermitage (28 years) is released and Glenmorangie Artisan Cask is launched in November.

2005 A 30 year old is launched.

2007 The entire range gets a complete makeover with 15 and 30 year olds being discontinued and the rest given new names as well as new packaging.

2008 An expansion of production capacity is started. Astar and Signet are launched.

2009 The expansion is finished and Sonnalta PX is released for duty free.

2010 Glenmorangie Finealta is released.

2011 28 year old Glenmorangie Pride is released.

2012 Glenmorangie Artein is released.

2013 Glenmorangie Ealanta is released.

2014 Companta, Taghta and Dornoch are released.

2015 Túsail and Duthac are released.

Tasting notes Glenmorangie Original 10 year old:

GS – The nose offers fresh fruits, butterscotch and toffee. Silky smooth in the mouth, mild spice, vanilla, and well-defined toffee. The fruity finish has a final flourish of ginger.

Duthac   Túsail   Taghta

Original 10 years old   Signet   Nectar D´Or

Glen Moray

[glen mur•ree]

Owner:
La Martiniquaise (COFEPP)

Region/district:
Speyside

Founded: **Status:** **Capacity:**
1897 Active (vc) 3 300 000 litres

Address: Bruceland Road, Elgin,
Morayshire IV30 1YE

Website: **Tel:**
glenmoray.com 01343 542577

The bulk of Glen Moray´s production is used by the owners for Label 5 and the blending of that whisky takes place at the distillery.

Graham Coull, the distillery manager, is also the master blender for Label 5 and recently he launched an addition to the range by way of Gold Heritage, a vatting of different whiskies aged up to 20 years. Label 5 is the 10th most popular blend in the world with 28 million bottles sold in 2014.

Since Glen Moray was acquired by French La Martiniquaise in 2008, the distillery has undergone a remarkable transformation from a small and relatively unknown distillery to one of the ten biggest in the industry.

Last year the distillery was expanded to 3.3 million litres and, by the end of 2015, they will be able to produce 6.5 million! The old Saladin maltings have already been demolished to make room for a new, 11 ton mash tun, 12 new washbacks and 6 more stills. The new mash tun is highly efficient with a 3 hour mash cycle compared to the previous one with a 7.5 hour cycle. After the expansion the equipment will consist of one mash tun, 21 stainless steel washbacks and 6 pairs of stills. And that is not where the buck stops. There is a possibility that the capacity will increase to 9 million litres in the near future. During 2014 a new effluent treatment plant was also built, as well as two new warehouses.

The distillery has been producing 7 days per week for many years and for 2015 that means 3,6 million litres of which 300,000 litres will be heavily peated (50ppm) spirit. Peated production is fairly new to the distillery and did not start until 2009.

The Glen Moray single malt core range consists of **Classic, Classic Port** and **Classic Peated** (new since spring 2015 and the first peated release from the distillery). All of them are without age statement. Furthermore, we can also find the **10 year old, 12 year old** and the **16 year old Chardonnay cask**. Recent limited releases include a **25 year old Port finish** and a **30 year old**.

History:

1897 West Brewery, dated 1828, is reconstructed as Glen Moray Distillery.

1910 The distillery closes.

1920 Financial troubles force the distillery to be put up for sale. Buyer is Macdonald & Muir.

1923 Production restarts.

1958 A reconstruction takes place and the floor maltings are replaced by a Saladin box.

1978 Own maltings are terminated.

1979 Number of stills is increased to four.

1996 Macdonald & Muir Ltd changes name to Glenmorangie plc.

1999 Three wood finishes are introduced - Chardonnay (no age) and Chenin Blanc (12 and 16 years respectively).

2004 Louis Vuitton Moët Hennessy buys Glenmorangie plc and a 1986 cask strength, a 20 and a 30 year old are released.

2005 The Fifth Chapter (Manager's Choice from Graham Coull) is released.

2006 Two vintages, 1963 and 1964, and a new Manager's Choice are released.

2007 New edition of Mountain Oak is released.

2008 The distillery is sold to La Martiniquaise.

2009 A 14 year old Port finish and an 8 year old matured in red wines casks are released.

2011 Two cask finishes and a 10 year old Chardonnay maturation are released.

2012 A 2003 Chenin Blanc is released.

2013 A 25 year old port finish is released.

2014 Glen Moray Classic Port Finish is released.

2015 Glen Moray Classic Peated is released.

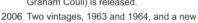

Classic Peated

Tasting notes Glen Moray 12 years old:

GS – Mellow on the nose, with vanilla, pear drops and some oak. Smooth in the mouth, with spicy malt, vanilla and summer fruits. The finish is relatively short, with spicy fruit.

Glen Ord

[glen <u>ord</u>]

Owner:
Diageo

Region/district:
Northern Highlands

Founded: **Status:**
1838   Active (vc)

Capacity:
11 000 000 litres

Address: Muir of Ord, Ross-shire IV6 7UJ

Website:
malts.com

Tel:
01463 872004 (vc)

If a distillery expands to more than double its size, one would think that the appearance would also change radically to look more like a modern, industrial building.

This, however, is not the case with Glen Ord distillery. The latest expansion, completed in early 2015, has been carried out in such a way that buildings being more than a century old, have been elegantly utilized to house the new equipment. Facing the street is the old stillhouse with six stills. Behind this we find the old kiln and malt storage which have been converted into a beautiful tunroom with 12 new wooden washbacks. To the rear side of this, the old Saladin maltings have been transformed into a new still room with eight stills in one row and the condensers on the opposite side. After the expansion, the complete set up of equipment comprises of two stainless steel mashtuns, each with a 12.5 ton mash. There are 22 wooden washbacks with a fermentation time of 75 hours and no less than 14 stills. The total capacity is now 11 million litres of alcohol which makes Glen Ord one of the five, biggest malt whisky distilleries in Scotland.

Part of the production at Glen Ord goes into Diageo's different blended whiskies, but more and more is being reserved for the Singleton of Glen Ord single malt. It has become the biggest success in terms of sales for many years in the Scotch whisky industry. The brand is reserved for the Asian market but the core range can also be found at the excellent visitor centre.

The core range is the **Singleton of Glen Ord 12, 15** and **18 year old**. In 2013 a new sub-range, The Singleton Reserve Collection, exclusive to duty free, was launched. The first release was **Singleton of Glen Ord Signature**, followed by **Trinité**, **Liberté** and **Artisan**. Limited releases include a **32** and **35 year old**, as well as a **cask strength**. The latest addition was **The Master's Casks 40 year old**, making it the oldest expression from the distillery so far. Only 999 bottles were released in spring 2015.

History:

1838   Thomas Mackenzie founds the distillery.

1855   Alexander MacLennan and Thomas McGregor buy the distillery.

1870   Alexander MacLennan dies and the distillery is taken over by his widow who marries the banker Alexander Mackenzie.

1877   Alexander Mackenzie leases the distillery.

1878   Alexander Mackenzie builds a new still house and barely manages to start production before a fire destroys it.

1896   Alexander Mackenzie dies and the distillery is sold to James Watson & Co. for £15,800.

1923   John Jabez Watson, James Watson's son, dies and the distillery is sold to John Dewar & Sons. The name changes from Glen Oran to Glen Ord.

1961   A Saladin box is installed.

1966   The two stills are increased to six.

1968   Drum maltings is built.

1983   Malting in the Saladin box ceases.

1988   A visitor centre is opened.

2002   A 12 year old is launched.

2003   A 28 year old cask strength is released.

2004   A 25 year old is launched.

2005   A 30 year old is launched as a Special Release from Diageo.

2006   A 12 year old Singleton of Glen Ord is launched.

2010   A Singleton of Glen Ord 15 year old is released in Taiwan.

2011   Two more washbacks are installed, increasing the capacity by 25%.

2012   Singleton of Glen Ord cask strength is released.

2013   Singleton of Glen Ord Signature, Trinité, Liberté and Artisan are launched.

2015   The Master's Casks 40 years old is released.

12 years old

Tasting notes Glen Ord 12 years old:

GS – Honeyed malt and milk chocolate on the nose, with a hint of orange. These characteristics carry over onto the sweet, easy-drinking palate, along with a biscuity note. Subtly drying, with a medium-length, spicy finish.

Glenrothes

[glen•roth•iss]

Owner:	**Region/district:**
The Edrington Group	Speyside
(the brand is owned by Berry Bros)	

Founded:	**Status:**	**Capacity:**
1878	Active	5 600 000 litres

Address: Rothes, Morayshire AB38 7AA

Website:	**Tel:**
theglenrothes.com	01340 872300

Usually it is the distillery owner that is also in possession of the single malt brand. At Glenrothes we have the unique situation of Edrington owning and operating the distillery, while the Glenrothes brand is in the hands of Berry Brothers.

This arrangement dates back to 2010 when the two companies decided to "swap" brands. Cutty Sark blended Scotch, which relied heavily on whisky from Glenrothes, was already established by BBR in 1923. In recent years the company realised that a high volume mega brand like Cutty didn't really fit in to their profile, something that Glenrothes single malt, owned by Edrington, definitely would do. Edrington, already selling Famous Grouse, were happy to take on another blend and a deal was concluded.

Glenrothes distillery is equipped with a 5.5 ton stainless steel full lauter mash tun. Twelve washbacks made of Oregon pine are in one room, whilst an adjacent modern tun room houses eight new stainless steel washbacks. Since 2012, the distillery has been working a 7-day week with 50 mashes per week. The magnificent, cathedral-like still house has five pairs of stills performing a very slow distillation. In 2015, the distillery will be producing 5.4 million litres of alcohol.

Some of the more recent vintages include **1988, 1998** and **2001**. Other core expressions include the **Select Reserve** and the **Alba Reserve**, both without age statement. A new addition to the range since 2014, is **Sherry Cask Reserve**, the first Glenrothes that has been matured 100% in first fill sherry casks. Spring 2015 saw the release of **Vintage Single Malt** which contains ten different vintages from 1989 to 2007 married together. The duty free range includes **Robur Reserve, Three Decades** and the exclusive **Oldest Reserve** which is a vatting of vintages from 1971 to 1998. Furthermore, there is the Manse Brae series with **Manse Reserve, Elder´s Reserve** and **Minister´s Reserve**. Finally, another new range called The Glenrothes Extraordinary Cask was introduced in 2012, with the third expression released in 2015 - a **1968 second fill sherry hogshead**.

History:
1878 James Stuart & Co. begins planning the new distillery with Robert Dick, William Grant and John Cruickshank as partners. Stuart has financial problems so Dick, Grant and Cruickshank terminate the partnership and continue the building.

1879 Production starts in May.

1884 The distillery changes name to Glenrothes-Glenlivet.

1887 William Grant & Co. joins forces with Islay Distillery Co. and forms Highland Distillers Company.

1897 A fire ravages the distillery.

1903 An explosion causes substantial damage.

1963 Expansion from four to six stills.

1980 Expansion from six to eight stills.

1989 Expansion from eight to ten stills.

1999 Edrington and William Grant & Sons buy Highland Distillers.

2002 Four single casks from 1966/1967 are launched.

2005 A 30 year old is launched together with Select Reserve and Vintage 1985.

2007 A 25 year old is released as a duty free item.

2008 1978 Vintage and Robur Reserve are launched.

2009 The Glenrothes John Ramsay, two vintages (1988 and 1998), Alba Reserve and Three Decades are released.

2010 Berry Brothers takes over the brand.

2011 Editor´s Casks are released.

2013 2001 Vintage and the Manse Brae range are released.

2014 Sherry Cask Reserve and 1969 Extraordinary Cask are released.

2015 Glenrothes Vintage Single Malt is released.

Vintage 2001

Tasting notes Glenrothes Select Reserve:

GS – The nose offers ripe fruits, spice and toffee, with a whiff of Golden Syrup. Faint wood polish in the mouth, vanilla, spicy and slightly citric. Creamy and complex. Slightly nutty, with some orange, in the drying finish.

1920 prohibition in America

The negative aspects of alcohol had been discussed during the entire 1800s in the USA and in 1851 a law was passed in Maine prohibiting sales of alcoholic beverages.

Four years later a further 12 states had followed suit. The temperance movement found this encouraging of course and they started working even harder for a nationwide prohibition. An even stronger lobby organization which strived towards prohibition was the Anti-Saloon League, founded in 1893. The increasing crime and immoral behaviour was blamed on immigrants in the larger cities and these immigrants frequented the many saloons operating there. Prohibition was ratified as the 18th Amendment in 1919 and went into effect on 17 January 1920. Manufacture, importation, sale and transport of alcohol (anything above 0.5%) was made illegal. On the other hand, did the act not prohibit consumption of alcohol and sales of, for example, whiskey for medicinal purposes was allowed. It is assumed that over a million gallons were consumed every year by way of these prescriptions.

Prohibition became a gold mine for criminals who saw the opportunity to supply thirsty consumers with illegal liquor, either as bootleggers, producing their own alcohol, or by smuggling products into the country. The alcohol was often offered to the public in clubs and bars that were called "speakeasy" and solely in New York alone, it is thought that anywhere from 30,000 to 100,000 speakeasies were operating during the 1920s. It is often said that Prohibition failed to reduce consumption but, as a matter of fact, figures dropped by 50% compared to pre-prohibition consumption and did not recover until the mid 1940s. However, spirits strengthened its position against beer. For the smugglers it was far more efficient and profitable to handle bottles of strong liquor than a much larger volume of beer.

Prohibition was a major blow to the legal alcohol industry in the USA, but also to many other groups which had indirectly been dependant on the business, like farmers. The negative economical consequences for the country became clearer, not least because the federal government lost a lot of tax revenue, and when the great depression hit the country in 1929, the downside of the law became apparent. Franklin D Roosevelt went to the polls in 1932 with the promise to repeal prohibition and on 5 December 1933, the 21st Amendment put an end to the dry years.

During these thirteen years, smuggling of spirits into the USA had been a lucrative business, not the least for the producers of Scotch whisky. Already during the years leading up to 1920, large quantities of Scotch whisky were exported across the Atlantic and a demand for the product had been established. When prohibition came into force, the companies in the UK redirected their shipments to countries and territories close to the USA. They carefully avoided any direct contact with criminal organisations, but sent the merchandise to a respectable customer in the Bahamas or in Canada. At these locations, a reloading took place onto boats steering towards the coastline of the USA, making sure that they stayed out of American territorial water. The final step involved the re-loading of the boxes onto small speed boats which brought the cargo to the mainland, simultaneously trying to avoid the US Coast Guard. The Scotch whisky producers made sure to deliver bottles and not whisky in bulk, which would have been easier to tamper with. Scotch whisky therefore received a quality stamp of approval from the consumers. The producers began to make customized whisky to suit the American´s taste for a whisky light in colour and flavour. Brands that are well-known today, like Cutty Sark and J&B, quickly took the lion´s share of the market.

What initially appeared to be a disastrous blow to the entire spirits industry, turned out to be a way in for the producers of Scotch to a market which, today, has become their most important.

Glen Scotia

[glen sko•sha]

Owner:
Loch Lomond Group
(majority owner Exponent)

Region/district:
Campbeltown

Founded: 1832
Status: Active (vc)
Capacity: 800 000 litres

Address: High Street, Campbeltown, Argyll PA28 6DS

Website:
glenscotia-distillery.co.uk

Tel:
01586 552288

Glen Scotia has progressed from being a neglected plant for decades to being a top priority distillery in just a few years.

Since the new owners, Exponent Private Equity, took over in spring 2014, large investments have been poured into Glen Scotia and its sister distillery, Loch Lomond. Glen Scotia has been thoroughly upgraded and renovated which includes three new wash backs, a new spirit safe, new roofs for the still house and tun room, upgrading of the stills and last, but not least – a visitor centre which incorporates a tasting area and a shop.

Apart from the two distilleries, the new owners acquired Glen Catrine Bonded Warehouse, which was established in 1974 and, today, is one of the largest bottling plants in Scotland with an annual production of 40 million bottles of whisky, vodka, gin, rum and brandy. Included in the £210m deal were also brands such as High Commissioner blended Scotch and Glen´s Vodka.

Glen Scotia is equipped with a traditional cast iron mash tun, nine washbacks made of stainless steel and one pair of stills. Fermentation time has varied from 48 hours up to five days, but starting as from 2015, the shortest fermentation time will be 70 hours. In order to achieve the desired character for Glen Scotia new make, the distillery has a very short middle cut (possibly the shortest in the industry). They start collecting the spirit at 71% and come off spirit at 68%. The production in 2015 will increase to 500,000 litres of pure alcohol, of which 12% is a combination of lightly peated (15ppm) and heavily peated (50ppm) spirit.

The core range, which used to consist of just a 12 year old, was completely revamped in 2012 with a 10, 12, 16, 18 and 21 year old. This range has, on the other hand, already been replaced in the spring of 2015 with three new expressions; **Double Cask** (matured in bourbon casks and with a 3-4 months finish in PX sherry), **15 year old** (American oak and a short finish in oloroso casks) and **Victoriana** which has been bottled at cask strength. A limited (197 bottles) **Distillery Edition** was also released towards the end of May.

History:

1832 The families of Stewart and Galbraith start Scotia Distillery.

1895 The distillery is sold to Duncan McCallum.

1919 Sold to West Highland Malt Distillers.

1924 West Highland Malt Distillers goes bankrupt and Duncan MacCallum buys back the distillery.

1928 The distillery closes.

1930 Duncan MacCallum commits suicide and the Bloch brothers take over.

1933 Production restarts.

1954 Hiram Walker takes over.

1955 A. Gillies & Co. becomes new owner.

1970 A. Gillies & Co. becomes part of Amalgated Distillers Products.

1979 Reconstruction takes place.

1984 The distillery closes.

1989 Amalgated Distillers Products is taken over by Gibson International and production restarts.

1994 Glen Catrine Bonded Warehouse Ltd takes over and the distillery is mothballed.

1999 The distillery re-starts under Loch Lomond Distillery supervision using labour from Springbank.

2000 Loch Lomond Distillers runs operations with its own staff from May onwards.

2005 A 12 year old is released.

2006 A peated version is released.

2012 A new range (10, 12, 16, 18 and 21 year old) is launched.

2014 A 10 year old and one without age statement are released - both heavily peated.

2015 A new range is released; Double Cask, 15 year old and Victoriana.

Tasting notes Glen Scotia Double Cask:

GS – The nose is sweet, with bramble and redcurrant aromas, plus caramel and vanilla. Smooth mouth-feel, with ginger, sherry and more vanilla. The finish is quite long, with spicy sherry and a final hint of brine.

Double Cask

Glen Spey

Photo: © Raymond MacDonald

[glen spey]

Owner:	**Region/district:**
Diageo	Speyside
Founded: **Status:**	**Capacity:**
1878 Active	1 400 000 litres

Address: Rothes, Morayshire AB38 7AU

Website:	**Tel:**
malts.com	01340 831215

The increasing demand for Scotch whisky towards the end of the 1800s was noted also south of the border and in 1887, W & A Gilbey became the first English company to buy a Scottish distillery.

But the famous gin producer, established in 1857, didn't settle at that. Eight years later, Strathmill was bought and in 1904 Knockando was acquired. Glen Spey distillery is situated in the middle of the town Rothes, at the foot of the hill where the ruins of Castle Rothes is to be found. This was the home of the Leslie family (Earl of Rothes) for four hundred years until 1662 when it was destroyed in a fire. Only a fragment of the outer walls remain on site, but many of the stones were used for building houses in Rothes. King Edward I was a guest at the castle in 1296 during his expedition to the north to quell the rebellious Scots. On his way back to London he brought The Stone of Destiny, on which Scottish kings had been crowned for centuries. It took exactly 700 years before it was returned to the Scots.

The single malt from Glen Spey has its biggest importance in the blend J&B, where it is one of the signature malts. Ten years ago J&B was the second most sold Scotch after Johnnie Walker, but has since lost more than a third of its sales and now finds itself in fifth place with 43 million bottles being sold during 2014.

Glen Spey is equipped with a semi-lauter mash tun, eight stainless steel washbacks and two pairs of stills, where the spirit stills are equipped with purifiers to obtain a lighter character of the spirit. The distillery is currently doing 24 mashes per week which means they only do short fermentations, 52 hours.

The core expression is the **12 year old Flora & Fauna** bottling. In 2010, two limited releases were made – a **1996 single cask** from new American Oak and a **21 year old** with maturation in ex-sherry American oak.

History:

1878 James Stuart & Co. founds the distillery which becomes known by the name Mill of Rothes.

1886 James Stuart buys Macallan.

1887 W. & A. Gilbey buys the distillery for £11,000 thus becoming the first English company to buy a Scottish malt distillery.

1920 A fire breaks out.

1962 W. & A. Gilbey combines forces with United Wine Traders and forms International Distillers & Vintners (IDV).

1970 The stills are increased from two to four.

1972 IDV is bought by Watney Mann which is then acquired by Grand Metropolitan.

1997 Guiness and Grand Metropolitan merge to form Diageo.

2001 A 12 year old is launched in the Flora & Fauna series.

2010 A 21 year old is released as part of the Special Releases and a 1996 Manager´s Choice single cask is launched.

12 years old

Tasting notes Glen Spey 12 years old:

GS – Tropical fruits and malt on the comparatively delicate nose. Medium-bodied with fresh fruits and vanilla toffee on the palate, becoming steadily nuttier and drier in a gently oaky, mildly smoky finish.

Glentauchers

[glen•tock•ers]

Owner:
Chivas Brothers
(Pernod Ricard)

Region/district:
Speyside

Founded: 1897

Status: Active

Capacity: 4 200 000 litroc

Address: Mulben, Keith, Banffshire AB55 6YL

Website: -

Tel: 01542 860272

Traditionally, malt whisky in Scotland has been distilled in pot stills, whereas column stills were used for grain whisky. Since a few years ago, there's even been a mandatory law which states that malt whisky can only be made "by batch distillation in pot stills".

Around 1910 a trial run was initiated to produce malt whisky in continuous stills at Glentauchers, Convalmore and Lochruan distilleries. All three distilleries were owned by James Buchanan and trials only lasted a few years. This was by no means the first time. Already in 1855, a Coffey still (aka known as continuous still or column still) was installed at Glenmavis distillery for production of single malt whisky. In more recent years, Loch Lomond distillery has done the same, but it was forced by the Scotch Whisky Association to label it as grain whisky and not malt. Glentauchers is one of Chivas Bros least known distilleries. Founded by whisky baron, James Buchanan, his Black & White blend was obviously the first to rely on Glentauchers for its character. As owners changed hands over the years, it became a signature malt for Teacher's and, today, it is an integral part of Ballantine's.

The distillery is equipped with a 12 tonnes stainless steel full lauter mash tun installed in 2007. There are six washbacks made of Oregon pine and three pairs of stills. A couple of years ago the production increased from six to seven days, which translates to 18 mashes per week and a total of 4 million litres per year. Most of the process at Glentauchers is done mechanically using traditional methods. The thought behind this is that new employees and trainees will be able to work here for a while to learn the basic techniques of whisky production.

There are no official bottlings of Glentauchers but Chivas Brothers have recently made an attempt to lift the whisky from obscurity. A range called Ballantine's 17 year old Signature Distillery Editions was launched in order to highlight the four signature malts of the famous blend. The Glentauchers version was released in 2014 and is reserved for markets in Asia.

History:

1897 James Buchanan and W. P. Lowrie, a whisky merchant from Glasgow, found the distillery.

1898 Production starts.

1906 James Buchanan & Co. takes over the whole distillery and acquires an 80% share in W. P. Lowrie & Co.

1915 James Buchanan & Co. merges with Dewars.

1923 Mashing house and maltings are rebuilt.

1925 Buchanan-Dewars joins Distillers Company Limited (DCL).

1930 Glentauchers is transferred to Scottish Malt Distillers (SMD).

1965 The number of stills is increased from two to six.

1969 Floor maltings is decommissioned.

1985 DCL mothballs the distillery.

1989 United Distillers (formerly DCL) sells the distillery to Caledonian Malt Whisky Distillers, a subsidiary of Allied Distillers.

1992 Production recommences in August.

2000 A 15 year old Glentauchers is released.

2005 Chivas Brothers (Pernod Ricard) become the new owner through the acquisition of Allied Domecq.

Gordon & MacPhail
Glentauchers 1990

Tasting notes Glentauchers 1991 G&M:

GS – Fresh and floral aromas, with sweet fruits and peppery peaches. Medium to full-bodied in the mouth, with cereal and sweet spice. The finish is medium to long.

Glenturret

[glen•turr•et]

Owner:
The Edrington Group

Region/district:
Southern Highlands

Founded: **Status:** **Capacity:**
1775 Active (vc) 340 000 litres

Address: The Hosh, Crieff, Perthshire PH7 4HA

Website: **Tel:**
thefamousgrouse.com 01764 656565

Glenturret is the spiritual home of Famous Grouse blended Scotch and boasts an excellent, recently refurbished Famous Grouse Experience visitor centre.

Sales figures for the Famous Grouse blend, the best selling Scotch in Scotland, has remained steady at around 36 million bottles annually since 2006. In May 2015 it was time for an overhaul of the packaging to make it more appealing and elegant. The reason, according to Edrington, was that consumers tend to move away from mid-level brands to either premium or value whiskies. The core range consists of The Famous Grouse, Black Grouse, Black Grouse Alpha, Snow Grouse and Naked Grouse. In 2014 the Famous Grouse 16 year old Double Matured was launched as the first of a series of limited edition blends.

Glenturret, considered to be the oldest working distillery in Scotland, is equipped with a stainless steel, open mash tun, the only one left in Scotland where the mash is stirred by hand and where the draff at the end of the process must be removed manually. Furthermore, there are eight Douglas fir washbacks with a minimum fermentation time of 48 hours and one pair of stills with vertical condensers. During 2015, the distillery will be doing 8-10 mashes per week which adds up to 190,000 litres of alcohol. The main part of this, approximately 126,000 litres, will be unpeated Glenturret, while the remaining 64,000 litres are made up of the heavily peated (80ppm in the barley) Ruadh Maor, which has been produced at the distillery since 2009 and is used for blended whisky.

There is only one official bottling in the core range, namely, the **10 year old**. In 2013, an **18 year old**, ex sherry butt was released exclusively at the visitor centre, and in July 2014, a further 1,800 bottles of a **Glenturret 1986** were released. During the same year, three bottlings, exclusive to the Whisky Shop chain, were released; **Sherry Edition**, **Peated Edition** and **Glenturret Triple Wood**. Finally, in May 2015, **Glenturret The Brock Malloy Edition** was launched. Named after two Glenturret still men in 1986, it was a single cask bottled at 47.1%.

History:

1775 Whisky smugglers establish a small illicit farm distillery named Hosh Distillery.

1818 John Drummond is licensee until 1837.

1826 A distillery in the vicinity is named Glenturret, but is decommissioned before 1852.

1852 John McCallum is licensee until 1874.

1875 Hosh Distillery takes over the name Glenturret Distillery and is managed by Thomas Stewart.

1903 Mitchell Bros Ltd takes over.

1921 Production ceases and the buildings are used for whisky storage only.

1929 Mitchell Bros Ltd is liquidated, the distillery dismantled and the facilities are used as storage for agricultural needs.

1957 James Fairlie buys the distillery and re-equips it.

1959 Production restarts.

1981 Remy-Cointreau buys the distillery and invests in a visitor centre.

1990 Highland Distillers takes over.

1999 Edrington and William Grant & Sons buy Highland Distillers for £601 million. The purchasing company, 1887 Company, is a joint venture between Edrington (70%) and William Grant (30%).

2002 The Famous Grouse Experience, a visitor centre costing £2.5 million, is inaugurated.

2003 A 10 year old Glenturret replaces the 12 year old as the distillery´s standard release.

2007 Three new single casks are released.

2013 An 18 year old bottled at cask strength is released as a distillery exclusive.

2014 A 1986 single cask is released.

2015 Glenturret - The Brock Malloy Edition is released.

10 years old

Tasting notes Glenturret 10 years old:

GS – Nutty and slightly oily on the nose, with barley and citrus fruits. Sweet and honeyed on the full, fruity palate, with a balancing note of oak. Medium length in the sweet finish.

Highland Park

[hi•land park]

| Owner: | Region/district: |
| The Edrington Group | Highlands (Orkney) |

| Founded: | Status: | Capacity: |
| 1790 | Active (vc) | 2 500 000 litres |

Address: Holm Road, Kirkwall, Orkney KW15 1SU

| Website: | Tel: |
| highlandpark.co.uk | 01856 874619 |

Highland Park single malt is one of the most respected Scotch whiskies and the distillery is known especially for its dedication to sherry casks.

The oak is left to air dry for four years in Spain and, thereafter, it is filled with sherry for two to three years. The owners spend £10m on wood alone every year. In spite of its reputation, sales figures remained pretty stable for several years. The reason being that there was a shortage of stock aged 10-15 years which, in turn, was caused by a somewhat irregular production towards the end of the 1990s. The supply has now come to meet the demand and during the last four years the sales have increased by almost 30% to 1.5 million bottles in 2014.

Highland Park is malting 30% of its malt themselves. The single malt is peated but done in a completely different way to that of the whiskies from Islay. The phenol specification is lower but it is the peat itself that sets Highland Park apart. Because of the lack of trees in Orkney, the local peat adds a sweet, heathery note to the character, as opposed to that of Islay's with their aroma of seaweed and iodine. There are five malting floors with a capacity of almost 36 tonnes of barley. The malt is dried for 18 hours using peat and the final 18 hours using coke. The phenol content is 30-40 ppm in its own malt and the bought malt from Simpson's is unpeated. The two varieties are always mixed before mashing.

The distillery is equipped with a semi-lauter mash tun, twelve Oregon pine washbacks with a fermentation time of between 50 and 80 hours, and two pairs of stills. The mash tun has a 12 tonnes capacity but is only filled to 50%. The plan for 2015 is to produce 2.1 million litres of alcohol.

The core range of Highland Park consists of **12, 15, 18, 21, 25, 30, 40** and **50 year old**. During the summer of 2014, Highland Park, like so many other producers, decided to launch an addition to the core range without an age statement. The name of the whisky is **Dark Origins**. The age of the different whiskies will change from batch to batch, but lately is has been between 11 and 15 years. Bottled at 46.8%, the recipe is made up of 80% first fill sherry casks, twice as many as for the standard 12 year old. The Highland Park range for duty free follows the same non age-statement trend, abandoning the vintages that, for years, were favoured by the owners. For duty free, the range will mirror the Viking heritage, which is of great importance to Orkney. **Svein, Einar** and **Harald** were released in spring 2013 and were followed by **Sigurd, Ragnvald** and **Thorfinn** in autumn that same year.

Highlighting the Vikings is also a range which comprises of four expressions called the Valhalla Collection. The first bottling was the 16 year old **Thor** in 2012, followed in 2013 by **Loki** (15 years) and in spring 2014 by the 15 year old **Freya**. The fourth and final instalment, **Odin**, was released in spring 2015. It is 16 years old and bottled at 55.8%. Unlike Freya, which was 100% bourbon matured, Odin is matured for 13 years in sherry hogsheads and the final three years in sherry butts.

History:

1798 David Robertson founds the distillery. The local smuggler and businessman Magnus Eunson previously operated an illicit whisky production on the site.

1816 John Robertson, an Excise Officer who arrested Magnus Eunson, takes over production.

1826 Highland Park obtains a license and the distillery is taken over by Robert Borwick.

1840 Robert's son George Borwick takes over but the distillery deteriorates.

1869 The younger brother James Borwick inherits Highland Park and attempts to sell it as he does not consider the distillation of spirits as compatible with his priesthood.

1876 Stuart & Mackay becomes involved and improves the business by exporting to Norway and India.

1895 James Grant (of Glenlivet Distillery) buys Highland Park.

1898 James Grant expands capacity from two to four stills.

1937 Highland Distilleries buys Highland Park.

1979 Highland Distilleries invests considerably in marketing Highland Park as single malt which increases sales markedly.

1986 A visitor centre, considered one of Scotland's finest, is opened.

1997 Two new Highland Park are launched, an 18 year old and a 25 year old.

1999 Highland Distillers are acquired by Edrington Group and William Grant & Sons.

2000 Visit Scotland awards Highland Park "Five Star Visitor Attraction".

History continued:

2005 Highland Park 30 years old is released. A 16 year old for the Duty Free market and Ambassador´s Cask 1984 are released.

2006 The second edition of Ambassador´s Cask, a 10 year old from 1996, is released. New packaging is introduced.

2007 The Rebus 20, a 21 year old duty free exclusive, a 38 year old and a 39 year old are released.

2008 A 40 year old and the third and fourth editions of Ambassador´s Cask are released.

2009 Two vintages and Earl Magnus 15 year are released.

2010 A 50 year old, Saint Magnus 12 year old, Orcadian Vintage 1970 and four duty free vintages are released.

2011 Vintage 1978, Leif Eriksson and 18 year old Earl Haakon are released.

2012 Thor and a 21 year old are released.

2013 Loki and a new range for duty free, The Warriors, are released.

2014 Freya and Dark Origins are released.

2015 Odin is released.

Tasting notes Highland Park 12 year old:

GS – The nose is fragrant and floral, with hints of heather and some spice. Smooth and honeyed on the palate, with citric fruits, malt and distinctive tones of wood smoke in the warm, lengthy, slightly peaty finish.

Tasting notes Highland Park Dark Origins:

GS – The nose offers bananas, milk chocolate-coated caramel and a hint of background soot. The palate is smooth, with dark chocolate, berries, spicy dry sherry and coal. The finish features black pepper, and is smoky, dry and lengthy.

Dark Origins

Einar

Odin

12 years old

18 years old

21 years old

Inchgower

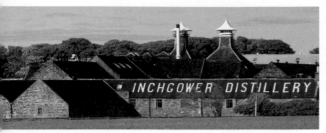

[inch•gow•er]

Owner:	**Region/district:**
Diageo	Speyside
Founded: **Status:**	**Capacity:**
1871 Active	3 200 000 litres

Address: Buckie, Banffshire AB56 5AB

Website:	**Tel:**
malts.com	01542 836700

That the owners of Inchgower, Alexander Wilson & Co, went bankrupt in 1936 might not have been so strange, as the previous years were amongst the toughest for the Scottish whisky industry.

The Wall Street crash in 1929 had led to a deep recession and in 1933 only two pot distilleries (Glenlivet and Glen Grant) and 13 grain distilleries were working in Scotland. The total production of malt whisky during that year was only 285,000 gallons, the lowest since 1824. But when Roosevelt became president of the USA in the same year, he pledged to abolish prohibition and, slowly but surely, the market for Scotch improved. Arthur Bell & Co had in the meanwhile kept ahead of the game and already acquired Blair Athol, as well as Dufftown in 1933. When they bought Inchgower in 1938 the amount of malt whisky produced was over 10 million gallons. For almost 50 years Inchgower was run by Bells until the Guinness Group acquired Bells for £356m. The distillery is today owned by Diageo and it continues to produce mainly for the blended industry, especially for Bells and Johnnie Walker.

After an extensive upgrade in 2012, the distillery is now equipped with a stainless steel semilauter mash tun, six wooden washbacks and 2 pairs of stills. The distillation process is tailored to produce a nutty spirit. This means cloudy worts, short fermentation (around 50 hours) and a quick distillation process to give as little copper contact as possible. In 2015, the plan is to do 19 mashes per week and 3 million litres of alcohol in the year. Most of the production is matured elsewhere, but there are also five dunnage and four racked warehouses on site.

Inchgower is situated on the south side of Moray Firth and is difficult to miss as it is situated just at the A98 near the small fishing port of Buckie. If one is driving from Elgin towards Banff, it is even easier to spot as the name appears on the roof. Besides the official **Flora & Fauna 14 year old**, there have also been a few limited bottlings of Inchgower single malt. In 2010, for example, a **single sherry cask** distilled in **1993**, was released.

History:

1871 Alexander Wilson & Co. founds the distillery. Equipment from the disused Tochineal Distillery, also owned by Alexander Wilson, is installed.

1936 Alexander Wilson & Co. becomes bankrupt and Buckie Town Council buys the distillery and the family's home for £1,600.

1938 The distillery is sold on to Arthur Bell & Sons for £3,000.

1966 Capacity doubles to four stills.

1985 Guinness acquires Arthur Bell & Sons.

1987 United Distillers is formed by a merger between Arthur Bell & Sons and DCL.

1997 Inchgower 1974 (22 years) is released as a Rare Malt.

2004 Inchgower 1976 (27 years) is released as a Rare Malt.

2010 A single cask from 1993 is released.

14 years old

Tasting notes Inchgower 14 years old:

GS – Ripe pears and a hint of brine on the light nose. Grassy and gingery in the mouth, with some acidity. The finish is spicy, dry and relatively short.

Jura

[joo•rah]

Owner:	**Region/district:**
Whyte & Mackay	Highlands (Jura)
(Emperador Inc)	
Founded: **Status:**	**Capacity:**
1810 Active (vc)	2 200 000 litres

Address: Craighouse, Isle of Jura PA60 7XT

Website:	**Tel:**
isleofjura.com	01496 820240

The incredible journey that Jura single malt started in 2010 continues. From having sales of about 350,000 bottles per year, the sales figures have now really escalated. During the last five years the brand has increased to 1.9 million bottles in 2014.

The high demand has furthermore presently placed Jura as the third most sold single malt in the UK after Glenfiddich and Glenmorangie. The distillery is lovely and the Isle of Jura with its 197 people and 4,000 deer is breathtakingly beautiful, but the quality of the whisky produced did not do justice to its ambience. The turning point admittedly came about 5 years ago, but the actual cause could be traced back to 1999. It was then found that there were too many casks in the warehouses that were of poor quality. They started to re-rack them into high quality bourbon casks and that, together with Master Blender Richard Paterson´s experienced nose, has resulted in a variety of exciting, new releases.

Jura distillery is equipped with one semi-lauter mash tun, six stainless steel washbacks with a fermentation time of 54 hours and two pairs of stills. Working a 7-day week since 2011, they will be doing 28 mashes per week during 2015, which will include one month of peated production (at 50ppm) producing 2.2 million litres of alcohol in the year.

The core range consists of **Origin** (10 years), **Diurach´s Own** (16 years), **Superstition** (with a part of peated Jura), as well as the peated **Prophecy**. Jura Elixir, a 12 year old matured in both American and European oak, was released as an exclusive for Sainsbury´s but is now available in other markets. Recent limited releases include the **1977 Juar**, the **30 year old Camas an Staca** and **Vintage 1984**. **Turas-Mara**, a duty free exclusive with no age statement, was released in 2013. The launch of a 40 year old, first announced for spring 2014, will now not be released before 2016. This year´s Jura Tastival bottling was an **18 year old**, bottled at 52% and matured in sparkling wine casks from Saumur in the Loire Valley.

History:

1810 Archibald Campbell founds a distillery named Small Isles Distillery.

1853 Richard Campbell leases the distillery to Norman Buchanan from Glasgow.

1867 Buchanan files for bankruptcy and J. & K. Orr takes over the distillery.

1876 Licence transferred to James Ferguson & Sons.

1901 Ferguson dismantles the distillery.

1960 Charles Mackinlay & Co. extends the distillery. Newly formed Scottish & Newcastle Breweries acquires Charles Mackinlay & Co.

1963 The first distilling takes place.

1985 Invergordon Distilleries acquires Charles Mackinlay & Co., Isle of Jura and Glenallachie from Scottish & Newcastle Breweries.

1993 Whyte & Mackay (Fortune Brands) buys Invergordon Distillers.

1996 Whyte & Mackay changes name to JBB (Greater Europe).

2001 The management buys out the company and changes the name to Kyndal.

2002 Isle of Jura Superstition is launched.

2003 Kyndal reverts back to its old name, Whyte & Mackay. Isle of Jura 1984 is launched.

2004 Two cask strengths (15 and 30 years old) are released in limited numbers.

2006 The 40 year old Jura is released.

2007 United Spirits buys Whyte & Mackay. The 18 year old Delmé-Evans and an 8 year old heavily peated expression are released.

2008 A series of four different vintages, called Elements, is released.

2009 The peated Prophecy and three new vintages called Paps of Jura are released.

2010 Boutique Barrels and a 21 year old Anniversary bottling are released.

2012 The 12 year old Jura Elixir is released.

2013 Camas an Staca, 1977 Juar and Turas-Mara are released.

2014 Whyte & Mackay is sold to Emperador Inc.

Tasting notes Jura 10 years old:

GS – Resin, oil and pine notes on the delicate nose. Light-bodied in the mouth, with malt and drying saltiness. The finish is malty, nutty, with more salt, plus just a wisp of smoke.

10 years old

Kilchoman

[kil•ho•man]

Owner:	Region/district:
Kilchoman Distillery Co.	Islay

Founded:	Status:	Capacity:
2005	Active (vc)	200 000 litres

Address: Rockside farm, Bruichladdich, Islay PA49 7UT

Website:	Tel:
kilchomandistillery.com	01496 850011

From its inception in 2005, Kilchoman has been a true barley to bottle distillery with a large part of the barley coming from fields which surround the distillery.

In June 2015, this was further reinforced when the company bought Rockside Farm which for ten years had delivered barley for its production and also happens to be the site on which the distillery is located. The farm was owned for more than 30 years by Mark and Rohaise French.

The distillery has its own floor maltings with a quarter of the barley requirements coming from fields that surround the distillery. The malt is peated from 20 to 25 ppm and the remaining malt (50 ppm) is bought from Port Ellen. Other equipment includes a stainless steel semi-lauter mash tun, five stainless steel washbacks (the last one installed in autumn 2015) and one pair of stills. The distillery is currently doing 8 mashes per week which translates to 170,000 litres of alcohol. With the addition of a fifth washback, they will be able to increase to 9 mashes and 200,000 litres in 2016. The owners sold 165,000 bottles in 2014 and the prediction is to sell 200,000 bottles in 2015.

The first release from the distillery, a 3 year old, was launched in 2009. During the ensuing years, a number of limited bottlings were released and it was not until 2012 that the distillery had a core expression – **Machir Bay**. Limited, but regular releases are the sherry matured **Loch Gorm** and **100% Islay**, both with new editions released in 2015. The latter is made from 100% barley grown on Islay. Recent, limited releases include a **10th anniversary bottle** in May 2015 which was a vatting of sherry and bourbon casks filled between 2005 and 2012 and a **Madeira cask** maturation in September. The special Feis Ile 2015 bottling from the distillery was a **7 year old** vatting of three bourbon barrels. In March 2014 it was time for the distillery's first duty free exclusive. This came in the form of **Coull Point**, which is around 5 years old, matured in bourbon casks and with a few weeks' finish in Oloroso sherry.

History:

2002 Plans are formed for a new distillery at Rockside Farm on western Islay.

2005 Production starts in June.

2006 A fire breaks out in the kiln causing a few weeks´ production stop but malting has to cease for the rest of the year.

2007 The distillery is expanded with two new washbacks.

2009 The first single malt, a 3 year old, is released on 9th September followed by a second release.

2010 Three new releases and an introduction to the US market. John Maclellan from Bunnahabhain joins the team as General Manager.

2011 Kilchoman 100% Islay is released as well as a 4 year old and a 5 year old.

2012 Machir Bay, the first core expression, is released together with Kilchoman Sherry Cask Release and the second edition of 100% Islay.

2013 Loch Gorm and Vintage 2007 are released.

2014 A 3 year old port cask matured and the first duty free exclusive, Coull Point, are released.

2015 The distillery celebrates its 10th anniversary.

Machir Bay

Tasting notes Kilchoman Machir Bay:

GS – A nose of sweet peat and vanilla, undercut by brine, kelp and black pepper. Filled ashtrays in time. A smooth mouth-feel, with lots of nicely-balanced citrus fruit, peat smoke and Germolene on the palate. The finish is relatively long and sweet, with building spice, chili and a final nuttiness.

Kininvie

[kin•in•vee]

Owner:
William Grant & Sons

Region/district:
Speyside

Founded: 1990

Status: Active

Capacity: 4 800 000 litres

Address: Dufftown, Keith, Banffshire AB55 4DH

Website:
-

Tel:
01340 820373

In 1990, the Scotch whisky industry had started to recover from the slump of the late 1970s and 1980s. At that time, younger consumers had turned away from whisky in favour of vodka and white wine.

Slowly, but surely, an interest for single malt was being nurtured. Owners of the big brands realised that malt whisky, necessary for blending purposes, could become scarce in the future if the new trend continued. For William Grant, producers of the best selling single malt in the world, Glenfiddich, increased capacity was vital to cope with the growing demand for both Glenfiddich, as well as the megablend, Grant´s. It was with this purpose in mind that Kininvie was built in 1990.

Kininvie distillery only consists of one still house which is made of white, corrugated metal and neatly tucked away behind Balvenie. It is equipped with a stainless steel full lauter mash tun which is placed next to Balvenie´s in the Balvenie distillery. Ten Douglas fir washbacks can be found in two separate rooms next to the Balvenie washbacks. Three wash stills and six spirit stills are furthermore all heated by steam coils. The distillery was silent in 2011/2012, but made 2.8 million litres in 2014 with further plans to do 21 mashes per week and 2.6 million litres during 2015. Kininvie malt whisky is frequently sold to other companies for blending purposes, under the name Aldundee.

The first time that Kininvie appeared as an official single malt bottling was in 2006, when a Hazelwood 15 year old was launched to celebrate the 105th birthday of Janet Sheed Roberts, the last, surviving grand-daughter of William Grant. In 2008 it was time for a Hazelwood 17 year old to celebrate her 107th birthday. It wasn't until autumn 2013 that Kininvie single malt was launched under its own name for the first time as a **23 year old**. This was followed by batch 2 and 3, both in limited numbers but available in select markets globally. A **17 year old** second expression, has also been released for duty free and, according to the owners, further bottlings can be expected.

History:

1990 Kininvie distillery is inaugurated on 26th June and the first distillation takes place 18th July.

2001 A bottling of blended whisky containing Kininvie malt is released under the name Hazelwood Centennial Reserve 20 years old.

2006 The first expression of a Kininvie single malt is released as a 15 year old under the name Hazelwood.

2008 In February a 17 year old Hazelwood Reserve is launched at Heathrow´s Terminal 5.

2013 A 23 year old Kininvie is launched in Taiwan.

2014 A 17 year old and batch 2 of the 23 year old are released.

BALVENIE & KININVIE DISTILLERIES

17 years old

Tasting notes Kininvie 17 years old:

GS – The nose offers tropical fruits, coconut and vanilla custard, with a hint of milk chocolate. Pineapple and mango on the palate, accompanied by linseed oil, ginger, and developing nuttiness. The finish dries slowly, with more linseed, plenty of spice, and soft oak.

Knockando

[nock•an•doo]

Owner: Diageo.	**Region/district:** Speyside
Founded: 1898	**Status:** Active	**Capacity:** 1 400 000 litres

Address: Knockando, Morayshire AB38 7RT

Website: malts.com	**Tel:** 01340 882000

Knockando single malt has had a strong position in France and Spain for a long time where over 600,000 bottles are sold annually. But it is as a key malt in the J&B blend that the whisky plays an important role.

Like Cutty Sark, the J&B blend was designed for a specific market – America, where the consumers preferred a whisky that was light in flavour and colour. Cutty Sark was first on the shelves as early as the beginning of the 1920s under the Prohibition. Charlie Guttman and Jake Culhane were distributors through their company, The Buckingham Corporation. Towards the end of the 1930s, the partners no longer saw eye-to-eye and Charlie Guttman decided to go solo but needed a strong brand. He turned to Justerini & Brooks in London and offered his services. The name of the whisky was shortened to J&B and Guttman began to challenge Cutty Sark for first place. It was only until 1970 before J&B reached its milestone – a year where no less than 31 million bottles were sold in USA alone.

The distillery is equipped with a small (4.4 tonnes), semi-lauter mash tun, eight Douglas fir washbacks and two pairs of stills. Knockando has always worked a five-day week with 16 mashes per week, 8 short fermentations (50 hours) and 8 long (100 hours). In 2015 this will mean a production of 1.4 million litres of alcohol. The spirit is tankered away to Auchroisk and Glenlossie, and some of the casks are returned to the distillery for maturation in two dunnage and two racked warehouses. Knockando´s nutty character, a result of the cloudy worts coming from the mash tun, has given it its fame. However, in order to balance the taste, the distillers also wish to create the typical Speyside floral notes by using boiling balls on the spirit stills to increase reflux.

The core range consists of **12 year old**, **15 year old Richly Matured**, **18 year old Slow Matured** and the **21 year old Master Reserve**. In 2011 a **25 year old** matured in first fill European oak was released as part of the Special Releases.

History:

1898 John Thompson founds the distillery. The architect is Charles Doig.

1899 Production starts in May.

1900 The distillery closes in March and J. Thompson & Co. takes over administration.

1903 W. & A. Gilbey purchases the distillery for £3,500 and production restarts in October.

1962 W. & A. Gilbey merges with United Wine Traders (including Justerini & Brooks) and forms International Distillers & Vintners (IDV).

1968 Floor maltings is decommissioned.

1969 The number of stills is increased to four.

1972 IDV is acquired by Watney Mann who, in its turn, is taken over by Grand Metropolitan.

1978 Justerini & Brooks launches a 12 year old Knockando.

1997 Grand Metropolitan and Guinness merge and form Diageo; simultaneously IDV and United Distillers merge to United Distillers & Vintners.

2010 A Manager´s Choice 1996 is released.

12 years old

Tasting notes Knockando 12 years old:

GS – Delicate and fragrant on the nose, with hints of malt, worn leather, and hay. Quite full in the mouth, smooth and honeyed, with gingery malt and a suggestion of white rum. Medium length in the finish, with cereal and more ginger.

Knockdhu

[nock•<u>doo</u>]

Owner:
Inver House Distillers
(Thai Beverages plc)

Region/district:
Highland

Founded: **Status:**
1893 Active

Capacity:
2 000 000 litres

Address: Knock, By Huntly, Aberdeenshire AB54 7LJ

Website:
ancnoc.com

Tel:
01466 771223

Knockdhu is the only malt distillery that was built by Distiller´s Company Ltd. (DCL) – a formation of distillers that had a profound effect on the Scotch whisky industry during the 20th century.

The company was formed in 1877 by six grain distilleries in the Lowlands in order to control the output of grain whisky so that the overproduction did not harm the market. By the 1920s, all the major whisky producers (like Walker´s, Buchanan´s and Dewar´s) had joined the company and started to dominate the market. In 1986, DCL was acquired by the brewing giant, Guinness, and the company was renamed United Distillers. Finally, by way of yet another merger, Diageo was born in 1997 and it is currently the world`s biggest spirits company.

Knockdhu distillery is equipped with a 5 tonnes stainless steel lauter mash tun, eight washbacks made of Oregon pine (with fermentation time now increased to 65 hours) and one pair of stills with worm tubs. For 2015 they plan to do 2 million litres of spirit of which 450,000 litres will be heavily peated (45ppm). The spirit is filled mainly into bourbon casks with an additional 15% of sherry butts.

The biggest markets for anCnoc single malt are UK, USA, Sweden and Germany and 180,000 bottles were sold in 2013. The core range consists of **12, 18, 22 years old** and the limited **Vintage 1975**. This was released in February 2015 and is the oldest expression yet to be released by the owner. In addition to that there is the peated range with **Rutter, Flaughter, Tushkar** and **Cutter**. An new expression in the range, **Peatlands** with a phenol content of 9ppm, was launched in autumn 2015. The peated expressions are all between 8 and 12 years old. Every year a new vintage is released and in early 2014, it was a **1999** which was replaced by a **Vintage 2000** in September of 2014. In 2012, a series of bottlings called **Peter Arkle Collection** was introduced for duty free, but these have now been discontinued. They were replaced in early 2015 by two new expressions; **Black Hill Reserve** and the peated (13.5ppm) **Barrow**. Both have matured in bourbon casks and are bottled at 46%.

History:

1893 Distillers Company Limited (DCL) starts construction of the distillery.

1894 Production starts in October.

1930 Scottish Malt Distillers (SMD) takes over production.

1983 The distillery closes in March.

1988 Inver House buys the distillery from United Distillers.

1989 Production restarts on 6th February.

1990 First official bottling of Knockdhu.

1993 First official bottling of anCnoc.

2001 Pacific Spirits purchases Inver House Distillers at a price of $85 million.

2003 Reintroduction of anCnoc 12 years.

2004 A 14 year old from 1990 is launched.

2005 A 30 year old from 1975 and a 14 year old from 1991 are launched.

2006 International Beverage Holdings acquires Pacific Spirits UK.

2007 anCnoc 1993 is released.

2008 anCnoc 16 year old is released.

2011 A Vintage 1996 is released.

2012 A 35 year old is launched.

2013 A 22 year old and Vintage 1999 are released.

2014 A peated range with Rutter, Flaughter, Tushkar and Cutter is introduced.

2015 Vintage 1975 and Peatlands are released as well as Black Hill Reserve and Barrow for duty free.

Tasting notes anCnoc 12 years old:

GS – A pretty, sweet, floral nose, with barley notes. Medium bodied, with a whiff of delicate smoke, spices and boiled sweets on the palate. Drier in the mouth than the nose suggests. The finish is quite short and drying.

12 years old

Lagavulin

[lah•gah•<u>voo</u>•lin]

Owner:
Diageo

Region/district:
Islay

Founded: **Status:**
1816   Active (vc)

Capacity:
2 460 000 litres

Address: Port Ellen, Islay, Argyll PA42 7DZ

Website:
malts.com

Tel:
01496 302749 (vc)

No one seems able to touch Laphroaig as Islay´s best selling malt within the foreseeable future. The battle for second place however, is a fiercely contested struggle between Bowmore and Lagavulin.

Bowmore had a few good years around 2008/2009, but the sales during the last three years have basically been unmoved. Lagavulin, on the other hand, has increased its sales by 30% to 1.7 million bottles sold in 2014. If this trend continues, Lagavulin will grab second place in a couple of years' time. Apart from this, the brand has a history of being the biggest on the island. The sales during 1998 were more than that of Laphroaig and Bowmore together. Thereafter the sales dipped quickly. The reason for this could be ascribed to the fact that two day shifts were in force at Lagavulin during a large part of the 1980s and, of course, this had consequences for the stock of Lagavulin 16 year old. Five-day shifts were only reintroduced first in 1991.

Lagavulin distillery was, for a period of time, owned by Peter Mackie, creator of the famous White Horse blend. He also acted as sales agent for Laphroaig but lost the agency in 1907. Disappointed and infuriated by this, he then decided to build a distillery on the site of Lagavulin that would produce a whisky identical to Laphroaig. The Malt Mill distillery started production in 1908, but, unfortunately, the whisky had little resemblance with Laphroaig single malt, and was only used for various blends until 1962 when it finally closed. Malt Mill was made famous in the 2012 Ken Loach movie, Angel´s Share, where a cask of Malt Mill single malt is discovered and offered at an auction. This is all fiction, but a few weeks after the film had been released, a bottle of Malt Mill new make from the last distillation was discovered at Lagavulin and this is now on display at the distillery. It is believed that there may be two or three bottlings of Malt Mill single malt in the hands of collectors.

The distillery is equipped with a 4.4 ton stainless steel full lauter mash tun, ten washbacks made of larch with a 55 hour fermentation cycle and two pairs of stills. Bourbon hogsheads are used almost without exception for maturation and all of the new production is stored on the mainland. There are only around 16,000 casks on Islay, split between warehouses at Lagavulin, Port Ellen and Caol Ila. The distillery is working 24/7 for 51 weeks and the volume for 2015 will be 2,45 million litres of alcohol.

The core range of Lagavulin is unusually limited and only consists of a **12 year old cask strength**, a **16 year old** and the **Distiller´s Edition**, a Pedro Ximenez sherry finish. Recent, limited releases include a **21 year old** from first fill sherry casks in 2012 and a **37 year old** distilled in 1976 and released in 2013. Autumn 2014 also saw the launch of a **triple matured Lagavulin** for sale at the visitor centre and for The Friends of the Classic Malts. The Islay Festival special release for 2015, bottled at 59,9%, was distilled in **1991** with a **triple maturation** in bourbon casks, PX sherry casks and, finally, old oak puncheons.

History:

1816   John Johnston founds the distillery.

1825   John Johnston takes over the adjacent distillery Ardmore founded in 1817 by Archibald Campbell and closed in 1821.

1835   Production at Ardmore ceases.

1837   Both distilleries are merged and operated under the name Lagavulin by Donald Johnston.

1852   The brother of the wine and spirits dealer Alexander Graham, John Crawford Graham, purchases the distillery.

1867   The distillery is acquired by James Logan Mackie & Co. and refurbishment starts.

1878   Peter Mackie is employed.

1889   James Logan Mackie passes away and nephew Peter Mackie inherits the distillery.

1890   J. L. Mackie & Co. changes name to Mackie & Co. Peter Mackie launches White Horse onto the export market with Lagavulin included in the blend. White Horse blended is not available on the domestic market until 1901.

1908   Peter Mackie uses the old distillery buildings to build a new distillery, Malt Mill, on the site.

1924   Peter Mackie passes away and Mackie & Co. changes name to White Horse Distillers.

1927   White Horse Distillers becomes part of Distillers Company Limited (DCL).

1930   The distillery is administered under Scottish Malt Distillers (SMD).

1952   An explosive fire breaks out and causes considerable damage.

1962   Malt Mills distillery closes and today it houses Lagavulin's visitor centre.

1974   Floor maltings are decommisioned and malt is bought from Port Ellen instead.

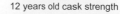

History continued:

1988 Lagavulin 16 years becomes one of six Classic Malts.

1998 A Pedro Ximenez sherry finish is launched as a Distillers Edition.

2002 Two cask strengths (12 years and 25 years) are launched.

2006 A 30 year old is released.

2007 A 21 year old from 1985 and the sixth edition of the 12 year old are released.

2008 A new 12 year old is released.

2009 A new 12 year old appears as a Special Release.

2010 A new edition of the 12 year old, a single cask exclusive for the distillery and a Manager´s Choice single cask are released.

2011 The 10th edition of the 12 year old cask strength is released.

2012 The 11th edition of the 12 year old cask strength and a 21 year old are released.

2013 A 37 year old and the 12th edition of the 12 year old cask strength are released.

2014 A triple matured for Friends of the Classic Malts and the 13th edition of the 12 year old cask strength are released.

Tasting notes Lagavulin 12 year old:

GS – Soft and buttery on the nose, with dominant, fruity, peat smoke, grilled fish and a hint of vanilla sweetness. More fresh fruit notes develop with the addition of water. Medium-bodied, quite oily in texture, heavily smoked, sweet malt and nuts. The finish is very long and ashy, with lingering sweet peat.

Distiller´s Edition   37 years old   FOCM

16 years old   12 years old cask strength

Laphroaig

[lah•froyg]

Owner:
Beam Suntory

Region/district:
Islay

Founded: 1815

Status: Active (vc)

Capacity: 3 300 000 litres

Address: Port Ellen, Islay, Argyll PA42 7DU

Website: laphroaig.com

Tel: 01496 302418

In 2015, the owners of Laphroaig could look back at 200 years of almost unbroken production. During this period of time, Laphroaig single malt has established itself as the most famous of peated whisky in the world with 3.2 million bottles sold in 2014.

From the founding until 1954, the distillery was owned by the same family (Johnston/Hunter), but it was the last in line, Ian Hunter, who played the biggest role in elevating Laphroaig to its current position. Not only did he double the capacity of the distillery, but was also able to increase the export of the whisky substantially. Hunter was also one of the first Scottish distillers to use bourbon casks for maturation and found that a mix of American and European oak enhanced the flavour of his whisky. Without heirs, Hunter left the distillery to his secretary and later manager, Bessie Williamson, who became a legend in her own right. During her years as owner, she expanded the distillery with another three stills and continued to promote the single malt, especially in America.

Laphroaig is equipped with a stainless steel full lauter mash tun and six stainless steel washbacks. The distillery uses an unusual combination of three wash stills and four spirit stills, all fitted with ascending lyne arms. Laphroaig is one of very few distilleries with its own maltings which produces 15% of its requirements. The balance comes from Port Ellen maltings or is imported from the mainland. Own malt and sourced malt are always blended before mashing. The own malt has a phenol specification of 40-60ppm, while bought malt lies between 35 and 45ppm. Laphroaig has one of the best visitor centres in the industry with tours at a variety of levels.

The core range consists of **Select** without age statement, **10 year old, 10 year old cask strength, Quarter Cask, Triple Wood, 18 year old** and a **25 year old**. There are four bottlings exclusive for duty free; **Laphroaig PX**, matured in hogsheads and quarter casks with a finish in Pedro Ximenez sherry casks, **Laphroaig QA Cask** with a double maturation in ex-bourbon barrels and new American oak, **Laphroaig An Cuan Mor**, also double matured and, finally, **Brodir**. The latter was first released in 2012 as a **13 year old** and sold exclusively on the Viking Line ferries between Sweden and Finland. A second, and more widely available version of Brodir, appeared in autumn 2014. This time it was a **14 year old** with a finish in ruby port pipes. The special **Feis Ile** bottling for 2015 was a **Cairdeas** without age statement, but probably around 12 years. The special feature was that it had been distilled in 2003, using 100% floor malted barley from its own maltings. Another three, limited releases appeared during 2015; the **15 year old**, which after having been replaced in 2009 by the 18 year old, made a temporary comeback with 72,000 bottles being released, a **21 year old** to celebrate the 21st anniversary of Friends of Laphroaig, as well as a **32 year old oloroso sherry** maturation which is to be launched in December.

History:

1815 Brothers Alexander and Donald Johnston found Laphroaig.

1836 Donald buys out Alexander and takes over operations.

1837 James and Andrew Gairdner found Ardenistiel a stone's throw from Laphroaig.

1847 Donald Johnston is killed in an accident in the distillery when he falls into a kettle of boiling hot burnt ale. The Manager of neigh-bouring Lagavulin, Walter Graham, takes over.

1857 Operation is back in the hands of the Johnston family when Donald's son Dougald takes over.

1860 Ardenistiel Distillery merges with Laphroaig.

1877 Dougald, being without heirs, passes away and his sister Isabella, married to their cousin Alexander takes over.

1907 Alexander Johnston dies and the distillery is inherited by his two sisters Catherine Johnston and Mrs. William Hunter (Isabella Johnston).

1908 Ian Hunter arrives in Islay to assist his mother and aunt with the distillery.

1924 The two stills are increased to four.

1927 Catherine Johnston dies and Ian Hunter takes over.

1928 Isabella Johnston dies and Ian Hunter becomes sole owner.

1950 Ian Hunter forms D. Johnston & Company

1954 Ian Hunter passes away and management of the distillery is taken over by Elisabeth "Bessie" Williamson, who was previously Ian Hunters PA and secretary. She becomes Director of the Board and Managing Director.

1967 Seager Evans & Company buys the distillery through Long John Distillery, having already acquired part of Laphroaig in 1962. The number of stills is increased from four to five.

History continued:

1972 Bessie Williamson retires. Another two stills are installed bringing the total to seven.

1975 Whitbread & Co. buys Seager Evans (now renamed Long John International) from Schenley International.

1989 The spirits division of Whitbread is sold to Allied Distillers.

1991 Allied Distillers launches Caledonian Malts. Laphroaig is one of the four malts included.

1994 HRH Prince Charles gives his Royal Warrant to Laphroaig. Friends of Laphroaig is founded.

1995 A 10 year old cask strength is launched.

2001 A 40 year old is released.

2004 Quarter Cask, a mix of different ages with a finish in quarter casks (i. e. 125 litres) is launched.

2005 Fortune Brands becomes new owner.

2007 A vintage 1980 (27 years old) and a 25 year old are released.

2008 Cairdeas, Cairdeas 30 year old and Triple Wood are released.

2009 An 18 year old is released.

2010 A 20 year old for French Duty Free and Cairdeas Master Edition are launched.

2011 Laphroaig PX and Cairdeas - The Ileach Edition are released. Triple Wood is moved to the core range and replaced in duty free by Laphroaig PX.

2012 Brodir and Cairdeas Origin are launched.

2013 QA Cask, An Cuan Mor, 25 year old cask strength and Cairdeas Port Wood Edition are released.

2014 Laphroaig Select and a new version of Cairdeas are released.

2015 A 21 year old, a 32 year old sherry cask and a new Cairdeas are released and the 15 year old is re-launched.

Tasting notes Laphroaig 10 year old:
GS – Old-fashioned sticking plaster, peat smoke and seaweed leap off the nose, followed by something a little sweeter and fruitier. Massive on the palate, with fish oil, salt and plankton, though the finish is quite tight and increasingly drying.

Select   Quarter Cask   15 years old

PX Cask   QA Cask

10 years old   Cairdeas 2015   An Cuan Mor

Linkwood

[link•wood]

Owner:
Diageo

Region/district:
Speyside

Founded: **Status:** **Capacity:**
1821　　　 Active　　 5 600 000 litres

Address: Elgin, Morayshire IV30 3RD

Website: **Tel:**
malts.com　　　　　　 01343 862000

Linkwood single malt has always had a good reputation among blenders. One of the owners has been especially skilful in marketing the distillery's whisky over the years.

The person in question was Innes Cameron who was the major stakeholder of Linkwood for the first three decades of the 1900s. Born in 1860, he began as a grocer's apprentice in Elgin but soon moved on to become a spirit merchant. Except for Linkwood, Cameron was also involved in Teaninich and Tamdhu. He was also the chairman of North of Scotland Malt Distiller's Association for the ten years leading up to his death in 1932.

Combining fresh, fruity, green/grassy flavours with a surprisingly powerful mouth feel – Linkwood incorporates the best from both worlds. To achieve the desired character, they try to get the wort as clear as possible, the fermentations are long (75 hours), the large stills are only filled just above the man door to enhance the copper contact, while the spirit stills are allowed to rest for a minimum of one hour between runs to allow for the regeneration of the copper. The old part of the distillery, which stopped producing in 1996, was equipped with worm tubs and had a slightly different character. Diageo's master blenders use Linkwood for a variety of the company's blends, especially for Johnnie Walker and White Horse.

On two occasions during 2011-2013, the distillery was expanded. The old distillery buildings facing Linkwood Road were demolished and an extension of the current still house, to house two of the stills and the tunroom, was constructed. The only original buildings from 1872 left standing are No. 6 warehouse and the redundant, old kiln with the pagoda roof. The set up of equipment now is one 12.5 ton full lauter mash tun, 11 wooden washbacks (80,000 litres but only filled with 57,000) and three pairs of stills. The two old stills were refurbished and moved into the new still house. Production during 2015 will be 5.1 million litres of alcohol.

The only official bottling is a **12 year old Flora & Fauna**.

History:

1821 Peter Brown founds the distillery.

1868 Peter Brown passes away and his son William inherits the distillery.

1872 William demolishes the distillery and builds a new one.

1897 Linkwood Glenlivet Distillery Company Ltd takes over operations.

1902 Innes Cameron, a whisky trader from Elgin, joins the Board and eventually becomes the major shareholder and Director.

1932 Innes Cameron dies and Scottish Malt Distillers takes over in 1933.

1962 Major refurbishment takes place.

1971 The two stills are increased by four. Technically, the four new stills belong to a new distillery referred to as Linkwood B.

1985 Linkwood A (the two original stills) closes.

1990 Linkwood A is in production again for a few months each year until 1996.

2002 A 26 year old from 1975 is launched as a Rare Malt.

2005 A 30 year old from 1974 is launched as a Rare Malt.

2008 Three different wood finishes (all 26 year old) are released.

2009 A Manager's Choice 1996 is released.

2013 Expansion of the distillery including two more stills.

12 years old

Tasting notes Linkwood 12 years old:

GS – Floral, grassy and fragrant on the nutty nose, while the slightly oily palate becomes increasingly sweet, ending up at marzipan and almonds. The relatively lengthy finish is quite dry and citric.

1963

Glenfiddich Straight Malt is released

The greater part of the 1900s belonged to the blends with some of the iconic brands having already been established in the late 1800s.

But it wasn't until after the second world war, that they monopolised the globe by selling millions of cases every year. Single malts on the other hand, were rarely heard of, apart from being an integral part of a blend. There were brands that were for sale, and had been for a long time, but to sell and to actively market are two different things. Nobody had come up with the idea of putting any effort into establishing single malt Scotch as a new category to reckon with. The idea was ludicrous and would surely only cannibalise on the sales of the important blends.

Two brothers were of a different opinion. In 1953, William Grant, grandson of the founder of Glenfiddich, died unexpectedly and his two sons, Alexander and Charles, became company directors aged 22 and 26 respectively. Of course, this was a gigantic responsibility for these two, young men but they would soon prove ready for the task. The two brothers quickly decided to divide the responsibilities of the company. Charles focused on the blended side of the business and, as such, his first task was to secure the company's need for grain whisky. Until the beginning of the 1960s, W Grant & Sons had relied mainly on deliveries of grain whisky from the mighty DCL. Charles didn't like the idea of being dependent on them, not least since DCL had their own blended brands which grew at the same pace as Grant's. It wouldn't be long, he thought, before Grant's would experience problems receiving deliveries. He decided that a grain distillery of his own would be the solution and in just nine months, he made sure that a new grain distillery was built in Girvan.

Alexander, or Sandy as he was fondly called, pursued the idea of selling and marketing Glenfiddich straight malt (although the term single malt didn't even exist at the time). In the meanwhile, Glenfiddich was enjoyed in pubs around Dufftown, but it wasn't sold to other parts of the country and neither was it available abroad. He packaged a 5 year old version in the same triangular bottle that had been used for Grants since 1957. This came to fruition in 1961 and he sensed that this could become huge. His competitors were sceptic, to say the least, and even more so when Sandy, in 1963, started to promote the new brand in a worldwide advertising campaign and using television as one of the means of advertising. Such a move was close to blasphemy in the conservative Scotch whisky business. The Glenfiddich Straight Malt was soon presented in a dark green bottle to clearly distinguish it from the Grants blend and the recipe was changed to an 8 year old with some older whiskies (12-13 years) included.

The Glenfiddich Straight Malt was launched in the USA in 1963 and it soon became a success story in other export markets as well. The other producers, although taking a very cautious approach to begin with, soon realised that there was money to be made from this "new" category. They followed suit, but not as fast as one would have thought. It lasted until 1978 when Macallan started to promote their single malt more actively and Glenmorangie followed suit in 1981. DCL were a little quicker with Cardhu which was allocated a marketing budget of £15.000 but, already in 1967, the marketing component ceased and the company didn't review its approach towards single malts until 1977.

A downturn in the global economy during the 1970s put a temporary hold to the expansion of single malts, but the actions taken by William Grants in 1963 paved the way for a renewed interest in the category from the late 1980s – an interest that today still shows no sign of diminishing.

Sandy Grant Gordon, right.

Loch Lomond

[lock low•mund]

Owner:
Loch Lomond Group
(majority owner Exponent)

Region/district:
Western Highlands

Founded: | **Status:** | **Capacity:**
1965 | Active | 5 000 000 litres

Address: Lomond Estate, Alexandria G83 0TL

Website:
lochlomonddistillery.com

Tel:
01389 752781

Loch Lomond distillery changed ownership in March 2014 when a management buy-in, backed by a private equity group, took place. The seller, at a price of £210m, was Sandy Bulloch who had built up his family business since the mid 1980s.

Also included in the deal were Glen Scotia distillery, the Glen Catrine Bonded Warehouse bottling plant and several brands. Soon after the acquisition, the new owners started to invest in the business. A major refurbishing of Glen Scotia including a visitor centre, upgrade of the bottling line in East Ayrshire and a revamp of the product ranges were all signs of how serious the owners were. Loch Lomond distillery also received its fair share of the investment when two new stills and three washbacks were added.

Loch Lomond distillery is equipped with one full lauter mash tun complemented by ten 25,000 litres and eleven 50,000 litres washbacks, all of which are made of stainless steel. The set-up of stills differs completely from any other distillery in Scotland. There are two, traditional, copper pot stills and six copper stills where the swan necks have been exchanged with rectifying columns. Furthermore, there is one Coffey still used for continuous distillation. And if this was not enough, an additional distillery with column stills producing grain whisky is housed in the same building. For the grain side of production there are twelve 100,000 litres and eight 200,000 litres washbacks. Its total capacity is 5 million litres of malt spirit and 18 million litres of grain.

The distillery used to produce a broad range of whiskies with two main brands – Loch Lomond and Inchmurrin. Further back in time, there were another 5-6 brands and several have contained peated whisky as well. Since spring 2015, when new bottlings were introduced, the range consists of **Loch Lomond Original Single Malt** without age statement and predominantly matured in bourbon casks, as well as two blends – the **Reserve** and the **Signature**. Furthemore, there are **Inchmurrin 12, 18** and a **Madeira Wood** without age statement, **Glengarry NAS** and **12 years old** and, finally, **Loch Lomond Single Grain**.

History:

1965 The distillery is built by Littlemill Distillery Company Ltd owned by Duncan Thomas and American Barton Brands.

1966 Production commences.

1971 Duncan Thomas is bought out.

1984 The distillery closes.

1985 Glen Catrine Bonded Warehouse Ltd buys Loch Lomond Distillery.

1987 The distillery resumes production.

1993 Grain spirits are also distilled.

1997 A fire destroys 300,000 litres of maturing whisky.

1999 Two more stills are installed.

2005 Inchmoan and Craiglodge as well as Inchmurrin 12 years are launched.

2006 Inchmurrin 4 years, Croftengea 1996 (9 years), Glen Douglas 2001 (4 years) and Inchfad 2002 (5 years) are launched.

2010 A peated Loch Lomond with no age statement is released as well as a Vintage 1966.

2012 New range for Inchmurrin released – 12, 15, 18 and 21 years.

2014 The distillery is sold to Exponent Private Equity. Organic versions of 12 year old single malt and single blend are released.

2015 Loch Lomond Original Single Malt is released together with a single grain and two blends, Reserve and Signature.

Tasting notes Loch Lomond Original:

GS – Initially earthy on the nose, with malt and subtle oak. The palate is rounded, with allspice, orange, lime, toffee, and a little smokiness. Barley, citrus fruits and substantial spiciness in the finish.

Loch Lomond Original

Longmorn

[long•morn]

Owner:
Chivas Brothers
(Pernod Ricard)

Region/district:
Speyside

Founded: | **Status:** | **Capacity:**
1894 | Active | 4 500 000 litres

Address: Longmorn, Morayshire IV30 8SJ

Website:
-

Tel:
01343 554139

Of the 15 malt distilleries operated by Chivas Brothers in Scotland, there are two where the owners have focused on the single malts, Glenlivet and Aberlour, both being in the top six of the world. The question is if a third distillery is about to join them.

Longmorn single malt is often referred to as a hidden gem or the whisky blender's favourite, and to the owners, Chivas Brothers, the whisky has become an integral part of several of their blends. Sales of Longmorn single malt today are minuscule and only one widely available expression exists. Recently, however, rumours have been floating around that a new range of single malts from the distillery could be in the offing.

Four years after John Duff had founded the distillery, he built another distillery just a few hundred metres from Longmorn which was called Longmorn 2 (later to become Benriach). The two plants worked as sister distilleries to such an extent that a private railroad was built to transport barley, peat, coal and casks between the two.

In 2012, Longmorn distillery was completely revamped and expanded. A new 8.5 tonnes Briggs full lauter mash tun replaced the old, traditional tun and seven of the eight, old stainless steel washbacks were moved to the new tun room and an additional three were installed. There are currently four pairs of stills, all fitted with sub-coolers and the wash stills now have external heat exchangers. The production capacity has also increased by 30% to 4.5 million litres.

Longmorn single malt is mainly used for some of the owner's top blends, especially Chivas Regal 18 year old and Royal Salute. In 2007 a **16 year old** replaced a 15 year old and this is still the only official core expression. Apart from that, a cask strength version can also be found at Chivas' visitor centres. The latest version is a **1999** bottled in 2014.

History:

1893 John Duff & Company, which founded Glenlossie already in 1876, starts construction. John Duff, George Thomson and Charles Shirres are involved in the company. The total cost amounts to £20,000.

1894 First production in December.

1897 John Duff buys out the other partners.

1898 John Duff builds another distillery next to Longmorn which is called Benriach (at times aka Longmorn no. 2). Duff declares bankruptcy and the shares are sold by the bank to James R. Grant.

1970 The distillery company is merged with The Glenlivet & Glen Grant Distilleries and Hill Thomson & Co. Ltd. Own floor maltings ceases.

1972 The number of stills is increased from four to six. Spirit stills are converted to steam firing.

1974 Another two stills are added.

1978 Seagrams takes over through The Chivas & Glenlivet Group.

1994 Wash stills are converted to steam firing.

2001 Pernod Ricard buys Seagram Spirits & Wine together with Diageo and Pernod Ricard takes over the Chivas group.

2004 A 17 year old cask strength is released.

2007 A 16 year old is released replacing the 15 year old.

2012 Production capacity is expanded.

16 years old

Tasting notes Longmorn 16 years old:

GS – The nose offers cream, spice, toffee apples and honey. Medium bodied in the mouth, with fudge, butter and lots of spice. The finish is quite long, with oak and late-lingering dry spices.

Macallan

[mack•al•un]

Owner:	**Region/district:**
Edrington Group	Speyside

Founded:	**Status:**	**Capacity:**
1824	Active (vc)	11.000 000 litres

Address: Easter Elchies, Craigellachie, Morayshire
AB38 9RX

Website:	**Tel:**
themacallan.com	01340 871471

In spring 2015 the main part of the Macallan Estate was turned into a construction-site and judging by the size of it, it looked like an entire village would be built. This is, of course, in line with the new distillery which is supposed to be ready by 2017.

The continued success for Macallan single malt (based on the 9,9 million bottles which were sold in 2014), has prompted the owners to build a new plant which will become Scotland´s biggest malt distillery with a capacity of 16 million litres. Located on the left side of the road leading up to the estate, the landmark building will be shaped like five hills with meadow grass on the roof. One part will house the new visitor centre, there will be three still houses, each with 4 wash stills and 8 spirit stills, and the fifth part will be the mash house with two mash tuns. The total cost of the project will be a staggering £100m!

Another project involving Macallan, although they are not involved financially, is the on-going construction of a £74m biomass combined heat and power plant. This is a joint venture between infrastructure development firm, John Laing, and the Green Investment Bank. The plant, due to be commissioned in March 2016, will power more than 20,000 homes, as well as providing 90% of the heat requirement for the distillery. Carbon emission savings from the plant will be the equivalent of taking 18,000 cars off the road.

Until the new distillery is ready, the production currently takes place in two separate plants. The number one plant holds one full lauter mash tun, 19 stainless steel washbacks (three were installed in July 2014), five wash stills and ten spirit stills. The number two plant is comprised of one semi-lauter mash tun, six wooden washbacks and three made of stainless steel, two wash stills and four spirit stills. The plan for 2015 is to do 73 mashes per week which means that they will end up with 11 million litres of alcohol – an all-time high for the distillery!

The core range for Macallan consists of the 1824 Series launched in 2012 – **Gold, Amber, Sienna** and **Ruby**. The whiskies are sold without age statement and have been matured in ex-sherry casks, but the origin of the oak and the age of the whisky will vary. Macallan´s two previous ranges, **Sherry Oak (12, 18, 25** and **30 year old)** and **Fine Oak (10, 12, 15, 17, 18, 21, 25** and **30 year old)** will gradually be phased out, but will still be available for purchase at selected markets for a fairly long period of time. In 2014, an extension of the range called 1824 Masters Series was launched. Four expressions, all sherry matured and bottled in Lalique decanters, have been included – **Rare Cask, Reflexion, No.6** and **M**. The Macallan duty free range holds four expressions; **Select Oak, Whisky Maker´s Edition, Estate Reserve** and **Oscuro**. A fifth expression, **Rare Cask Black**, is due for release in October 2015. Finally, there is also **The Fine & Rare** range – which showcases vintages from 1926 to 1990.

History:

1824 The distillery is licensed to Alexander Reid under the name Elchies Distillery.

1847 Alexander Reid passes away and James Shearer Priest and James Davidson take over.

1868 James Stuart takes over the licence. He founds Glen Spey distillery a decade later.

1886 James Stuart buys the distillery.

1892 Stuart sells the distillery to Roderick Kemp from Elgin. Kemp expands the distillery and names it Macallan-Glenlivet.

1909 Roderick Kemp passes away and the Roderick Kemp Trust is established to secure the family's future ownership.

1965 The number of stills is increased from six to twelve.

1966 The trust is reformed as a private limited company.

1968 The company is introduced on the London Stock Exchange.

1974 The number of stills is increased to 18.

1975 Another three stills are added, now making the total 21.

1984 The first official 18 year old single malt is launched.

1986 Japanese Suntory buys 25% of Macallan-Glenlivet plc stocks.

1996 Highland Distilleries buys the remaining stocks. 1874 Replica is launched.

1999 Edrington and William Grant & Sons buys Highland Distilleries (where Edrington, Suntory and Remy-Cointreau already are shareholders) for £601 million. They form the 1887 Company which owns Highland Distilleries with 70% held by Edrington and 30% by William Grant & Sons (excepting the 25% share held by Suntory).

History continued:

2000 The first single cask from Macallan (1981) is named Exceptional 1.

2001 A new visitor centre is opened.

2002 Elegancia replaces 12 year old in the duty-free range. 1841 Replica, Exceptional II and Exceptional III are also launched.

2003 1876 Replica and Exceptional IV, single cask from 1990 are released.

2004 Exceptional V, single cask from 1989 is released as well as Exceptional VI, single cask from 1990. The Fine Oak series is launched.

2005 New expressions are Macallan Woodland Estate, Winter Edition and the 50 year old.

2006 Fine Oak 17 years old and Vintage 1975 are launched.

2007 1851 Inspiration and Whisky Maker´s Selection are released as a part of the Travel Retail range. 12 year old Gran Reserva is launched in Taiwan and Japan.

2008 Estate Oak and 55 year old Lalique are released.

2009 Capacity increased by another six stills. The Macallan 1824 Collection, a range of four duty free expressions, is launched. A 57 year old Lalique bottling is released.

2010 Oscuro is released for Duty Free.

2011 Macallan MMXI is released for duty free.

2012 Macallan Gold, the first in the new 1824 series, is launched.

2013 Amber, Sienna and Ruby are released.

2014 1824 Masters Series (with Rare Cask, Reflexion and No. 6) is released.

2015 Rare Cask Black is released.

Tasting notes Macallan Gold:

GS – The nose offers apricots and peaches, fudge and a hint of leather. Medium-bodied, with malt, walnuts and spices on the palate. Quite oaky in the medium-length finish.

Tasting notes Macallan Amber:

GS – Sweet sherry, malt, spicy fudge and a hint of cinnamon on the nose. Sherry and malt carry over from the nose to the palate, with ginger emerging in time. Christmas cake flavours and Jaffa orange in the relatively lengthy finish.

Tasting notes Macallan 12 year old sherry oak:

GS – The nose is luscious, with buttery sherry and Christmas cake characteristics. Rich and firm on the palate, with sherry, elegant oak and Jaffa oranges. The finish is long and malty, with slightly smoky spice.

Amber

Sienna

Ruby

Fine Oak 15   Select Oak   Sherry Oak 12

Gold

No. 6   Rare Cask   Reflexion

Macduff

[mack•duff]

Owner:
John Dewar & Sons Ltd
(Bacardi)

Region/district:
Highlands

Founded: **Status:** **Capacity:**
1960 Active 3 340 000 litres

Address: Banff, Aberdeenshire AB45 3JT

Website: **Tel:**
lastgreatmalts.com 01261 812612

A lot of effort has recently gone into re-launching the whisky from the distillery as a single malt. Yet, there can be no doubt about the role that the Macduff distillery plays in the company – mainly to make the signature malt for the William Lawson´s blend.

The brand was registered in 1889 by an Irish blending company called E. & J. Burke. The whisky is named after William Lawson who became export manager of the company one year before the blend was launched. Sales increased steadily, but it wasn´t until Martini & Rossi took over in 1963 that volumes began to skyrocket. Today it is the 7th biggest Scotch in the world, not least in Russia where it was launched five years ago and is already selling 1 million cases per year. The standard version without age statement was complemented with a 13 year old in 2013. In autumn 2014, William Lawson took the same route as several brands have done recently. They produced a flavoured spirit drink based on whisky and launched William Lawson´s Super Spiced which is infused with vanilla and spices.

The distillery is equipped with a stainless steel semi-lauter mash tun, nine washbacks made of stainless steel and the rather unusual set-up of five stills (two wash stills and three spirit stills). The fifth still was installed in 1990. In order to fit the stills into the still room, the lyne arms on four of the stills are bent in a peculiar way and on one of the wash stills it is U-shaped. In 2015 the distillery will be doing 26 mashes per week for 48 weeks producing 3.4 million litres of alcohol.

Official bottlings from Macduff have always been made under the name Glen Deveron. When a completely new range of bottlings was launched in September 2015, the name had changed to The Deveron. The core range now consists of a **10 year old**, exclusive to France, as well as a **12** and **18 year old**. There are also plans for a possible **25 year old** in the future. For duty free, a new range was launched in 2013 under the name The Royal Burgh Collection encompassing a **16**, a **20** and a **30 year old**.

History:

1960 The distillery is founded by Marty Dyke, George Crawford and Brodie Hepburn (who is also involved in Tullibardine and Deanston). Macduff Distillers Ltd is the name of the company.

1965 The number of stills is increased from two to three.

1967 Stills now total four.

1972 William Lawson Distillers, part of General Beverage Corporation which is owned by Martini & Rossi, buys the distillery from Glendeveron Distilleries.

1990 A fifth still is installed.

1993 Bacardi buys Martini Rossi (including William Lawson) and eventually transfered Macduff to the subsidiary John Dewar & Sons.

2013 The Royal Burgh Collection (16, 20 and 30 years old) is launched for duty free.

2015 A new range is launched - 10, 12 and 18 years old.

12 years old

Tasting notes The Deveron 12 years old:

GS – Soft, sweet and fruity on the nose, with vanilla, ginger, and apple blossom. Medium-bodied, gently spicy, with butterscotch and Brazil nuts. Caramel contrasts with quite dry spicy oak in the finish.

Mannochmore

[man•och•moor]

Owner:
Diageo

Region/district:
Speyside

Founded: **Status:** **Capacity:**
1971 Active 6 000 000 litres

Address: Elgin, Morayshire IV30 8SS

Website: **Tel:**
malts.com 01343 862000

Mannochmore is associated with the release of the weird and infamous Loch Dhu single malt. It was distilled at Mannochmore in 1986 and released in 1996 as a 10 year old.

Already in 1997 it was withdrawn from the market. The whisky was almost black (Loch Dhu means black lake in Gaelic) and the official explanation for the colour was that it had been matured in double-charred casks. Most people that have tasted it rather think that it's due to a hefty addition of caramel. It did not get many good reviews but, over the years, it has achieved cult status in certain circuits and the few bottles that still show up are sold for hundreds of pounds.

Two miles south of Elgin lies the Glenlossie/Mannochmore complex. It is a busy site with 14 warehouses holding 250,000 casks from many of Diageo's 28 distilleries. There is also a dark grains plant processing pot ale into cattle feed, as well as a newly installed biomass burner converting draff from 20 nearby distilleries into steam, which will power the whole site. And, of course, there are two distilleries with Glenlossie having been built almost a hundred years before its younger sister, Mannochmore. Five years ago, the two distilleries produced a combined total of 5 million litres – a figure which today, thanks largely to a powerful expansion and switching to 7-day weeks, has almost doubled to 9 million litres. Mannochmore has now outgrown the older sibling and accounts for two thirds of the capacity. Since summer 2013 the distillery is equipped with an 11 ton Briggs full lauter mash tun, eight wooden washbacks and another eight external made of stainless steel and four pairs of stills. All the wooden washbacks date from 1971 when the distillery was built and are due for replacement within the next two years. The 8 new washbacks will also be made of larch.

The core range of Mannochmore is just a **12 year old Flora & Fauna**. In 2009, a limited **18 year old** was released and in 2010 it was time for a sherry matured single cask from **1998**.

History:

1971 Scottish Malt Distillers (SMD) founds the distillery on the site of their sister distillery Glenlossie. It is managed by John Haig & Co. Ltd.

1985 The distillery is mothballed.

1989 In production again.

1992 A Flora & Fauna series 12 years old becomes the first official bottling.

2009 An 18 year old is released.

2010 A Manager's Choice 1998 is released.

2013 The number of stills is increased to four.

Tasting notes Mannochmore 12 years old:

GS – Perfumed and fresh on the light, citric nose, with a sweet, floral, fragrant palate, featuring vanilla, ginger and even a hint of mint. Medium length in the finish, with a note of lingering almonds.

12 years old

Miltonduff

[mill•ton•duff]

Owner:
Chivas Brothers
(Pernod Ricard)

Region/district:
Speyside

Founded: 1824   **Status:** Active   **Capacity:** 5 800 000 litres

Address: Miltonduff, Elgin, Morayshire IV30 8TQ

Website:
-

Tel:
01343 547433

Even if the distance between Miltonduff and Glenburgie is more than 10 kilometres, one can still speak of them as being sister distilleries. The common link since way back is the Ballantine's blend.

It was through the self-made Canadian billionaire, Harry C Hatch, that both distilleries ended up in the same company in 1936, Hiram Walker, together with the Ballantine's brand. Hatch was a very sharp and pragmatic businessman who, in a skilful way, used the prohibition in the USA to build his business. Hiram Walker Ltd, based in Canada, had been owned by the Walker family for many years. They were American citizens and when US prohibition came into force in 1920, they faced a problem in supplying American consumers with Canadian liquor. Harry Hatch had no such scruples, so he bought the company. His attitude was clear; "The legislation does not prevent us from exporting at all. It prevents somebody from over there from importing. There's a difference." Harry Hatch subsequently died in 1946 aged 62.

Miltonduff distillery is equipped with an 8 tonne full lauter mash tun with a copper dome, 16 stainless steel washbacks with a fermentation time of 56 hours and six, large stills. The lyne arms are all sharply descending allowing for very little reflux. This makes for a rather robust and oily newmake in contrast to the lighter and more floral Glenburgie. Together, the two malts form the backbone of Ballantine's – the second biggest Scotch in the world after Johnnie Walker.

The most recent official bottling of Miltonduff is a **1997, 15 year old**, which was released in Chivas Brothers' cask strength series and is available only at Chivas' visitor centres. In 2013, a special Miltonduff version of the 17 year old Ballantine's was launched in its Signature Distillery Editions range. The idea is to highlight the four signature malts of the world's number 2 blended Scotch. The first two (Scapa and Glenburgie) were released in 2012 and the fourth and final one (Glentauchers) came in 2014.

History:

1824 Andrew Peary and Robert Bain obtain a licence for Miltonduff Distillery. It has previously operated as an illicit farm distillery called Milton Distillery but changes name when the Duff family buys the site it is operating on.

1866 William Stuart buys the distillery.

1895 Thomas Yool & Co. becomes new part-owner.

1936 Thomas Yool & Co. sells the distillery to Hiram Walker Gooderham & Worts. The latter transfers administration to the newly acquired subsidiary George Ballantine & Son.

1964 A pair of Lomond stills is installed to produce the rare Mosstowie.

1974 Major reconstruction of the distillery.

1981 The Lomond stills are decommissioned and replaced by two ordinary pot stills, the number of stills now totalling six.

1986 Allied Lyons buys 51% of Hiram Walker.

1987 Allied Lyons acquires the rest of Hiram Walker.

1991 Allied Distillers follow United Distillers' example of Classic Malts and introduce Caledonian Malts in which Tormore, Glendro-nach and Laphroaig are included in addition to Miltonduff. Tormore is later replaced by Scapa.

2005 Chivas Brothers (Pernod Ricard) becomes the new owner through the acquisition of Allied Domecq.

- CHIVAS BROTHERS -
CASK STRENGTH EDITION
NON CHILL-FILTERED
SPEYSIDE SINGLE MALT SCOTCH WHISKY
Miltonduff
16 YEARS   MD 16 004
Bottled straight from the cask at **52.9**
6 March 1998   6 August 2014   50cl
PRODUCT OF SCOTLAND

1998 16 years old

Tasting notes Miltonduff 16 years old:

GS – Vanilla and pine on the nose, with soft toffee and a hint of cinnamon. Zesty orange, mixed nuts and cocoa on the palate. The finish is long, nutty and slighlty citric.

Mortlach

[mort•lack]

Owner:	**Region/district:**
Diageo	Speyside
Founded: **Status:**	**Capacity:**
1823 Active	3 800 000 litres

Address: Dufftown, Keith, Banffshire AB55 4AQ

Website:	**Tel:**
malts.com	01340 822100

Mortlach as a brand was re-launched in a major way in spring/summer 2014 and it was also announced that the distillery would be expanded, doubling its capacity, to cope with future demand.

With autumn, however, came another statement from the owners. The expansion, together with two other projects at Clynelish and Teaninich, were put on hold. It was not said straight out, but the reason was probably the decline in sales that Diageo, and most of the other producers, experienced during 2014. It seems likely though, that all the expansions that were put on hold, will go ahead, the only question is when that will happen. This doesn't mean, however, that there is no construction work going on at the distillery. From early May it was shut down for 20 weeks to carry out some major works; replacing 4 washbacks, installing a larger yeast tank in preparation for an extended distillery, replacing necks, shoulders and lye pipes on some of the stills, etc. This means only 2 million litres will be produced during 2015.

The distillery is equipped with a 12 tonnes full lauter mash tun, six washbacks made of larch and six stills in various sizes, all of them attached to worm tubs for cooling the spirit vapours. There are three wash stills and three spirit stills where the No. 3 pair acts as a traditional double distillation. The low wines from wash stills No. 1 and 2 are directed to the remaining two spirit stills according to a certain distribution. In one of the spirit stills, called Wee Witchie, the charge is redistilled twice and, with all the various distillations taken into account, it could be said that Mortlach is distilled 2.8 times.

The **16 year old Flora and Fauna** bottling, has now been replaced with a new range; **Rare Old** with no age statement and matured mostly in first fill bourbon casks, but also in ex sherry casks, **Special Strength**, exclusive to duty free, similar to Rare Old but bottled at the higher strength of 49%, **18 year old** matured in a combination of first fill sherry casks and re-fill American oak and **25 year old**, predominantly matured in re-fill American oak.

History:
- 1823 The distillery is founded by James Findlater.
- 1824 Donald Macintosh and Alexander Gordon become part-owners.
- 1831 The distillery is sold to John Robertson for £270.
- 1832 A. & T. Gregory buys Mortlach.
- 1837 James and John Grant of Aberlour become part-owners. No production takes place.
- 1842 The distillery is now owned by John Alexander Gordon and the Grant brothers.
- 1851 Mortlach is producing again after having been used as a church and a brewery for some years.
- 1853 George Cowie joins and becomes part-owner.
- 1867 John Alexander Gordon dies and Cowie becomes sole owner.
- 1895 George Cowie Jr. joins the company.
- 1897 The number of stills is increased from three to six.
- 1923 Alexander Cowie sells the distillery to John Walker & Sons.
- 1925 John Walker becomes part of Distillers Company Limited (DCL).
- 1964 Major refurbishment.
- 1968 Floor maltings ceases.
- 1996 Mortlach 1972 is released as a Rare Malt.
- 1998 Mortlach 1978 is released as a Rare Malt.
- 2004 Mortlach 1971, a 32 year old cask strength is released.
- 2014 Four new bottlings are released - Rare Old, Special Strength, 18 year old and 25 year old.

Rare Old

Tasting notes Mortlach Rare Old:
GS – Fresh and fruity on the nose, majoring in peaches and apricots. Parma violets, milk chocolate, and finally caramel. Fruit carries over from the nose to the nutty palate, with cinnamon spice. The finish is relatively long and spicy.

Oban

[oa•bun]

Owner:
Diageo

Region/district:
Western Highlands

Founded: **Status:**
1794   Active (vc)

Capacity:
870 000 litres

Address: Stafford Street, Oban, Argyll PA34 5NH

Website:
malts.com

Tel:
01631 572004 (vc)

The fans of Oban single malt, and there are many, have not been spoilt for choice of new releases. It was therefore a surprise in spring 2015 when Oban Little Bay was launched.

The whisky, which doesn't carry an age statement, has been matured in three types of casks; sherry casks made of European oak, re-fill casks with new oak cask ends and re-fill American oak hogsheads. It was first released in the USA, which has always been the primary market for Oban, with a global distribution planned for later in the year.

With its location in the middle of town and it being surrounded by other buildings, there is no place for expansion of the distillery. With the increased demand (Oban single malt now sells more than 1 million bottles per year), the distillery crew has been working hard to increase the output using the available equipment. The 870,000 litres distilled in 2014 was an all time high but, due to re-placement of some of the equipment in January 2015, production will be 820,000 litres for 2015.

The distillery is equipped with a 6.5 tonnes traditional stainless steel mash tun with rakes, four washbacks made of European larch and one pair of stills. Attached to the stills is a rectangular, stainless steel, double worm tub to condensate the spirit vapours. One washback will fill the wash still twice. However, the character of Oban single malt is dependent on long fermentations (110 hours), hence they can only manage six mashes per week, giving five long and one short fermentation. The distillery boasts one of the best visitor centres in the business with 35,000 visitors every year.

The core range consists of four expressions – the new **Little Bay**, a **14 year old**, an **18 year old** exclusive for USA and a **Distiller's Edition** with a montilla fino sherry finish. In 2010 a distillery exclusive bottling, available only at the distillery, was released. It was finished in fino sherry casks and has no age statement. In autumn 2013 a **21 year old** bottled at 58,5% was launched as part of the annual Special Releases.

History:

1793 John and Hugh Stevenson found the distillery.

1820 Hugh Stevenson dies.

1821 Hugh Stevenson's son Thomas takes over.

1829 Bad investments force Thomas Stevenson into bankruptcy. His eldest son John takes over.

1830 John buys the distillery from his father's creditors for £1,500.

1866 Peter Cumstie buys the distillery.

1883 Cumstie sells Oban to James Walter Higgins who refurbishes and modernizes it.

1898 The Oban & Aultmore-Glenlivet Co. takes over with Alexander Edwards at the helm.

1923 The Oban Distillery Co. owned by Buchanan-Dewar takes over.

1925 Buchanan-Dewar becomes part of Distillers Company Limited (DCL).

1931 Production ceases.

1937 In production again.

1968 Floor maltings ceases and the distillery closes for reconstruction.

1972 Reopening of the distillery.

1979 Oban 12 years is on sale.

1988 United Distillers launches Classic Malts and Oban 14 year old is included.

1998 A Distillers' Edition is launched.

2002 The oldest Oban (32 years) so far is launched.

2004 A 1984 cask strength is released.

2009 Oban 2000, a single cask, is released.

2010 A no age distillery exclusive is released.

2013 A limited 21 year old is released.

2015 Oban Little Bay is released.

Oban Little Bay

Tasting notes Oban 14 years old:

GS – Lightly smoky on the honeyed, floral nose. Toffee, cereal and a hint of peat. The palate offers initial cooked fruits, becoming spicier. Complex, bittersweet, oak and more gentle smoke. The finish is quite lengthy, with spicy oak, toffee and new leather.

Pulteney

[poolt•ni]

Owner:
Inver House Distillers
(Thai Beverages plc)

Region/district:
Northern Highlands

Founded: **Status:** **Capacity:**
1826   Active (vc)   1 800 000 litres

Address: Huddart St, Wick, Caithness KW1 5BA

Website: **Tel:**
oldpulteney.com   01955 602371

Pulteney together with Oban and Glen Garioch, is the most urban distillery in Scotland. It lies right in the middle of Wick and is surrounded by buildings on all sides.

The distillery itself is charmingly compact with very little room for expansion, should that ever be required in the future. The sales of Old Pulteney (the name of their single malt) has increased steadily and is now at about 550,000 bottles per year.

The distillery is equipped with a stainless steel semi-lauter mash tun clad with wood and with a copper canopy. There are six washbacks from the 1920s all made by Corten steel and are due to be replaced by stainless steel in 2016. Fermentation time is a mix of short (50 hours) and long (110 hours). The wash still, equipped with a huge boil ball and a very thick lye pipe, is quaintly chopped off at the top. An interesting detail is the spirit safe which was previously used at the long since closed Glen Flagler distillery. Both stills use stainless steel worm tubs for condensing the spirit. The plan for 2015 is to produce 1.5 million litres of alcohol.

The core range of Old Pulteney is made up of **12, 17, 21** and **35 year old**. To complement the bottlings with age statements, a new, un chillfiltered expression with no age statement was introduced in 2013 – **Old Pulteney Navigator**. During that same year, the owners started to release a series of whiskies, all destined for duty free and named after lighthouses that are scattered all around Wick. **Noss Head** is matured in ex-bourbon American oak, **Duncansby Head** is a mix of ex-bourbon and ex-sherry, while **Pentland Skerries** is matured in Spanish ex-sherry casks. The latter was replaced in 2015 by **Dunnet Head**. A limited **Vintage 1990**, which was released in 2013, will be replaced by a **1989 Vintage** towards the end of 2015, together with the second release of the **35 year old**. Finally, there are two casks at the distillery where visitors can fill their own bottles; a **1989 refill bourbon** and a **2000 first fill bourbon**.

History:

1826 James Henderson founds the distillery.

1920 The distillery is bought by James Watson.

1923 Buchanan-Dewar takes over.

1930 Production ceases.

1951 In production again after being acquired by the solicitor Robert Cumming.

1955 Cumming sells to James & George Stodart, a subsidiary to Hiram Walker & Sons.

1958 The distillery is rebuilt.

1959 The floor maltings close.

1961 Allied Breweries buys James & George Stodart Ltd.

1981 Allied Breweries changes name to Allied Lyons.

1995 Allied Domecq sells Pulteney to Inver House Distillers.

1997 Old Pulteney 12 years is launched.

2001 Pacific Spirits (Great Oriole Group) buys Inver House at a price of $85 million.

2004 A 17 year old is launched.

2005 A 21 year old is launched.

2006 International Beverage Holdings acquires Pacific Spirits UK.

2010 WK499 Isabella Fortuna is released.

2012 A 40 year old and WK217 Spectrum are released.

2013 Old Pulteney Navigator, The Lighthouse range (3 expressions) and Vintage 1990 are released.

2014 A 35 year old is released.

2015 Dunnet Head is released for travel retail.

12 years old

Tasting notes Old Pulteney 12 years old:

GS – The nose presents pleasingly fresh malt and floral notes, with a touch of pine. The palate is comparatively sweet, with malt, spices, fresh fruit and a suggestion of salt. The finish is medium in length, drying and decidedly nutty.

Royal Brackla

[royal brack•lah]

Owner:
John Dewar & Sons
(Bacardi)

Region/district:
Highlands

Founded:
1812

Status:
Active

Capacity:
4 000 000 litres

Address: Cawdor, Nairn, Nairnshire IV12 5QY

Website:
lastgreatmalts.com

Tel:
01667 402002

For most of the 20th century, the whisky from Royal Brackla has been used for blended whisky, especially Dewar's. The gentle and fruity character of the single malt makes it a perfect part of any blend.

A number of related factors during production work together to create its distinguished flavour; clear wort, long fermentations (70 hours), long foreshots (30 minutes), a slow distillation and ascending lyne arms on the stills to create as much reflux as possible during the distillation. Almost the entire output is used by the owners for their blends and only a hard-to-find 10 year old has been available as single malt. This, however, changed in autumn 2015 when a completely new range was launched. This release wasn't entirely problem free. When Dewar's bought the distillery from Diageo in 1998, no maturing whisky was included in the deal and for the older expressions, stock had to be bought from other dealers.

In 1835 the distillery was the first to be given a royal warrant by King William IV and the whisky became known as "The King's Own Whisky". More success followed in 1860 when Andrew Usher, whose company was also a partner in the distillery, used Royal Brackla malt in his (and thereby Scotland's) first blended whisky.

Royal Brackla is equipped with a 12.5 tonnes full lauter mash tun from 1997. There are six wooden washbacks and another two made of stainless steel which have been placed outside. Finally, there are also two pairs of stills. At the moment the distillery is running at full capacity, which means 17 mashes per week and 4 million litres of alcohol per year.

The new core range, replacing the 10 year old, consists of a **12, 16** and **21 year old** and there are also plans for a possible release of a **30 year old** in the future. Recent, limited releases include a **25 year old** and a **35 year old**. The latter, launched in April 2014 at Changi airport in Singapore, is the oldest official Royal Brackla ever.

History:

1812 The distillery is founded by Captain William Fraser.

1835 Brackla becomes the first of three distilleries allowed to use 'Royal' in the name.

1852 Robert Fraser & Co. takes over the distillery.

1898 The distillery is rebuilt and Royal Brackla Distillery Company Limited is founded.

1919 John Mitchell and James Leict from Aberdeen purchase Royal Brackla.

1926 John Bisset & Company Ltd takes over.

1943 Scottish Malt Distillers (SMD) buys John Bisset & Company Ltd and thereby acquires Royal Brackla.

1966 The maltings closes.

1970 Two stills are increased to four.

1985 The distillery is mothballed.

1991 Production resumes.

1993 A 10 year old Royal Brackla is launched in United Distillers' Flora & Fauna series.

1997 UDV spends more than £2 million on improvements and refurbishing.

1998 Bacardi–Martini buys Dewar's from Diageo.

2004 A new 10 year old is launched.

2014 A 35 year old is released for Changi airport in Singapore.

2015 A new range is released; 12, 16 and 21 year old.

12 years old

Tasting notes Royal Brackla 12 years old:

GS – Warm spices, malt and peaches in cream on the nose. The palate is robust, with spice and mildly smoky soft fruit. Quite lengthy in the finish, with citrus fruit, mild spice and cocoa powder.

Royal Lochnagar

[royal loch•nah•gar]

Owner:		Region/district:
Diageo		Eastern Highlands

Founded:	Status:	Capacity:
1845	Active (vc)	500 000 litres

Address: Crathie, Ballater, Aberdeenshire AB35 5TB

Website:	Tel:
malts.com	01339 742700

Of the 28 malt distilleries owned by Diageo, Royal Lochnagar is definitely the smallest! At the same time it is the home to Diageo's Malt Advocate courses, where employees, during a five day course, can learn about all aspects of whisky.

Set in beautiful surroundings with Royal Deeside and the imposing Lochnagar mountain situated to the south and Balmoral, the Queen's summer residence, just a stone's throw to the north, the single malt makes an important contribution to Diageo's key blend in Korea – namely Windsor. The Korean market has been sluggish in recent years and in 2014 a decision was made to introduce a new version of Windsor, technically not a whisky, but a spirit drink. It is called W Ice by Windsor and is meant to attract new customer groups. Made from Scotch whisky (not least Lochnagar), ingredients such as pine, dates and dried fig essence have been added. The spirit has also been chillfiltered at minus 8 degrees Celsius and diluted to 35% alcohol. Sales figures have been positive and the new brand is now competing with Golden Blue, the market leader in this new segment.

The distillery is equipped with a 5.4 ton open, traditional stainless steel mash tun. Fermentation takes place in three wooden washbacks, with short fermentations of 60 hours and long ones of 106 hours. The two stills are quite small with a charge in the wash still of 6,100 litres and 4,000 litres in the spirit still. The cooling of the spirit vapours takes place in cast iron worm tubs. The whole production is filled on site with around 1,000 casks being stored in the only warehouse and the rest being sent to Glenlossie. Four mashes per week during 2015 will result in 450,000 litres of pure alcohol.

The core range consists of the **12 year old** and **Selected Reserve**. The latter is a vatting of selected casks, usually around 18-20 years of age. There is also a **Distiller's Edition** with a second maturation in Muscat casks. In 2013 a limited expression for Friends of the Classic Malts was released with no age statement and **triple matured** (American oak refill, charred American oak hogsheads and European oak refill).

History:

1823   James Robertson founds a distillery in Glen Feardan on the north bank of River Dee.

1826   The distillery is burnt down by competitors but Robertson decides to establish a new distillery near the mountain Lochnagar.

1841   This distillery is also burnt down.

1845   A new distillery is built by John Begg, this time on the south bank of River Dee. It is named New Lochnagar.

1848   Lochnagar obtains a Royal Warrant.

1882   John Begg passes away and his son Henry Farquharson Begg inherits the distillery.

1896   Henry Farquharson Begg dies.

1906   The children of Henry Begg rebuild the distillery.

1916   The distillery is sold to John Dewar & Sons.

1925   John Dewar & Sons becomes part of Distillers Company Limited (DCL).

1963   A major reconstruction takes place.

2004   A 30 year old cask strength from 1974 is launched in the Rare Malts series (6,000 bottles).

2008   A Distiller's Edition with a Moscatel finish is released.

2010   A Manager's Choice 1994 is released.

2013   A triple matured expression for Friends of the Classic Malts is released.

WASHBACK No.3

ROYAL LOCHNAGAR
HIGHLAND SINGLE MALT
SCOTCH WHISKY
12

12 years old

Tasting notes Royal Lochnagar 12 years old:

GS – Light toffee on the nose, along with some green notes of freshly-sawn timber. The palate offers a pleasing and quite complex blend of caramel, dry sherry and spice, followed by a hint of liquorice before the slightly scented finish develops.

Scapa

[ska•pa]

Owner:
Chivas Brothers
(Pernod Ricard)

Region/district:
Highlands (Orkney)

Founded: **Status:** **Capacity:**
1885 Active 1 300 000 litres

Address: Scapa, St Ola, Kirkwall, Orkney KW15 1SE

Website: **Tel:**
scapamalt.com 01856 876585

Always in the shadow of Highland Park and often referred to as "the other Orkney whisky", it now seems that Scapa is finally about to find its place in the spotlight.

For the first time in the distillery´s 130 year old history, Scapa is now open for visitors. Many have walked down the road from Kirkwall to Scapa Flow, only to be met by a sign saying: "Sorry - no visitors". On 27th of April 2015 the sign was removed and a visitor centre with a shop was opened, thus affording visitors the opportunity to see a unique distillery and buy the single malt. And it seems like this is only the first step of getting Scapa more recognition in the whisky industry. Ever since Pernod Ricard assumed ownership in 2005, the distillery has only been working 3-4 days per week. From 2015 the production has been increased to 7 days, which means around 1 million litres of alcohol per year.

The equipment at Scapa distillery consists of a 2.9 ton semi-lauter mash tun with a copper dome, eight washbacks (four made of Corten steel and four of stainless steel) and two stills. Due to the increased production, fermentation time is now down to 52 hours from the previous 160 hours. The wash still is only one of two surviving Lomond stills in the industry. Instead of a swan neck, a Lomond still has a straight tube with adjustable plates. The idea was that, by modifying the position of the plates, you could make different types of spirits with the same still. Today, however, the plates of the Scapa Lomond still have been removed. The distillery was in danger of being closed down in 2000 but the owners at the time, Allied Domecq, decided to invest £2 million to refurbish it. There are three dunnage and three racked warehouses on site.

For many years, the Scapa core range was just the **16 year old** but the line-up was expanded in September 2015 with **Scapa Skiren**. Matured in first fill bourbon, it doesn´t carry an age statement and is bottled at 40%. Skiren means glittering bright skies in the old norse language and the new expression will first be released in the UK and in France. There is also a **14 year old cask strength** distilled in 2000, sold exclusively at Chivas´ visitor centres.

History:

1885 Macfarlane & Townsend founds the distillery with John Townsend at the helm.

1919 Scapa Distillery Company Ltd takes over.

1934 Scapa Distillery Company goes into voluntary liquidation and production ceases.

1936 Production resumes.

1936 Bloch Brothers Ltd (John and Sir Maurice) takes over.

1954 Hiram Walker & Sons takes over.

1959 A Lomond still is installed.

1978 The distillery is modernized.

1994 The distillery is mothballed.

1997 Production takes place a few months each year using staff from Highland Park.

2004 Extensive refurbishing takes place at a cost of £2.1 million. Scapa 14 years is launched.

2005 Production ceases in April and phase two of the refurbishment programme starts. Chivas Brothers becomes the new owner.

2006 Scapa 1992 (14 years) is launched.

2008 Scapa 16 years is launched.

2015 The distillery opens for visitors and Scapa Skiren is launched.

Tasting notes Scapa 16 years old:

GS – The nose offers apricots and peaches, nougat and mixed spices. Pretty, yet profound. Medium-bodied, with caramel and spice notes in the mouth. The finish is medium in length and gingery, with fat, buttery notes emerging at the end.

Scapa Skiren

Speyburn

[spey•burn]

Owner: **Region/district:**
Inver House Distillers   Speyside
(Thai Beverages plc)

Founded: **Status:**   **Capacity:**
1897   Active   4 200 000 litres

Address: Rothes, Aberlour, Morayshire AB38 7AG

Website: **Tel:**
speyburn.com   01340 831213

Speyburn is the bestselling brand in the Inver House range of single malts with more than 500,000 bottles sold in 2014. For decades it has been one of the top ten malts in USA.

The brand's success in USA can probably be explained by the fact that Inver House had American owners for the first 24 years. Publicker Industries founded the company in 1964 and then sold it to Standard Brands in 1979. A management buyout in 1988, however, ended the American ownership. The connections that were established during these years led to an agreement in 1993 with Barton Brands to distribute Speyburn single malt on the American market. The agreement with Barton expired in 2009 and Speyburn is now sold through a subsidiary of the current owners, Thai Beverages.

An impressive expansion of the distillery commenced during 2014 and was completed in spring of 2015. The expansion had cost £4m and included a new, 6 ton stainless steel mash tun which was fitted into the existing tun room, thus replacing the old cast iron mash tun. Four of the six wooden washbacks were kept but they have also expanded with no less than 15 washbacks made of stainless steel. These new washbacks were placed in a converted dunnage warehouse leaving the distillery with just the one warehouse. Finally, the existing wash still was converted to a spirit still of exactly the same shape as the other one, while a new and much larger wash still was installed. The two spirit stills are connected to a worm tub while the wash still is fitted with a shell and tube condenser. The fermentation time has also been lengthened from the original 48 hours to a minimum of 72 hours. The result of the expansion was a doubling of the capacity to 4.2 million litres.

The core range of Speyburn single malt is the **10 year old** and **Bradan Orach** without age statement. There is also a limited **25 year old**. A few years ago a community on the internet called Clan Speyburn was created for the fans of the brand and in 2013 the first of the Clan casks for members was released – a **1975 PX sherry single cask**.

History:

1897   Brothers John and Edward Hopkin and their cousin Edward Broughton found the distillery through John Hopkin & Co. They already own Tobermory. The architect is Charles Doig. Building the distillery costs £17,000 and the distillery is transferred to Speyburn-Glenlivet Distillery Company.

1916   Distillers Company Limited (DCL) acquires John Hopkin & Co. and the distillery.

1930   Production stops.

1934   Productions restarts.

1962   Speyburn is transferred to Scottish Malt Distillers (SMD).

1968   Drum maltings closes.

1991   Inver House Distillers buys Speyburn.

1992   A 10 year old is launched as a replacement for the 12 year old in the Flora & Fauna series.

2001   Pacific Spirits (Great Oriole Group) buys Inver House for $85 million.

2005   A 25 year old Solera is released.

2006   Inver House changes owner when International Beverage Holdings acquires Pacific Spirits UK.

2009   The un-aged Bradan Orach is introduced for the American market.

2012   Clan Speyburn is formed.

2014   The distillery is expanded.

10 years old

Tasting notes Speyburn 10 years old:

GS – Soft and elegant on the spicy, nutty nose. Smooth in the mouth, with vanilla, spice and more nuts. The finish is medium, spicy and drying.

Speyside

[spey•side]

Owner:	**Region/district:**
Speyside Distillers Co.	Speyside
Founded: **Status:**	**Capacity:**
1976 Active	600 000 litres

Address: Glen Tromie, Kingussie, Inverness-shire
PH21 1NS

Website:	**Tel:**
speysidedistillery.co.uk	01540 661060

John Harvey McDonough, whose ancestors´ involvement in the whisky business dates back to the 1770s, had been buying whisky from Speyside distillery ever since he launched his Spey brand 20 years ago.

During the 1990s, Harvey McDonough was working in the spirits business in Taiwan and the networking that he had created there, has now made Spey the third biggest single malt in Taiwan. Increased sales prompted the need for their own distillery and this opportunity presented itself in 2012, when the company bought Speyside distillery. In line with Diageo´s Haig Club and David Beckham, Speyside Distillers chose an exfootballer, Michael Owen, in 2014 to be the leading figure-head of the brand.

Speyside produces on a small-scale and is set in beautiful surroundings. The distillery is equipped with a 4 ton semi-lauter mash tun, four stainless steel washbacks with a fermentation time of 48 hours and one pair of stills. In 2015 they will be working a 6 day week with a total production of 600,000 litres of alcohol. There are also plans to build a second distillery in Rothiemurches near Aviemore.

A new core range of Spey single malt was launched during 2014/2015; **Tenné** (with a 6 month finish in tawny port), **12 year old** (6 months finish in new American oak), **18 year old** (sherry matured), **Chairman´s Choice** and **Royal Choice**. The latter two are multi-vintage marriages from both American and European oak. Destined for export markets is also **Black Burn** without age statement. Limited releases during 2015 include **Lord Byron**, bourbon matured with a tawny port finish and two Michael Owen releases for Asia – **Golden Choice** (a mix of American and European oak) and **1412** (bourbon matured with a finish in both sherry and port casks). Finally, as a replacement for the discontinued, famous black whisky Cu Duhb, **Beinn Dubh** was released. The new owners also produce blended whisky for selected export markets such as Scotch Guard and Glen Hood.

History:

1956 George Christie buys a piece of land at Drumguish near Kingussie.

1957 George Christie starts a grain distillery near Alloa.

1962 George Christie (founder of Speyside Distillery Group in the fifties) commissions the drystone dyker Alex Fairlie to build a distillery in Drumguish.

1986 Scowis assumes ownership.

1987 The distillery is completed.

1990 The distillery is on stream in December.

1993 The first single malt, Drumguish, is launched.

1999 Speyside 8 years is launched.

2000 Speyside Distilleries is sold to a group of private investors including Ricky Christie, Ian Jerman and Sir James Ackroyd.

2001 Speyside 10 years is launched.

2012 Speyside Distillers is sold to Harvey´s of Edinburgh.

2014 A new range, Spey from Speyside Distillery, is launched (NAS, 12 and 18 year old).

2015 The range is revamped again. New expressions include Tenné, 12 years old and 18 years old.

12 years old

Tasting notes Spey 12 years old:

GS – Malt and white pepper on the nose, with a mildly savoury background. The palate features vanilla, orange, hazelnuts and cloves. Black pepper and lively oak in the medium-length finish.

1988 launch of The Classic Malts

The origins of The Classic Malts of Scotland sit with the attempts of the Distillers Company Limited (DCL) to enter the nascent single malt whisky market in 1985/86 – a market which had begun with Glenfiddich in 1963.

Although still far from convinced that single malts were a viable proposition for future growth, and concerned to protect the reputation of their blended Scotch whiskies, DCL launched the Ascot Cellar (thus named because DCL's home trade offices were in Ascot). The Ascot Cellar was principally conceived as an on-trade offering of a mixed case of individual malts – four single malt whiskies plus two vatted malts. The single malts were Lagavulin 12, Talisker 10, Rosebank 8 and Linkwood 12. The other bottles were The Strathconon 12 and Glenleven 12. In 1987, DCL "merged" with Guinness. The architect of this major consolidation was the subsequently disgraced Sir Ernest Saunders. Saunders had promised his "co-conspirators", and the City more generally, that the newly formed company would deliver great things. Considerable pressure was placed on its new marketing teams to show that they could deliver for this now marketing-led drinks giant. One outcome was to be The Classic Malts of Scotland.

The two people responsible for the creation of The Classic Malts were Mike Collings, a well-travelled drinks marketeer with a background in wine; and Roy MacMillan, a Scot with a family background in the whisky business. These two saw the opportunity to explode the single malt whisky category by offering non-specialist on-trade and off-trade accounts the opportunity to become a "malt whisky specialist" overnight through the simple but brilliant mechanic of a mixed case of six bottles. In selecting the six whiskies to make up The Classic Malts, a "regionality" concept was used. Although people argued then and later about the different Scotch whisky regions, for the convenience of the six-case model, these were the regions and whiskies that were chosen:

West Highland (Oban), Islands (Talisker), Islay (Lagavulin), Highlands (Dalwhinnie), Lowlands (Glenkinchie) and Speyside (Cragganmore).

Choosing the whiskies to match this concept was in some cases not difficult: Oban, for example, was the only brand that qualified for the West Highland region - loosely, a bastardised version of Campbeltown. For more congested regions such as Speyside, the choice was more complex. Whilst there was a strong desire to ensure that the six chosen liquids were highly differentiated and spoke strongly of their regional origins, other factors, such as the attractiveness of the distillery and the ease of access by road for tourists, were taken into account. But another point that should perhaps have been taken a little more seriously was the available inventory for the distilleries finally selected. Such was the unforeseen success of The Classic Malts that within less than 10 years of launch, sales had far exceeded initial estimates and in the case of at least one brand, supply was effectively capped.

The Classic Malts were first launched in duty-free channels and domestic launches followed in Europe and USA, around 1989/90. The two main marketing mechanics were the Classic Malts plinth – which can still be found in use, in its original form, in on-trade premises around the world and can be considered one of the most effective point-of-sale assets ever developed by United Distillers – and the regional whisky map of Scotland. The map was initially deployed in stand-out installations in travel retail channels, and framed prints can still be found in bars, pubs and shops around the world. In addition to this cutting edge technology was deployed to create a VHS video featuring Michael Jackson discussing each of the six bottlings.

The reception given by the trade was overwhelmingly positive, not only from specialists but also from restaurants and pubs. The success also inspired other producers and more and more single malts became available to the consumers. The Classic Malts was the stepping stone to the huge interest in single malts we see today.

Springbank

[spring•bank]

Owner:	Region/district:
Springbank Distillers (J & A Mitchell)	Campbeltown

Founded.	Status:	Capacity:
1828	Active (vc)	750 000 litres

Address: Well Close, Campbeltown, Argyll PA28 6ET

Website:	Tel:
springbankdistillers.com	01586 551710

In the "good old days" there used to be five official Scotch whisky regions – Highlands, Speyside (originally a sub-category of the Highlands), Lowlands, Islay and Campbeltown.

Nobody questioned the existence of Campbeltown, a town that once had close to 30 distilleries operating. But in 1998 only one was left – Springbank – and the SWA (Scotch Whisky Association) decided that one distillery can't constitute a region. Shortly after that, Glen Scotia was re-opened but two weren't enough for SWA. Then Hedley Wright, the owner of Springbank, opened up Glengyle making the number of distilleries equal to that of the Lowlands, and Campbeltown was once again able to function as a region on its own.

The distillery is equipped with an open cast iron mash tun, six washbacks made of Scandinavian larch, one wash still and two spirit stills. The wash still is unique in Scotland, as it is fired by both an open oil-fire and internal steam coils. Ordinary condensers are used to cool the spirit vapours, except in the first of the two spirit stills, where a worm tub is used. Springbank is also the only distillery in Scotland that malts its entire need of barley using own floor maltings.

Springbank produces three distinctive single malts with different phenol contents in the malted barley. Springbank is distilled two and a half times (12-15ppm), Longrow is distilled twice (50-55 ppm) and Hazelburn is distilled three times and unpeated. In 2015 a total of 150,000 litres will be produced of which 10% is Longrow and 10% Hazelburn.

The core range is Springbank **10, 15** and **18 year old**, a **12 year old cask strength** and a **21 year old** (limited, but with yearly batches). Longrow is represented by **Longrow**, the **18 year old** and the **Longrow Red**. The 2014 edition of the latter was a port cask finish and it was followed up in 2015 by a whisky that got its final character from New Zealand pinot noir casks. Finally, there is Hazelburn where the core range consists of a **10 year old** (which was released in summer 2014) and a **12 year old**. Recent, limited editions have included **Springbank 9 year old** with a 5 year old second maturation in **Gaja Barolo** casks, **Springbank 17 year old sherry wood** and **Hazelburn Rundlets & Kilderkins** (a 10 year old matured in small casks of 68 and 82 litres). Towards the end of 2014, we saw the release of a rare **25 year old** Springbank (with a second edition in October 2015), matured in both bourbon and sherry and married together for a few months in port casks. At the same time, a **Springbank Green 12 year old** was released, made of organic barley and matured in bourbon casks. A second release (a **13 year old**) was made in October 2015, this time matured in sherry casks. Both "green" releases are one-offs and organic barley hasn't been used for malting since 2002. Finally, a limited number of single cask releases are made every year for various markets, for example, a **12 year old port wood** Springbank was destined for the UK market in April of 2015.

History:

1828 The Reid family, in-laws of the Mitchells (see below), founds the distillery.

1837 The Reid family encounters financial difficulties and John and William Mitchell buy the distillery.

1897 J. & A. Mitchell Co Ltd is founded.

1926 The depression forces the distillery to close.

1933 The distillery is back in production.

1960 Own maltings ceases.

1969 J. & A. Mitchell buys the independent bottler Cadenhead.

1979 The distillery closes.

1985 A 10 year old Longrow is launched.

1987 Limited production restarts.

1989 Production restarts.

1992 Springbank takes up its maltings again.

1997 First distillation of Hazelburn.

1998 Springbank 12 years is launched.

1999 Dha Mhile (7 years), the world's first organic single

2000 A 10 year old is launched.

2001 Springbank 1965 'Local barley' (36 years), 741 bottles, is launched.

2002 Number one in the series Wood Expressions is a 12 year old with five years on Demerara rum casks. Next is a Longrow sherry cask (13 years). A relaunch of the 15 year old replaces the 21 year old.

2004 Springbank 10 years 100 proof is launched as well as Springbank Wood Expression bourbon, Longrow 14 years old, Springbank 32 years old and Springbank 14 years Port Wood.

2005 Springbank 21 years, the first version of Hazel burn (8 years) and Longrow Tokaji Wood Ex pression are launched.

Springbank 15 years   Hazelburn 12 years   Springbank 21 years

History continued:

2006 Longrow 10 years 100 proof, Springbank 25 years, Springbank 9 years Marsala finish, Springbank 11 years Madeira finish and a new Hazelburn 8 year old are released.

2007 Springbank Vintage 1997 and a 16 year old rum wood are released.

2008 The distillery closes temporarily. Three new releases of Longrow - CV, 18 year old and 7 year old Gaja Barolo.

2009 Springbank Madeira 11 year old, Springbank 18 year old, Springbank Vintage 2001 and Hazelburn 12 year old are released.

2010 Springbank 12 year old cask strength and a 12 year old claret expression together with new editions of the CV and 18 year old are released. Longrow 10 year old cask strength and Hazelburn CV are also new.

2011 Longrow 18 year old and Hazelburn 8 year old Sauternes wood expression are released.

2012 New releases include Springbank Rundlets & Kilderkins, Springbank 21 year old and Longrow Red.

2013 Longrow Rundlets & Kilderkins, a new edition of Longrow Red and Springbank 9 year old Gaja Barolo finish are released.

2014 Hazelburn Rundlets & Kilderkins, Hazelburn 10 year old and Springbank 25 years old are launched.

2015 New releases include Springbank Green 12 years old and a new edition of the Longrow Red.

Springbank Green 12

Tasting notes Springbank 10 years old:

GS – Fresh and briny on the nose, with citrus fruit, oak and barley, plus a note of damp earth. Sweet on the palate, with developing brine, nuttiness and vanilla toffee. Long and spicy in the finish, coconut oil and drying peat.

Tasting notes Longrow NAS:

GS – Initially slightly gummy on the nose, but then brine and fat peat notes develop. Vanilla and malt also emerge. The smoky palate offers lively brine and is quite dry and spicy, with some vanilla and lots of ginger. The finish is peaty with persistent, oaky ginger.

Tasting notes Hazelburn 12 years old:

GS – A highly aromatic nose, featuring nutty toffee, sherry, dried fruits and dark chocolate. The palate is rich and spicy, with cocoa, coffe, ginger and sweet notes of caramel and orange marmalade. Long and spicy in the finish, with more caramel, coffee and chocolate.

Springbank 10 years old   Longrow

Strathisla

[strath•eye•la]

Owner:
Chivas Bros (Pernod Ricard)

Region/district:
Speyside

Founded: **Status:**
1700   Active (vc)

Capacity:
2 450 000 litres

Address: Seafield Avenue, Keith,
Banffshire AB55 5BS

Website: **Tel:**
maltwhiskydistilleries.com   01542 783044

Strathisla is the oldest distillery in Speyside and next year it can celebrate its 230th anniversary. It is probably one of the most photographed distilleries in Scotland and encapsulates what most people think a distillery should look like.

In 1830 the distillery was bought by William Longmore, a great man and benefactor in the town of Keith. For 110 years it was owned by him and his descendants when in 1940, George "Jay" Pomery managed to get control of the company and the distillery. He was a man with a dubious reputation and within ten years he managed to bring the historical distillery to the verge of bankruptcy. He sold all the output from the distillery to clients in London, but tax authorities eventually found out that the companies were fictitious and he was charged with tax evasion. The company and distillery were declared bankrupt, but Seagram's/Chivas Brothers came to the rescue and its owner, Sam Bronfman, bought the company in 1950. Strathisla has ever since served as the public figure-head for the Chivas Regal and Royal Salute blends. The latter was first released in 1953 and the range has since then been expanded with the 38 year old Stone of Destiny in 2005, the 62 Gun Salute in 2010 and Royal Salute Tribute to Honour in 2013. Of the last one, only 21 bottles were released at £150,000 a piece.

The distillery is equipped with a 5 tonnes traditional mash tun with a raised copper canopy, seven washbacks made of Oregon pine and three of larch – all with a 54 hour fermentation. There are two pairs of stills in a cramped, but very charming still room. The wash stills are of lantern type with descending lyne arms and the spirit stills have boiling balls and the lyne arms are slightly ascending. The spirit produced at Strathisla is piped to nearby Glen Keith distillery for filling or to be tankered away. A small amount is stored on site in two racked and one dunnage warehouse.

Two official bottlings exist – the **12 year old** and a **cask strength** (currently from 1995 and bottled in 2014) which is sold only at the distillery.

History:

1786 Alexander Milne and George Taylor found the distillery under the name Milltown, but soon change it to Milton.

1823 MacDonald Ingram & Co. purchases the distillery.

1830 William Longmore acquires the distillery.

1870 The distillery name changes to Strathisla.

1880 William Longmore retires and hands operations to his son-in-law John Geddes-Brown. William Longmore & Co. is formed.

1890 The distillery changes name to Milton.

1940 Jay (George) Pomeroy acquires majority shares in William Longmore & Co. Pomeroy is jailed as a result of dubious business transactions and the distillery goes bankrupt in 1949.

1950 Chivas Brothers buys the run-down distillery at a compulsory auction for £71,000 and starts restoration.

1951 The name reverts to Strathisla.

1965 The number of stills is increased from two to four.

1970 A heavily peated whisky, Craigduff, is produced but production stops later.

2001 The Chivas Group is acquired by Pernod Ricard.

Tasting notes Strathisla 12 years old:

GS – Rich on the nose, with sherry, stewed fruits, spices and lots of malt. Full-bodied and almost syrupy on the palate. Toffee, honey, nuts, a whiff of peat and a suggestion of oak. The finish is medium in length, slightly smoky and a with a final flash of ginger.

12 years old

Strathmill

[strath•mill]

Owner:
Diageo

Region/district:
Speyside

Founded: 1891
Status: Active
Capacity: 2 600 000 litres

Address: Keith, Banffshire AB55 5DQ

Website: malts.com
Tel: 01542 883000

For the current owners, Diageo, Strathmill distillery is a true workhorse, producing malt whisky for the megablend J&B. Very little is being bottled as single malt.

But it hasn't always been that way. The famous gin producer W. & A. Gilbey bought the distillery in 1895. They already owned Glen Spey and were later (in 1904) to buy Knockando as well. They used Strathmill for blends – Spey Royal for instance – while marketing the single malt for export markets at the same time. In the Australian newspaper of December 1905, The Manawatu Times, one could read the following. "Of eight of the most popular whiskies submitted for analysis, the Western Australian Government analyst in his annual report to Parliament pronounces Strathmill to be the most genuine matured malt whisky."

Strathmill is neatly tucked away right next to the River Isla that flows through Keith. It is also Isla that supplies the cooling water for the distillery while the process water for the mashing comes from a borehole on the site. In October 2014, severe flooding caused the river to burst its banks and three people were trapped inside the distillery before fire-fighters could rescue them.

Since last year, Strathmill has gone from a 5-day production week to 7 days, which means that the number of mashes per week has increased to 13 with 2.6 million litres being produced in the year. The equipment consists of a 9 ton stainless steel semi-lauter mash tun and six stainless steel washbacks with a 76 hour fermentation period. There are two pairs of stills and Strathmill is one of a select few distilleries still using purifiers on the spirit stills. This device is mounted between the lyne arm and the condenser and acts as a mini-condenser, allowing the lighter alcohols to travel towards the condenser and forcing the heavier alcohols to go back into the still for another distillation. The result is a lighter and fruitier spirit. In Strathmill's case, both purifiers and condensers are fitted to the outside of the still house.

The only official bottling is the **12 year old Flora & Fauna**, but a limited **25 year old** was launched in 2014 as part of the Special Releases.

History:

1891 The distillery is founded in an old mill from 1823 and is named Glenisla-Glenlivet Distillery.

1892 The inauguration takes place in June.

1895 The gin company W. & A. Gilbey buys the distillery for £9,500 and names it Strathmill.

1962 W. & A. Gilbey merges with United Wine Traders (including Justerini & Brooks) and forms International Distillers & Vintners (IDV).

1968 The number of stills is increased from two to four and purifiers are added.

1972 IDV is bought by Watney Mann which later the same year is acquired by Grand Metropolitan.

1993 Strathmill becomes available as a single malt for the first time since 1909 as a result of a bottling (1980) from Oddbins.

1997 Guinness and Grand Metropolitan merge and form Diageo.

2001 The first official bottling is a 12 year old in the Flora & Fauna series.

2010 A Manager's Choice single cask from 1996 is released.

2014 A 25 year old is released.

12 years old

Tasting notes Strathmill 12 years old:

GS – Quite reticent on the nose, with nuts, grass and a hint of ginger. Spicy vanilla and nuts dominate the palate. The finish is drying, with peppery oak.

Talisker

[tal•iss•kur]

Owner:		Region/district:
Diageo		Highlands (Skye)

Founded:	Status:	Capacity:
1830	Active (vc)	2 700 000 litres

Address: Carbost, Isle of Skye,
Inverness-shire IV47 8SR

Website:	Tel:
malts.com	01478 614308 (vc)

Last year we wrote that Diageo had set up a goal a couple of years ago, and that Talisker should become one of the Top 10 single malts in terms of sales.

In 2014 it finally happened, when Talisker surpassed both Bowmore and Cardhu to be placed number ten with 2.3 million bottles sold. The main driving force has, of course, been the 10 year old bottling but recent expressions like Storm and Dark Storm, have quickly reached strong sales figures and, as it seems, without cannibalising on sales of the 10 year old.

Talisker is not as heavily peated as some of its cousins on Islay with a phenol specification of 18-20 ppm in the barley which gives a phenol content of 5-7 ppm in the new make. The distillery is equipped with a stainless steel lauter mash tun with a capacity of 8 tonnes, eight washbacks made of Oregon pine and five stills (two wash stills and three spirit stills) all of them connected to wooden wormtubs. The wash stills are equipped with a special type of purifiers, using the colder outside air, and have a u-bend in the lyne arm. The purifiers and the peculiar bend of the lyne arms allow for more copper contact and increase the reflux during distillation. The fermentation time is quite long (65-75 hours) and the middle cut from the spirit still is collected between 76% and 65% which, together with the phenol specification, gives a medium peated spirit. Production in 2015 will be around 18 mashes per week which adds up to 2.7 million litres of alcohol.

For a distillery to be situated on an island, can sometimes prove to be a daunting task, especially when it comes to the supply of cooling water. One example is Tobermory on Mull which has been forced to close on a few occasions in the past due to drought and poor water supply. To ensure that Talisker wouldn't find itself in the same situation, the owners installed a cooling system in 2014 where sea water is used. The water is pumped from approximately 8 metres below the surface into the wormtubs. Once it has been through the worms, it goes through a heat exchanger and the cooled water goes back into the worm tubs. A similar system was recently installed at Caol Ila on Islay.

The range of Talisker single malts has been given plenty of attention in recent years and with the new expressions from 2013, Talisker's core range now consists of **Storm** without age statement, **10, 18, 25** and **30 year old**, **Distiller's Edition** with an Amoroso sherry finish, **Talisker 57° North** which is released in small batches, and **Port Ruighe** (pronounced Portree after the main town on Isle of Skye) with a finish in ruby port casks. In spring 2015, a further addition to the core range was made by way of **Talisker Skye**. It doesn't carry any age statement and has been matured in toasted and refill American oak casks. Bottled at 45.8%, it is a slightly softer version than Storm and the 10 year old. There is also **Dark Storm**, the peatiest Talisker so far, which is exclusive to duty free. Recent limited releases include a **27 year old** which was launched in the autumn of 2013 and a triple matured bottling for **Friends of the Classic Malts** in 2014.

History:

1830 Hugh and Kenneth MacAskill, sons of the local doctor, found the distillery.

1848 The brothers transfer the lease to North of Scotland Bank and Jack Westland from the bank runs the operations.

1854 Kenneth MacAskill dies.

1857 North of Scotland Bank sells the distillery to Donald MacLennan for £500.

1863 MacLennan experiences difficulties in making operations viable and puts the distillery up for sale.

1865 MacLennan, still working at the distillery, nominates John Anderson as agent in Glasgow.

1867 Anderson & Co. from Glasgow takes over.

1879 John Anderson is imprisoned after having sold non-existing casks of whisky.

1880 New owners are now Alexander Grigor Allan and Roderick Kemp.

1892 Kemp sells his share and buys Macallan Distillery instead.

1894 The Talisker Distillery Ltd is founded.

1895 Allan dies and Thomas Mackenzie, who has been his partner, takes over.

1898 Talisker Distillery merges with Dailuaine-Glenlivet Distillers and Imperial Distillers to form Dailuaine-Talisker Distillers Company.

1916 Thomas Mackenzie dies and the distillery is taken over by a consortium consisting of, among others, John Walker, John Dewar, W. P. Lowrie and Distillers Company Limited (DCL).

1928 The distillery abandons triple distillation.

1960 On 22nd November the distillery catches fire and substantial damage occurs.

1962 The distillery reopens after the fire.

History continued:

1972 Own malting ceases.

1988 Classic Malts are introduced, Talisker 10 years included. A visitor centre is opened

1998 A new stainless steel/copper mash tun and five new worm tubs are installed. Talisker is launched as a Distillers Edition with an amoroso sherry finish.

2004 Two new bottlings appear, an 18 year old and a 25 year old.

2005 To celebrate the 175th birthday of the distillery, Talisker 175th Anniversary is released. The third edition of the 25 year old cask strength is released.

2006 A 30 year old and the fourth edition of the 25 year old are released.

2007 The second edition of the 30 year old and the fifth edition of the 25 year old are released.

2008 Talisker 57° North, sixth edition of the 25 year old and third edition of the 30 year old are launched.

2009 New editions of the 25 and 30 year old are released.

2010 A 1994 Manager´s Choice single cask and a new edition of the 30 year old are released.

2011 Three limited releases - 25, 30 and 34 year old.

2012 A limited 35 year old is released.

2013 Four new expressions are released – Storm, Dark Storm, Port Ruighe and a 27 year old.

2014 A bottling for the Friends of the Classic Malts is released.

2015 Talisker Skye is released.

Port Ruighe

Storm

Skye

Tasting notes Talisker 10 years old:

GS – Quite dense and smoky on the nose, with smoked fish, bladderwrack, sweet fruit and peat. Full-bodied and peaty in the mouthy; complex, with ginger, ozone, dark chocolate, black pepper and a kick of chilli in the long, smoky tail.

Tasting notes Talisker Storm:

GS – The nose offers brine, burning wood embers, vanilla, and honey. The palate is sweet and spicy, with cranberries and blackcurrants, while peat-smoke and black pepper are ever-present. The finish is spicy, with walnuts, and fruity peat.

57° North

Dark Storm

10 years old

18 years

Distiller´s Edition

Tamdhu

[tam•doo]

Owner:
Ian Macleod Distillers

Region/district:
Speyside

Founded: **Status:** **Capacity:**
1896 Active 4 000 000 litres

Address: Knockando, Aberlour,
Morayshire AB38 7RP

Website: **Tel:**
tamdhu.com 01340 872200

Throughout its existence, Tamdhu has been working as a low-key distillery, supplying malt for blends. It was not until Ian Macleod Distillers took over in 2011 that they made a serious attempt at launching their single malt.

Robertson & Baxter, (the predecessor for today's Edrington) was involved in founding the distillery in 1896, took over ownership two years later and didn't sell again until 2011. William Alexander Robertson, blender and broker, who died just one month after Tamdhu started producing, was one of the great whisky profiles of the 19th century and he supplied whisky for big blends such as White Horse and Vat 69. The Robertson name lives on through The Robertson Trust, which not only controls Edrington, but which annually donates £15m to different charitable projects around the UK.

The distillery is equipped with an 11.85 tonne semilauter mash tun, nine Oregon pine washbacks and three pairs of stills. The tun room is unusually spacious with high ceilings, yellow tiles on the walls and large windows. There are four dunnage warehouses, one racked and five palletised with another four being built in 2016. The oldest cask in stock dates back to 1961. During 2015 the production target is 2.5 million litres of alcohol.

When the new owners restarted production at the distillery in 2012, a decision was taken not to use the old Saladin maltings. This type of malting is still used by some commercial maltsters but Tamdhu was the last distillery to use them. There are a total of 10 boxes and the capacity used to be 14,000 tonnes per year. The plant is in a surprisingly good state with all the equipment left intact, and with a little bit of refurbishing it would be possible for it to become operational again.

The core range of Tamdhu consists of a **10 year old** matured in first and second fill sherry casks, as well as the recently released **Tamdhu Batch Strength**. This was launched in spring of 2015, bottled at 58.8%, un chill-filtered and without colouring.

History:

1896 The distillery is founded by Tamdhu Distillery Company, a consortium of whisky blenders with William Grant as the main promoter. Charles Doig is the architect.

1897 The first casks are filled in July.

1898 Highland Distillers Company, which has several of the 1896 consortium members in managerial positions, buys Tamdhu Distillery Company.

1911 The distillery closes.

1913 The distillery reopens.

1928 The distillery is mothballed.

1948 The distillery is in full production again in July.

1950 The floor maltings is replaced by Saladin boxes when the distillery is rebuilt.

1972 The number of stills is increased from two to four.

1975 Two stills augment the previous four.

1976 Tamdhu 8 years is launched as single malt.

2005 An 18 year old and a 25 year old are released.

2009 The distillery is motballed.

2011 The Edrington Group sells the distillery to Ian Macleod Distillers.

2012 Production is resumed.

2013 The first official release from the new owners – a 10 year old.

2015 Tamdhu Batch Strength is released.

10 years old

Tasting notes Tamdhu 10 years old:

GS – Soft sherry notes, new leather, almonds, marzipan and a hint of peat on the nose. Very smooth and drinkable, with citrus fruit, gentle spice and more sweet sherry on the palate. Persistent spicy leather, with a sprinkling of black pepper in the finish.

Tamnavulin

AMNAVULIN
DISTILLERY

[tam•na•<u>voo</u>•lin]

Owner:	**Region/district:**
Whyte & Mackay (Emperador)	Speyside

Founded:	**Status:**	**Capacity:**
1966	Active	4 000 000 litres

Address: Tomnavoulin, Ballindalloch,
Banffshire AB3 9JA

Website:	**Tel:**
-	01807 590285

Tamnavulin has been one of few distilleries still using washbacks made of Corten steel. That, however, is no longer the case. In summer of 2015 the remaining four were replaced by stainless steel.

Traditionally, washbacks have been made of wood, Oregon pine or Siberian larch in particular. In the 1950s several distilleries started to install washbacks made of Corten steel but, since that material is subject to pitting, an extremely localized form of corrosion, stainless steel soon became the preferred type, even though it was much more expensive. Whether or not the material in the washbacks has an effect on the spirit is often discussed. Some argue that wooden washbacks, being more difficult to clean, always has a natural bacterial flora contributing to the flavour. Be that as it may, few distilleries dare change from wood to steel or vice versa for fear of changing the character of the spirit.

Tamnavulin distillery is equipped with a full lauter mash tun with 10.7 tonnes capacity, nine washbacks made of stainless steel (five of which were installed in summer 2015) with a fermentation time of 48 hours and three pairs of stills. Two racked warehouses (10 casks high) on site have a capacity of 35,000 casks with the oldest ones dating back to 1967, but several of the casks are from other distilleries. During 2015, the owners will be doing 16 mashes per week which equates to around 3.5 million litres in the year. From 2010 to 2013, part of the yearly production (around 5%) was heavily peated with a phenol specification in the barley of 55ppm.

Almost the entire production goes to blended whiskies, Whyte & Mackay in particular. The only recent official bottling of Tamnavulin was a 12 year old which has now been discontinued. There is, however, a 12 year old single malt called Ben Bracken which has been distilled at Tamnavulin and is mainly sold in Lidl supermarkets.

History:

1966 Tamnavulin-Glenlivet Distillery Company, a subsidiary of Invergordon Distillers Ltd, founds Tamnavulin.

1993 Whyte & Mackay buys Invergordon Distillers.

1995 The distillery closes in May.

1996 Whyte & Mackay changes name to JBB (Greater Europe).

2000 Distillation takes place for six weeks.

2001 Company management buy out operations for £208 million and rename the company Kyndal.

2003 Kyndal changes name to Whyte & Mackay.

2007 United Spirits buys Whyte & Mackay. Tamnavulin is opened again in July after having been mothballed for 12 years.

2014 Whyte & Mackay is sold to Emperador Inc.

Tasting notes Tamnavulin 12 years old:

GS – Delicate and floral on the nose, with light malt and fruit gums. Light to medium bodied, fresh, malty and spicy on the palate, with a whiff of background smoke. The finish is medium in length, with lingering spice, smoke, and notes of caramel.

TAMNAVULIN

*Single Malt
From Scotch Whisky*

12

12 years old

Teaninich

[tee•ni•nick]

Owner:	**Region/district:**
Diageo	Northern Highlands
Founded: **Status:**	**Capacity:**
1817 Active	9 800 000 litres

Address: Alness, Ross-shire IV17 0XB

Website:	**Tel:**
malts.com	01349 885001

Teaninich distillery, which lies in the rather un-romantic Teaninich Industrial Estate just south of Alness village, was, in the 1970s, one of the largest distilleries in Scotland with a capacity of 6 million litres.

In July 2014 Teaninich distillery stopped production. The reason was not a lack of demand for Teaninich single malt – to the contrary, it is a vital part of several Johnnie Walker blends. Instead, the site was due for a huge expansion. The original setup with ten washbacks and six stills was doubled and the new capacity will reach nearly 10 million litres per year. Three of the existing wash stills were altered into spirit stills so that the old stillhouse would house all six spirit stills, while a new house was built for the new wash stills. Teaninich has a very unusual mashing technique. It is the only Scottish distillery using a mash filter instead of a mash tun. The malt is ground into fine flour without husks in an Asnong hammer mill. Once the grist has been mixed with water, the mash passes through a Meura 2001 mash filter and the wort is collected. Water is added for a second time in the filter and a second run of mash is obtained. The procedure is repeated three times until a washback is filled. A brand new filter with a 14 tonnes mash has been installed to cope with the larger volumes. Apart from this, the production cycle will remain the same as before, with 75 hours of fermentation and a slow distillation to achieve Teaninich's green/oily character. Commissioning of the new equipment began in April 2015 and was still ongoing in August.

There were also plans to build a completely new distillery on the same grounds as Teaninich. With a total of 16 stills and a capacity of 13 million litres of pure alcohol, this would have become Diageo's second mega distillery after Roseisle. In autumn 2014, however, Diageo announced that the plans had been postponed, possibly due to reports of declining demand for Scotch whisky in certain markets.

The only official core bottling is the **10 year old** in the Flora & Fauna series. In autumn of 2009, a Teaninich **1996 single cask** was released in the new range Manager's Choice.

History:

1817 Captain Hugh Monro, owner of the estate Teaninich, founds the distillery.

1831 Captain Munro sells the estate to his younger brother John.

1850 John Munro, who spends most of his time in India, leases Teaninich to the infamous Robert Pattison from Leith.

1869 John McGilchrist Ross takes over the licence.

1895 Munro & Cameron takes over the licence.

1898 Munro & Cameron buys the distillery.

1904 Robert Innes Cameron becomes sole owner of Teaninich.

1932 Robert Innes Cameron dies.

1933 The estate of Robert Innes Cameron sells the distillery to Distillers Company Limited.

1970 A new distillation unit with six stills is commissioned and becomes known as the A side.

1975 A dark grains plant is built.

1984 The B side of the distillery is mothballed.

1985 The A side is also mothballed.

1991 The A side is in production again.

1992 United Distillers launches a 10 year old Teaninich in the Flora & Fauna series.

1999 The B side is decommissioned.

2000 A mash filter is installed.

2009 Teaninich 1996, a single cask in the new Manager's Choice range is released.

2014 Another six stills and eight washbacks are installed and the capacity is doubled.

2015 The distillery is expanded with six new stills and the capacity is doubled.

10 years old

Tasting notes Teaninich 10 years old:

GS – The nose is initially fresh and grassy, quite light, with vanilla and hints of tinned pineapple. Mediumbodied, smooth, slightly oily, with cereal and spice in the mouth. Nutty and slowly drying in the finish, with pepper and a suggestion of cocoa powder notes.

Tobermory

[tow•bur•mo•ray]

Owner:
Burn Stewart Distillers
(Distell Group Ltd)

Region/district:
Highland (Mull)

Founded: 1798
Status: Active (vc)
Capacity: 1 000 000 litres

Address: Tobermory, Isle of Mull, Argyllsh. PA75 6NR

Website:
tobermorydistillery.com

Tel:
01688 302647

The distillery was known as Tobermory from 1798 until its closure in 1930. In 1972, when the distillery was re-opened, the new owners decided to name the distillery Ledaig Distillery.

Seven years later it was time for yet another change of ownership when Stewart Jowett took over. During the sporadic production years of his ownership, he continued to use the name Ledaig. However, to confuse matters even further, he also launched both a blended whisky and a vatted malt under the name Tobermory. Nine years after Burn Stewart´s take-over in 1993, a decision was taken to use Tobermory as the brand for unpeated malts and Ledaig was reserved for all peated expressions with a phenol content of 30-40 ppm. Production from 1972 to 1993 was intermittent and of a hugely varying quality. The first task for Master Blender, Ian MacMillan was therefore to separate casks worthy of being bottled as single malts, while the rest was being used for blends.

The distillery is equipped with a traditional 5 ton cast iron mash tun, four wooden washbacks with a fermentation time of 50 to 90 hours and two pairs of stills. Two of the stills were replaced in August 2014. The owner´s plans are to do 8 mashes per week and 750,000 litres of alcohol in 2015 with a 50/50 split between Ledaig and Tobermory.

The core range from Tobermory distillery is the **10 and 15 year old Tobermory** and the **10 and 18 year old Ledaig**. The latter, finished in oloroso casks, was released in spring 2015 For a while, there has been talk of a couple of very old expressions from the distillery and the first of them was released in spring 2015 – the **42 year old Ledaig Dùsgadh**. It is the oldest bottling ever of Ledaig and only 500 bottles were produced. In autumn 2015, it was time for yet another **42 year old** – this time a **Tobermory**. Finally, there are two bottlings available at the distillery – a **19 year old Tobermory** and a **16 year old Ledaig**, both with a second maturation for over 10 years in PX sherry casks.

History:

1798 John Sinclair founds the distillery.

1837 The distillery closes.

1878 The distillery reopens.

1890 John Hopkins & Company buys the distillery.

1916 Distillers Company Limited (DCL) takes over John Hopkins & Company.

1930 The distillery closes.

1972 A shipping company in Liverpool and the sherrymaker Domecq buy the buildings and embark on refurbishment. When work is completed it is named Ledaig Distillery Ltd.

1975 Ledaig Distillery Ltd files for bankruptcy and the distillery closes again.

1979 The estate agent Kirkleavington Property buys the distillery, forms a new company, Tobermory Distillers Ltd and starts production.

1982 No production. Some of the buildings are converted into flats and some are rented to a dairy company for cheese storage.

1989 Production resumes.

1993 Burn Stewart Distillers buys Tobermory for £600,000 and pays an additional £200,000 for the whisky supply.

2002 Trinidad-based venture capitalists CL Financial buys Burn Stewart Distillers for £50m.

2005 A 32 year old from 1972 is launched.

2007 A Ledaig 10 year old is released.

2008 A limited edition Tobermory 15 year old is released.

2013 Burn Stewart Distillers is sold to Distell Group Ltd. A 40 year old Ledaig is released.

2015 Ledaig 18 years and 42 years are released together with Tobermory 42 years.

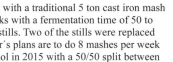

Tasting notes Tobermory 10 years old:

GS – Fresh and nutty on the nose, with citrus fruit and brittle toffee. A whiff of peat. Medium-bodied, quite dry on the palate with delicate peat, malt and nuts. Medium finish with a hint of mint and a slight citric tang.

Tasting notes Ledaig 10 years old:

GS – The nose is profoundly peaty, sweet and full, with notes of butter and smoked fish. Bold, yet sweet on the palate, with iodine, soft peat and heather. Developing spices. The finish is medium to long, with pepper, ginger, liquorice and peat.

10 years old

Tomatin

[to•mat•in]

Owner:	**Region/district:**
Tomatin Distillery Co	Highland
(Takara Shuzo Co., Kokubu & Co., Marubeni Corp.)	

Founded:	**Status:**	**Capacity:**
1897	Active (vc)	5 000 000 litres

Address: Tomatin, Inverness-shire IV13 7YT

Website:	**Tel:**
tomatin.com	01463 248144 (vc)

Tomatin´s transition from a producer of bulk whisky to a well-respected single malt brand has been a remarkable one.

Ten years ago, around 100,000 bottles were sold. In 2014 the number had grown to 420,000 with USA proving to be a fast growing market. The distillery has furthermore achieved great success, with the new sub-range of peated whiskies – Cù Bòcan.

The distillery is equipped with one 8 tonne stainless steel mash tun, 12 stainless steel washbacks with a fermentation time from 54 to 108 hours and six pairs of stills (only four of the spirit stills are used). There is an additional mash tun dating back to the time that the distillery was the largest in Scotland. A part of its wall has been cut away which gives the visitors a great opportunity to look into a mash tun and learn about the process. The goal is to produce 2.5 million litres in 2015, of which 100,000 litres will be peated at 30-35ppm.

The core range of single malts consists of **Legacy** (without age statement), **12, 18** and **30 year old**. Included are also **Cask Strength, 14 year old port finish** and a **Vintage 1988**. The Cask Strength (bottled at 57.5%) is the first of its kind in the core range and has been matured in a mix of bourbon and oloroso sherry casks. Recent limited releases include the **Tomatin Cuatro Series**, consisting of four 12 year old whiskies, distilled on the same day, matured in American oak for 9 years and then transferred to four different types of sherry butts for three years. In summer 2015, **Contrast** was launched – a pack of two bottles (bourbon and sherrymatured) with whiskies from six vintages (1973 to 2006). In 2013 the first official release of a peated Tomatin occurred. It was sold under the name **Cù Bòcan** which now has become a stand-alone brand. The range has been expanded and now includes a **1989 Vintage, Cù Bòcan Sherry, Cù Bòcan Virgin Oak** and **Cù Bòcan Bourbon**. The entire Tomatin range is due for a re-branding next year, at which time a travel retail range will also be launched.

History:
1897 The distillery is founded by Tomatin Spey Distillery Company.
1906 Production ceases.
1909 Production resumes through Tomatin Distillers.
1956 Stills are increased from two to four.
1958 Another two stills are added.
1961 The six stills are increased to ten.
1974 The stills now total 23 and the maltings closes.
1985 The distillery company goes into liquidation.
1986 Takara Shuzo Co. and Okara & Co., buy Tomatin through Tomatin Distillery Co.
1998 Okara & Co is liquidated and Marubeni buys out part of their shareholding.
2004 Tomatin 12 years is launched.
2005 A 25 year old and a 1973 Vintage are released.
2006 An 18 year old and a 1962 Vintage are launched.
2008 A 30 and a 40 year old as well as several vintages from 1975 and 1995 are released.
2009 A 15 year old, a 21 year old and four single casks (1973, 1982, 1997 and 1999) are released.
2010 The first peated release - a 4 year old exclusive for Japan.
2011 A 30 year old and Tomatin Decades are released.
2013 Cù Bòcan, the first peated Tomatin, is released.
2014 14 year old port finish, 36 year old, Vintage 1988, Tomatin Cuatro, Cù Bòcan Sherry Cask and Cù Bòcan 1989 are released.
2015 Cask Strength and Cù Bòcan Virgin Oak are released.

12 years old

Tasting notes Tomatin 12 years old:
GS – Barley, spice, buttery oak and a floral note on the nose. Sweet and medium-bodied, with toffee apples, spice and herbs in the mouth. Medium-length in the finish, with sweet fruitiness.

Tomintoul

[tom•in•towel]

Owner: **Region/district:**
Angus Dundee Distillers   Speyside

Founded: **Status:**   **Capacity:**
1965   Active   3 300 000 litres

Address: Ballindalloch, Banffshire AB37 9AQ

Website: **Tel:**
tomintouldistillery.co.uk   01807 590274

Tomintoul distillery became operational in 1965 and, therefore, celebrated its 50th anniversary this year. The first 35 years were eventful, but there was no focus on the distillery's single malts.

No less than six different owners came and went, one of them being Lonhro which was led by the controversial Tiny Rowlands. He was the son of a German adventurer and during WWII, he got involved in the Hitler Youth. After the war he was interned on the Isle of Man and then immigrated to South Rhodesia where he began to create a fortune for himself. With considerably more focus, the owners since 2000, Angus Dundee, has taken on the task of managing the distillery. Admittedly, they use the bulk part of the whisky for their many blends, but they have also put in a lot of work in establishing Tomintoul as a single malt of note.

Tomintoul distillery is equipped with a 11.6 tonnes semi lauter mash tun, six washbacks, all made of stainless steel and with a fermentation time of 54 hours, as well as two pairs of stills. There are currently 15 mashes per week, which means that capacity is used to its maximum, and the six racked warehouses have a storage capacity of 116,000 casks. The malt used for mashing is lightly peated, but every year a small batch of heavily peated spirit (55ppm) is produced. On the site there is also a blend centre with 14 large blending vats.

The core range consists of a **10, 14, 16** and **21 year old** (all kosher certified). There are also two **12 year olds** finished in **sherry** and **port** respectively. The peaty side of Tomintoul is represented by **Peaty Tang** (a vatting of 4-5 year old peated Tomintoul and 8 year old unpeated Tomintoul) and, as a standalone range, **Old Ballantruan** and **Old Ballantruan 10 year old**. The recent years' limited releases include a **1976 Vintage** (the latest bottling from 2013) and a **Single Cask** (the most recent being from **1981**). The highlight of 2015 was **Five Decades**, launched in August to celebrate the 50th anniversary. The bottle contains whiskies from 1965, 1975, 1985, 1995 and 2005. Towards the end of the year there was also the release of a **40 year old**.

History:

1965 The distillery is founded by Tomintoul Distillery Ltd, which is owned by Hay & MacLeod & Co. and W. & S. Strong & Co.

1973 Scottish & Universal Investment Trust, owned by the Fraser family, buys the distillery. It buys Whyte & Mackay the same year and transfers Tomintoul to that company.

1974 The two stills are increased to four and Tomintoul 12 years is launched.

1978 Lonhro buys Scottish & Universal Investment Trust.

1989 Lonhro sells Whyte & Mackay to Brent Walker.

1990 American Brands buys Whyte & Mackay.

1996 Whyte & Mackay changes name to JBB (Greater Europe).

2000 Angus Dundee plc buys Tomintoul.

2002 Tomintoul 10 year is launched.

2003 Tomintoul 16 years is launched.

2004 Tomintoul 27 years is launched.

2005 A young, peated version called Old Ballantruan is launched.

2008 1976 Vintage and Peaty Tang are released.

2009 A 14 year old and a 33 year old are released.

2010 A 12 year old Port wood finish is released.

2011 A 21 year old, a 10 year old Ballantruan and Vintage 1966 are released.

2012 Old Ballantruan 10 years old is released.

2013 A 31 year old single cask is released.

2015 Five Decades and a 40 year old are released.

10 years old

Tasting notes Tomintoul 10 years old:

GS – A light, fresh and fruity nose, with ripe peaches and pineapple cheesecake, delicate spice and background malt. Medium-bodied, fruity and fudgy on the palate. The finish offers wine gums, mild, gently spiced oak, malt and a suggestion of smoke.

Tormore

[tor•more]

Owner:	**Region/district:**
Chivas Bros (Pernod Ricard)	Speyside
Founded: **Status:**	**Capacity:**
1958 Active	4 400 000 litres

Address: Tormore, Advie, Grantown-on-Spey, Morayshire PH26 3LR

Website: **Tel:**
tormoredistillery.com 01807 510244

Tormore single malt may have been re-launched recently, but the main task for the distillery is to produce malt whisky for the Long John blend.

This legendary whisky, established in 1909, has seen better days but still managed to sell 4 million bottles last year. The brand was acquired by the American company, Schenley Industries, in 1956 and shortly thereafter, the company decided to build Tormore distillery in Speyside. The founder and main owner of Schenley, Lewis Rosenstiel, made it big during prohibition as a dealer in spirits and wines and, over the years, he managed to create a position for his company as the largest distiller and importer of spirits (after Seagram's) in the USA. He was also one of the most controversial profiles in the business, having been indicted, but never convicted, for bootlegging. It wasn´t until after his death in 1976, that the truth about Rosenstiel´s connections with Mafia members such as Frank Costello and Meyer Lansky really became known.

Tormore looks like no other distillery in Scotland and those who have travelled the A95 in the middle of Speyside will probably agree. The thought was to build a showpiece distillery, where the famous architect, Sir Albert Richardson, was hired and no expenses were spared. Positioned right next to the road, with its white washed walls, green copper roofs and extraordinary barred windows, its view is stupendous! Following an upgrade in early 2012, Tormore is now equipped with one stainless steel full lauter mash tun, 11 stainless steel washbacks and four pairs of stills. Tormore single malt is known for its fruity and light character which is achieved by a clear wort, a slow distillation and by using purifiers on all the stills.

Nearly everything that is produced at Tormore goes into a variety of blended Scotch but, since March 2014, there is a **14 year old** bottled at 43% and a **16 year old**, un chill-filtered and bottled at 48%. Both have been matured in American oak. The new version was first introduced in France and will later be rolled out to other markets.

History:

1958 Schenley International, owners of Long John, founds the distillery.

1960 The distillery is ready for production.

1972 The number of stills is increased from four to eight.

1975 Schenley sells Long John and its distilleries (including Tormore) to Whitbread.

1989 Allied Lyons (to become Allied Domecq) buys the spirits division of Whitbread.

1991 Allied Distillers introduce Caledonian Malts where Miltonduff, Glendronach and Laphroaig are represented besides Tormore. Tormore is later replaced by Scapa.

2004 Tormore 12 year old is launched as an official bottling.

2005 Chivas Brothers (Pernod Ricard) becomes new owners through the acquisition of Allied Domecq.

2012 Production capacity is increased by 20%.

2014 The 12 year old is replaced by two new expressions - 14 and 16 year old.

14 years old

Tasting notes Tormore 14 years old:

GS – Vanilla, butterscotch, summer berries and light spice on the nose. Milk chocolate and tropical fruit on the smooth palate, with soft toffee. Lengthy in the finish, with a sprinkling of black pepper.

Tullibardine

[tully•<u>bar</u>•din]

Owner:
Picard Vins & Spiritueux

Region/district:
Highlands

Founded: 1949
Status: Active (vc)
Capacity: 3 000 000 litres

Address: Blackford, Perthshire PH4 1QG

Website: tullibardine.com
Tel: 01764 682252

Since the complete revamp of the Tullibardine range of single malts two years ago, the company has, slowly but surely, started to establish itself in more and more parts of the world.

A total of 120,000 bottles were sold in 32 countries during 2014, with USA as one of the most important markets. The new strategy seems to pay off and the latest financial statements reflect an increase in profit of more than 70%. The distillery will now become the spiritual home to the famous brand, Highland Queen. It was first launched in 1893 by Roderick Macdonald, who 15 years later, took over Glenmorangie distillery. The brand had its heyday in the 1970s when it was sold all over the world. In 2008 it was taken over by the Picard family. The range now consists of three blends and six single malts, the oldest being a 40 year old.

The equipment at Tullibardine consists of a 6 ton stainless steel semi-lauter mash tun, nine stainless steel washbacks with a minimum fermentation of 52 hours and two pairs of stills. In 2015 the owners expect to produce 2.5 million litres of alcohol. The adjacent retail park (which was closed in 2014) has now been taken over by the distillery, transforming it into a bottling hall, a laboratory and more warehouses.

The whole range of Tullibardine single malts was completely revamped in 2013 and now consists of a bourbon matured core expression without age statement, **Sovereign**, three wood finishes, **225 Sauternes finish, 228 Burgundy finish** and **500 Sherry finish**. Finally, there are two older bottlings – a **20 year old** and a **25 year old**. All three wood finishes have received a second maturation for 12 months in other casks. The Sauternes casks came from Ch Suduiraut, the Burgundy casks previously held red wine from Chassagne Montrachet (owned by Picard) and for the Sherry finish, mainly PX casks were used. A spectacular bottling was released in summer 2015, as the first in a new range called Custodian Collection. It was a **60 year old** distilled in 1952 and since the whisky was at risk of dropping below the legal 40%, it was filled into a glass container already in 2012.

History:

1949 The architect William Delmé-Evans founds the distillery.

1953 The distillery is sold to Brodie Hepburn.

1971 Invergordon Distillers buys Brodie Hepburn Ltd.

1973 The number of stills increases to four.

1993 Whyte & Mackay (owned by Fortune Brands) buys Invergordon Distillers.

1994 Tullibardine is mothballed.

1996 Whyte & Mackay changes name to JBB (Greater Europe).

2001 JBB (Greater Europe) is bought out from Fortune Brands by management and changes name to Kyndal (Whyte & Mackay from 2003).

2003 A consortium buys Tullibardine for £1.1 million. The distillery is in production again by December.

2005 Three wood finishes from 1993, Port, Moscatel and Marsala, are launched together with a 1986 John Black selection.

2006 Vintage 1966, Sherry Wood 1993 and a new John Black selection are launched.

2007 Five different wood finishes and a couple of single cask vintages are released.

2008 A Vintage 1968 40 year old is released.

2009 Aged Oak is released.

2011 Three vintages (1962, 1964 and 1976) and a wood finish are released. Picard buys the distillery.

2013 A completely new range is launched – Sovereign, 225 Sauternes, 228 Burgundy, 500 Sherry, 20 year old and 25 year old.

2015 A 60 year old Custodian Collection is released.

Sovereign

Tasting notes Tullibardine Sovereign:

GS – Floral on the nose, with new-mown hay, vanilla and fudge. Fruity on the palate, with milk chocolate, brazil nuts, marzipan, malt, and a hint of cinnamon. Cocoa, vanilla, a squeeze of lemon and more spice in the finish.

Construction work at Annandale Distillery four years ago. Production started in November 2014.

New
distilleries

This section is reserved for distilleries
that were established in the last decade and the vast majority
of them have not yet released any whisky.
It is an interesting group of distilleries ranging from industrial giants
like Roseisle and Ailsa Bay, producing more than 10 million litres
every year to the likes of Abhainn Dearg and Strathearn,
craft distilleries aiming at a yearly production of 20-30,000 litres.
This chapter is bound to grow in the future as there are at least another
20 distilleries planned in Scotland. Most of them are presented in
the chapter The Year That Was under New Distilleries.

Wolfburn

[wolf•burn]

Owner:	**Region/district:**
Aurora Brewing Ltd.	Northern Highlands

Founded:	**Status:**	**Capacity:**
2013	Active	125 000 litres

Address: Henderson Park, Thurso,
Caithness KW14 7XW

Website:	**Tel:**
wolfburn.com	01847 891051

This is now the most northerly distillery on the Scottish mainland (thereby displacing Pulteney in Wick). The distillery is situated in an industrial area on the outskirts of Thurso and actually doesn't resemble a distillery as we are used to seeing it.

There are three, large, newly constructed buildings of which one is the distillery, while the other two are warehouses. The owners have chosen a site that is situated 350 metres from the ruins of the old Wolfburn Distillery which was founded in 1821 and subsequently closed down during the latter half of the 19th century. Construction work commenced in August 2012 and the first newmake came off the stills at the end of January 2013. From the very onset, before the equipment was ordered, the company hired Shane Fraser as the distillery manager. Shane had previously worked at Lochnagar and Oban and had for the last seven years been distillery manager at Glenfarclas. From the early design stages of the project, he was involved in determining the character of the future whisky.

The distillery is equipped with a 1.1 ton semi-lauter stainless steel mash tun with a copper canopy, three stainless steel washbacks with a fermentation time of 70-92 hours, holding 5,500 litres each, one wash still (5,500 litres) and one spirit still (3,600 litres). Each mash takes about 6 hours and the run in the spirit still is 10 minutes of foreshots, 2 hours of spirit cut and 2 hours on feints. Wolfburn uses a mix of casks: approximately one third of the spirit is laid down in ex-bourbon quarter casks, a further third is in ex-bourbon hogsheads as well as barrels, and the final third is laid down in ex-sherry butts.

The main part of the malt is unpeated and the intention is to create a smooth whisky. However, in 2014, there were 2 months production of lightly peated (10 ppm) spirit and this will increase during 2015 to 4 months. The first peated release will be in 2017, but the first bottling from the distillery which is expected in early 2016 will, nevertheless, have a smoky profile. The reason for this is that quarter casks from an Islay distillery have been used for the maturation. More than 2,000 casks have been filled since its inception and the owners expect to produce 125,000 litres of pure alcohol in 2015. The two warehouses were expanded during 2015 with a third in which a bottling plant will also be built.

Strathearn

[strath•earn]

Owner:
Tony Reeman-Clark

Region/district:
Southern Highlands

Founded: **Status:**
2013 Active

Capacity:
c 30 000 litres

Address: Bachilton Farm Steading, Methven PH1 3QX

Website:
strathearndistillery.com

Tel:
01738 840 100

This is something as unique as Scotland´s first micro-distillery. Admittedly, Abhainn Dearg on the Isle of Lewis potentially has the same capacity, but the stills at Strathearn are considerably smaller.

It is the brainchild of Tony Reeman-Clark and in early 2013 they received planning permission from the local council. Six months later, the distillery was up and running. It is situated at the old Bachilton Farm, a couple of miles west of Methven near Perth. Gin production was started at the beginning of August 2013 and the first whisky was filled into casks in October. The distillery uses the Maris Otter barley which was abandoned by other distillers years ago due to the low yield. Reeman-Clark prefers it though, because of the flavours that it contributes and he has also been experimenting with the original bere barley as used in the 1800s.

All the equipment is fitted into one room and consists of a stainless steel mash tun and two stainless steel washbacks (2,000 litres) with a fermentation time of 4-5 days. Furthermore, we also find a 1,000 litre wash still and a 500 litre spirit still. Both stills are of the Alambic type, made in Por-

tugal and with vertical tube copper condensers. When they are producing gin, they simply detach the lyne arm and mount a copper basket to the still to hold the botanicals. For maturation, the owners use 50-100 litre casks made of a variety of wood; virgin French oak, virgin American oak and ex-sherry casks.

Reeman-Clark and his team of three (distillery manager Stuart MacMillan, distiller Zak Shenfield and brewer Liam Pennycook) have also been experimenting with other types of wood like chestnut, mulberry and cherry. According to the rules laid out by Scotch Whisky Association, spirit matured in anything other than oak, cannot be called whisky. This problem has been solved by labelling the content Uisge Beatha – the ancient name for Scotch whisky. In an attempt to show how whisky would have been drunk a couple of hundred years ago, the company released this spirit, which had matured in 30-litre casks for only 28 days, in spring 2015. On the whisky side, both peated (35ppm) and un-peated whisky will be produced and the first bottlings are expected to appear in September 2016. So far, apart from the Uisge Beatha, four different gins have been released; Heather Rose, Classic, Oaked Highland and Homecoming Scotland 2014.

Kingsbarns

[kings•barns]

Owner:		**Region/district:**
Wemyss family		Lowlands
Founded:	**Status:**	**Capacity:**
2014	Active (vc)	600 000 litres

Address: East Newhall Farm, Kingsbarns, St Andrews KY16 8QE

Website:	**Tel:**
kingsbarnsdistillery.com	07717 754053

The plans for this distillery near St Andrews in Fife, were already drafted in 2008 and it has been a long journey until the goal was reached.

The idea, initiated by Greg Ramsay and Doug Clement, was to restore a dilapidated farmhouse from the late 18th century into a modern distillery. To advise them, they also had support in Bill Lark, known as the godfather of modern whisky-making in Australia. Planning permission was finally received in March 2011 while the initiators battled to find funding. The turning point came in September 2012 when the Scottish government awarded a grant of £670,000. This, in turn, led to the Wemyss family agreeing to inject £3m into the project and becoming the new owners. The family owns and operates the bottling company, Wemyss Malts, and adding a distillery of their own to the portfolio will, of course, facilitate the whisky business in the future.

Construction began in June 2013 and the distillery was officially opened on 30th November 2014 on St Andrew´s Day. Commissioning of the distillery began in January 2015 with the first casks being filled early in March. The distillery is equipped with a 1.5 ton stainless steel mash tun, four 7,500 litre stainless steel washbacks with a fermentation time of 55-72 hours, one 7,500 litre wash still and one 4,500 litre spirit still. Local barley will be used to produce a fruity, Lowland style whisky and predominantly, first fill bourbon barrels will be used for maturation, together with some sherry butts. The plan is to produce 140,000 litres of alcohol, although there is capacity for larger production in the future.

Peter Holroyd, with experience in the brewing business, became the distillery manager, while the founder of the project and the one who tirelessly fought to implement it, Doug Clement, is the visitor centre manager. There are currently three different types of tours for visitors to choose from. Since Kingsbarns is newly established, they can of course not offer the visitors a taste of their own produce. On the other hand, Wemyss Malts, as an independent bottler, offers an extensive range of whiskies from single malts to blends that can either be enjoyed at the distillery or bought in the shop.

Wemyss is a family-owned company based in Edinburgh, founded in 2005. The family owns another three companies in the field of wine and gin.

Ballindalloch

[bal•lin•da•lock]

Owner:
Ballindalloch Estate

Region/district:
Speyside

Founded:
2014

Status:
Active (vc)

Capacity:
100 000 litres

Address: Ballindalloch, Banffshire AB37 9AA

Website:
ballindallochdistillery.com

Tel:
01807 500 331

In the heart of Speyside, the owners of Ballindalloch Castle at the time, decided in 2012 to turn a steading from 1820 into a whisky distillery.

The previous proprietors of the castle were also involved in the whisky industry from 1923 to 1965, when they owned part of Cragganmore distillery, not far away from the castle. The old farm building was meticulously renovated with attention given to every little detail and the result is an amazingly beautiful distillery which can be seen from the A95 between Aberlour and Grantown-on-Spey.

Ballindalloch distillery takes its water from the nearby Garline Springs and all the barley (currently Concerto) is grown on the Estate. All of the distillery equipment are gathered on the second floor which makes it easy for visitors to get a good view of the production. The equipment consists of an extraordinary 1 ton semi lauter, copper clad mash tun with a copper dome. There are four washbacks made of Oregon pine with two short (65 hours) fermentations per week and three long (115 hours). Finally there is a 5,000 litre lantern-shaped wash still and a 3,600 litre spirit

still with a reflux ball. Both stills are connected to two wooden worm tubs for cooling the spirit vapours. The distillery is run by three persons only and, with no automation or computers. Distillery manager is Charlie Smith who has more than 40 years´ experience within the industry, some of which was as manager at Talisker. The distillery came on stream in September 2014 and was officially opened in May 2015 by Prince Charles. To start with, the distillery will be working 5 days a week, making 100,000 litres of alcohol. The idea is to produce a robust and bold whisky, enhanced not least by the use of worm tubs. Initially, they will fill the spirit into ex-bourbon and ex-sherry casks, but experiments with other types of casks are also in the pipeline. The one warehouse on site, filled up quickly and in May 2015 they started to move casks into a second warehouse close to Glenfarclas distillery.

The distillery is open for visitors by appointment and the stylish tasting rooms have been decorated with many objects from the family´s castle. There is also the opportunity to take part in The Art of Whisky Making experience, which means spending a day with the crew and learning about whisky from mashing to warehousing.

Ardnamurchan

[ard•ne•mur•ken]

Owner: **Region/district:**
Adelphi Distillery Ltd   Western Highlands

Founded: **Status:**   **Capacity:**
2014   Active (vc)   500 000 litres

Address: Glenbeg, Ardnamurchan, Argyll PH36 4JG

Website: **Tel:**
adelphidistillery.com   01972 500 285

The success for the independent bottler, Adelphi Distillery, has forced the owners to build their own distillery.

The chosen site is Glenbeg on the Ardnamurchan peninsula, just north of Isle of Mull, which makes it the most westerly distillery on mainland Scotland. It is a stunning location, on the shores of Loch Sunart, overlooking the Atlantic. Most of the buildings were completed by August 2013, the equipment started to arrive in the autumn and, finally on 11th July 2014, the distillery came on stream. The distillery was officially opened two weeks later by Princess Anne.

The distillery is equipped with a 2 tonne semi lauter mash tun made of stainless steel with a copper canopy, four oak washbacks and three made of stainless steel, a wash still (10,000 litres) and a spirit still (6,000 litres). The four wooden washbacks are very special because they were retrieved in France, where they had been used as cognac vats, were dismantled and then rebuilt at Ardnamurchan. Minimum fermentation time is 72 hours. The production started with 2 mashes per week and has now increased to five, which means around 150,000 litres per year with the intention of moving to 300,000 litres sometime soon. Two different styles of whisky will be produced; peated and unpeated. For the peated spirit, the barley has a phenol specification of 30-35ppm. The ultimate goal is to be self-sufficient with malted barley, but the plans for their own malting floors have now been postponed.

The owners have put in a lot of effort into creating a distillery whose environmental footprint is as small as possible. All the power and heat requirements for the distillery come from local renewables – the river that provides the distillery's cooling water has a hydro-electricity generator and the biomass boiler is fuelled by woodchip from the local forestry.

Adelphi Distillery is named after a distillery which closed in 1902. The company is owned by Keith Falconer and Donald Houston, who have recruited Alex Bruce from the wine trade to act as Marketing Director. Their whiskies are always bottled at cask strength, uncoloured and un chillfiltered. Adelphi bottles around 50 casks a year.

Annandale

[ann•an•dail]

Owner:
Annandale Distillery Co.

Region/district:
Lowlands

Founded: 2014
Status: Active (vc)
Capacity: 500 000 litres

Address: Northfield, Annan, Dumfriesshire DG12 5LL

Website:
annandaledistillery.co.uk

Tel:
01461 204816

In 2010 Professor David Thomson and his wife, Teresa Church, obtained consent from the local council for the building of the new Annandale Distillery in Dumfries and Galloway in the south-west of Scotland.

The old one had been producing since 1836 and was owned by Johnnie Walker from 1895 until it closed down in 1918. From 1924 to 2007, the site was owned by the Robinson family, who were famous for their Provost brand of porridge oats. The old distillery was used as a drying plant for the oats. David Thomson began the restoration of the site in June 2011 with the two, old sandstone warehouses being restored to function as two-level dunnage warehouses. The distillery was in a poor condition and the mash house, the tun room and the still house were completely reconstructed. The total cost, including restoration, construction and new equipment amounted to £10.5m.

Entering the production area of the new distillery is like walking into a beautiful village church. First you run into the 2.5 ton semi-lauter mash tun with an elegant copper dome. Then, with three wooden washbacks (a fermentation time of 72 hours) on each side, you are guided up to the two spirit stills (4,000 litres). Once you have reached them, you find the wash still (12,000 litres) slightly hidden behind a wall. The capacity is 500,000 litres per annum but, to begin with, they will only work one shift, which means 6 mashes per week and 250,000 litres. So far, the casks used for maturation have been first fill bourbon barrels from Buffalo Trace and second fill from Woodford Reserve but there are plans to also fill sherry butts during 2015.

The first cask was filled on 15 November 2014 and both unpeated and peated (45ppm) whisky is distilled. The unpeated version will be called Man o' Words inspired by the poet, Robert Burns, while the peated version will be named Man o' Sword, named after King Robert the Bruce. In spring 2015, the distillery decided to put up the first 100 casks that were filled for sale. Prices vary, but cask number one has a price tag of £1m, while barrel number eight, the Chinese lucky number, will cost £888,000. The old maltings, with the kiln and original pagoda roof, have been turned into an excellent visitor centre and the owners hope to attract 50,000 visitors to the distillery in the future.

Daftmill

[daf•mil]

Owner: Francis Cuthbert	**Region/district:** Lowlands

Founded: 2005	**Status:** Active	**Capacity:** c 65 000 litres

Address: By Cupar, Fife KY15 5RF

Website: daftmill.com	**Tel:** 01337 830303

Permission was granted in 2003 for a steading at Daftmill Farmhouse in Fife, just a few miles west of Cupar and dating back to 1655, to be converted into a distillery.

The first distillation was on 16th December 2005 and around 20,000 litres are distilled in a year. A little less was produced in 2013 when a new boiler was fitted. It is run as a typical farmhouse distillery. The barley is grown on the farm and they also supply other distilleries. Of the total 800 tonnes that Francis Cuthbert harvests in a year, around 100 tonnes are used for his own whisky. The malting is done without peat at Crisp's in Alloa. The equipment consists of a one tonne semi-lauter mash tun with a copper dome, two stainless steel washbacks with a fermentation between 72 and 100 hours and one pair of stills with slightly ascending lyne arms. The equipment is designed to give a lot of copper contact, a lot of reflux. The wash still has a capacity of 3,000 litres and the spirit still 2,000 litres.

Francis Cuthbert's aim is to do a light, Lowland style whisky similar to Rosebank. In order to achieve this they have very short foreshots (five minutes) and the spirit run starts at 78% to capture all of the fruity esters and already comes off at 73%. The spirit is filled mainly into ex-bourbon casks, always first fill, but there are also a few sherry butts in the two dunnage warehouses.

Taking care of the farm obviously prohibits Francis from producing whisky full time. His silent season is during spring and autumn when work in the fields take all of his time. Whisky distillation is therefore reserved for June-August and November-February. The whisky in the warehouse is now 9 years old and whisky enthusiasts have been asking for years when the first bottling is due. Francis however doesn't seem to be in a hurry or as he himself puts it "patience is a virtue".

Abhainn Dearg

[aveen jar•rek]

Owner: Mark Tayburn	**Region/district:** Islands (Isle of Lewis)

Founded: 2008	**Status:** Active	**Capacity:** c 20 000 litres

Address: Carnish, Isle of Lewis, Na h-Eileanan an Iar HS2 9EX

Website: abhainndearg.co.uk	**Tel:** 01851 672429

In September 2008, spirit flowed from a newly constructed distillery in Uig on the island of Lewis in the Outer Hebrides.

This was the first distillery on the island since 1840 when Stornoway distillery was closed. The Gaelic name of the new distillery is Abhainn Dearg which means Red River, and the founder and owner is Mark "Marko" Tayburn who was born and raised on the island. Part of the distillery was converted from an old fish farm while some of the buildings are new. There are two 500 kg mash tuns made of stainless steel and two 7,500 litre washbacks made of Douglas fir with a fermentation time of 4 days. The wash still has a capacity of 2,112 litres and the spirit still 2,057 litres. Both have very long necks and steeply descending lye pipes leading out into two wooden worm tubs. Both bourbon and sherry casks are used for maturation. The plan was to use 100% barley grown on Lewis and in 2013 the first 6 tonnes of Golden Promise (15% of the total requirement) were harvested. In 2015, the owner reported that all the barley needed for the production, now came from the island.

The first release from the distillery was The Spirit of Lewis (matured for a short time in sherry casks) in 2010 and the first single malt was a limited release (2,011 bottles) of a 3 year old in October 2011, followed up by a cask strength version (58%) in 2012. The owners currently still offer Spirit of St Lewis and the 3 year old single malt and it seems the plan is to await the release of a 10 year old in 2018.

Ailsa Bay

[ail•sah bey]

Owner:	**Region/district:**
William Grant & Sons	Lowlands
Founded: **Status:**	**Capacity:**
2007 Active	12 000 000 litres
Address: Girvan, Ayrshire KA26 9PT	
Website:	**Tel:**
-	01465 713091

Commisioned in September 2007, it only took nine months to build this distillery on the same site as Girvan Distillery near Ayr on Scotland's west coast.

Until recently, it was equipped with a 12,5 tonne full lauter mash tun, 12 washbacks made of stainless steel and eight stills. In August 2013 however, it was time for a major expansion when yet another mash tun, 12 more washbacks and eight more stills were commissioned, doubling the capacity to 12 million litres of alcohol.

Each washback will hold 50,000 litres and fermentation time is 60 hours for the heavier styles and 72 hours for the lighter "Balveniestyle". The stills are made according to the same standards as Balvenie's and two of the wash stills and two of the spirit stills have stainless steel condensers instead of copper. That way, they have the possibility

of making batches of a more sulphury spirit if desired. A unique feature is the octangular spirit safe which sits between the two rows of stills. Each side corresponds to one specific still. To increase efficiency and to get more alcohol, high gravity distillation is used. The wash stills are heated using external heat exchangers but they also have interior steam coils. The spirit stills are heated by steam coils. In 2015, the distillery will be producing 11.5 million litres of alcohol.

Four different types of spirit are produced. The most common is a light and rather sweet spirit. Then there is a heavy, sulphury style and two peated with the peatiest having a malt specification of 50ppm. A majority of the casks (60-70%) used for maturation, are refill bourbon casks and the rest is made up of first fill bourbon and sherry casks. The production is destined to become a part of Grant's blended Scotch which is currently the fourth most popular Scotch in the world with 56 million bottles sold in 2013.

Roseisle

[rose•eyel]

Owner:	**Region/district:**
Diageo	Highlands
Founded: **Status:**	**Capacity:**
2009 Active	12 500 000 litres
Address: Roseisle, Morayshire IV30 5YP	
Website:	**Tel:**
-	01343 832100

Roseisle distillery is located on the same site as the already existing Roseisle maltings just west of Elgin. The distillery has won several awards for its ambition towards sustainable production.

The distillery is equipped with two stainless steel mash tuns with a 12.5 tonne charge each. There are 14 huge (115,500 litres) stainless steel washbacks and 14 stills with the wash stills being heated by external heat exchangers while the spirit stills are heated using steam coils. The spirit vapours are cooled through copper condensers but on three spirit stills and three wash stills there are also stainless steel condensers attached, that you can switch to for a more sulphury spirit. The plan for 2015 is to do 21 mashes per week and a total of 10,6 million litres of alcohol.

The total cost for the distillery was £40m and how to use the hot water in an efficient way was very much a focal

point from the beginning. For example, Roseisle is connected by means of two long pipes with Burghead maltings, 3 km north of the distillery. Hot water is pumped from Roseisle and then used in the seven kilns at Burghead and cold water is then pumped back to Roseisle. The pot ale from the distillation will be piped into anaerobic fermenters to be transformed into biogas and the dried solids will act as a biomass fuel source. The biomass burner on the site, producing steam for the distillery, covers 72% of the total requirement. Furthermore, green technology has reduced the emission of carbon dioxide to only 15% of an ordinary, same-sized distillery.

So far no substantial quantities of peated spirit has been distilled at Roseisle and during 2014 they will concentrate on producing a whisky with a light Speyside character. The fermentation time for this style is 90-100 hours and for a heavier style it is 50-60 hours.

Eden Mill

[eden mill]

Owner: Paul Miller		**Region/district:** Lowlands
Founded: 2014	**Status:** Active (vc)	**Capacity:** 80 000 litres
Address: St Andrews, Fife, KY16 0UU		
Website: edenmill.com		**Tel:** 01334 834038

In 2012, Paul Miller, the former Molson Coors sales director, with a background in the whisky industry (Diageo and Glenmorangie), opened up the successful Eden Brewery in Guardbridge 3 miles west of St Andrews.

The site was an old paper mill and only 50 metres away, there was a distillery called Seggie which was operative between 1810 and 1860 and owned by the Haig family. As an extension of the brewery, Paul decided to build a distillery called Eden Mill Distillery. The distillery, with a capacity of 80,000 litres per year, will mainly produce malt whisky, but gin is also on the map. The distillery is equipped with two wash stills and one spirit still of the alambic type. Made by Hoga in Portugal, all three stills are of the same size – 1,000 litres. Eden Mill is the first combined brewery and distillery in Scotland – a combination which

has proven so successful, especially in the USA. The brewery/distillery also has a visitor centre which already attracts 20,000 visitors a year.

Whisky production started in November 2014 and, waiting for the first whisky to be ready for release, Miller has already released three, young malt spirits – St Andrews Day, Hogmanay and Robert Burns Day, all of them being sold out by now. Different varieties of gin, on the other hand, have been launched continuously since the start, which include Hop, Love, Golf and Oak Gin. Of the whisky production, around 10,000 litres a year will be reserved for the distillery's Private Cask Owners' Club, where customers can buy anything from octaves to hogsheads of whisky.

In spite of the comparatively small size of the operation, Eden Mill Brewery and Distillery employs around 25 people.

Dalmunach

[dal•moo•nack]

Owner: Chivas Brothers		**Region/district:** Speyside
Founded: 2015	**Status:** Active	**Capacity:** 10 000 000 litres
Address: Carron, Banffshire AB38 7QP		
Website: -		**Tel:** -

One of the newest distilleries in Scotland, and one of the most beautiful, has been built on the site of the former Imperial distillery.

Imperial distillery was inaugurated in 1898 and was then owned by DCL (later to become Diageo) from 1916 until 2005, when Chivas Brothers took over. It was out of production for 60% of the time until 1998 when it was mothballed. The owners probably never planned to use it for distillation again as it was put up for sale in 2005 to become available as residential flats. Soon after, it was withdrawn from the market and, in 2012, a decision was taken to tear down the old distillery and build a new. Demolition of the old distillery began in 2013 and by the end of that year, nothing was left, except for the old warehouses.

Construction on the new Dalmunach distillery started in 2013 and it was commissioned in October 2014. The ex-

ceptional and stunning distillery is equipped with a 12 ton Briggs full lauter mash tun, 16 stainless steel washbacks and 4 pairs of stills. The stills are positioned in a circle with a hexagonal spirit safe in the middle. The distillery, which cost £25m to build, has a capacity of 10 million litres and was officially opened in June 2015 by Nicola Sturgeon, First Minister of Scotland.

Glasgow

[glas•go]

Owner:	Region/district:
Liam Hughes, Ian McDougall	Lowlands

Founded:	Status:	Capacity:
2015	Active	100 000 litres

Address. 204 West George St, Glasgow G2 4QY

Website:	Tel:
glasgowdistillery.com	0141 4047191

There's been keen competition in the past year to see who builds what has been dubbed the first new distillery in Glasgow in more than one hundred years.

On the one hand there is Tim Morrison, the owner of AD Rattray, and his son, Andrew, who plan a distillery in the Ocean's Docks on the banks of the River Clyde. In the meanwhile, Liam Hughes and Ian McDougall from The Glasgow Distillery Company, backed up by Asian investors, have been working away on building another distillery at the Hillington Business Park and they finally beat the Morrison's to it. The first whisky was distilled in February 2015 and while awaiting that it will become at least three years old, they have already launched Makar Gin from own production. Whisky, however, is also being offered by the company. Prometheus, a sourced single malt where the distillery is not revealed, was released in April 2015. It is a 26

year old, slightly peated sherry maturation from Speyside. The next release, a 27 year old, is due early in 2016.

The distillery is equipped with a one ton mash tun, four wash backs (5,400 litres each), one 2,500 litre wash still, one 1,400 litre spirit still and one 450 litre gin still. The plan is to produce around 100,000 litres of alcohol for the first few years and then double that production.

The fact that this would have been the first malt whisky distillery in Glasgow in more than 100 years, isn't entirely true. In 1958, Strathclyde grain distillery was founded and within that distillery, Kinclaith malt distillery was constructed which operated until 1975. Strathclyde is still a working distillery. At one time there were 40 distilleries working in Glasgow, but only five malt distilleries (Adelphi, Camlachie, Dundashill, Provanmill and Yoker) and one grain distillery (Port Dundas) made it into the 1900s before they were closed.

Harris

[har•ris]

Owner:	Region/district:
Isle of Harris Distillers Ltd.	Islands (Isle of Harris)

Founded:	Status:	Capacity:
2015	Active (vc)	230 000 litres

Address: Tarbert, Isle of Harris, Na h-Eileanan an Iar HS3 3DJ

Website:	Tel:
harrisdistillery.com	01859 502098

Almost ten years ago, Anderson Bakewell had conjured up an idea that has now resulted in a distillery which has come to fruition on the Isle of Harris.

Bakewell, who has been connected to the island for more than 40 years, acquired the services of Simon Erlanger for the company's benefit at an early stage. Erlanger, a former marketing director for Glenmorangie, is now MD of the new distillery, while Bakewell is chairman of the company. Construction started in 2014 and it seems that distillation will commence in end of September 2015. The total cost for the whole project is £11.4m, but this sum probably also covers the cost for barley and casks until the first whisky is ready to be bottled. The distillery, located in Tarbert, is the second distillery after Abhainn Dearg on Lewis to be located in the Outer Hebrides.

The equipment consists of a 1.2 tonne semi lauter mash tun made of stainless steel but clad with American oak, 5 washbacks (6,000 litres each) made of Oregon pine, one 7,000 litre wash still and a 5,000 litre spirit still - both with descending lyne arms and made in Italy. The style of the whisky, which will be called Hearach (the Gaelic word for a person living on Harris), will be medium peated with a phenol specification in the barley of 12-14ppm. The first product to be released though, will be a gin, available only from the distillery and online. Apart from traditional gin botanicals, local ingredients will also be used such as sugar kelp.

Although the annual production capacity is 230,000 litres, the plan is to start with less than 100,000 litres in the first full year. With the new distillery, 20 new jobs will be created on the island, some of whom will be working in the distillery visitor centre.

Distilleries per owner

c = closed, d = demolished, mb = mothballed, dm = dismantled

Diageo
Auchroisk
Banff (d)
Benrinnes
Blair Athol
Brora (c)
Caol Ila
Cardhu
Clynelish
Coleburn (dm)
Convalmore (dm)
Cragganmore
Dailuaine
Dallas Dhu (c)
Dalwhinnie
Dufftown
Glen Albyn (d)
Glendullan
Glen Elgin
Glenesk (dm)
Glenkinchie
Glenlochy (d)
Glenlossie
Glen Mhor (d)
Glen Ord
Glen Spey
Glenury Royal (d)
Inchgower
Knockando
Lagavulin
Linkwood
Mannochmore
Millburn (dm)
Mortlach
North Port (d)
Oban
Pittyvaich (d)
Port Ellen (dm)
Rosebank (c)
Roseisle
Royal Lochnagar
St Magdalene (dm)
Strathmill
Talisker
Teaninich

Pernod Ricard
Aberlour
Allt-a-Bhainne
Braeval
Caperdonich (d)
Dalmunach
Glenallachie
Glenburgie
Glen Keith
Glenlivet
Glentauchers
Glenugie (dm)
Imperial (d)
Inverleven (d)
Kinclaith (d)
Lochside (d)
Longmorn

Miltonduff
Scapa
Strathisla
Tormore

Edrington Group
Glenrothes
Glenturret
Highland Park
Macallan

Inver House (Thai Beverage)
Balblair
Balmenach
Glen Flagler (d)
Knockdhu
Pulteney
Speyburn

John Dewar & Sons (Bacardi)
Aberfeldy
Aultmore
Craigellachie
Macduff
Royal Brackla

William Grant & Sons
Ailsa Bay
Balvenie
Glenfiddich
Kininvie
Ladyburn (dm)

Whyte & Mackay (Emperador)
Dalmore
Fettercairn
Jura
Tamnavulin

Morrison Bowmore (Suntory)
Auchentoshan
Bowmore
Glen Garioch

Burn Stewart Distillers (Distell)
Bunnahabhain
Deanston
Tobermory

Benriach Distillery Co.
Benriach
Glendronach
Glenglassaugh

Loch Lomond Group
Glen Scotia
Littlemill (d)
Loch Lomond

Beam Suntory
Ardmore
Laphroaig

J & A Mitchell
Glengyle
Springbank

Glenmorangie Co. (LVMH)
Ardbeg
Glenmorangie

Angus Dundee Distillers
Glencadam
Tomintoul

Ian Macleod Distillers
Glengoyne
Tamdhu

Campari Group
Glen Grant

Isle of Arran Distillers
Arran

Signatory
Edradour

Tomatin Distillery Co.
Tomatin

J & G Grant
Glenfarclas

Rémy Cointreau
Bruichladdich

David Prior
Bladnoch (c)

Gordon & MacPhail
Benromach

La Martiniquaise
Glen Moray

Ben Nevis Distillery Ltd (Nikka)
Ben Nevis

Picard Vins & Spiritueux
Tullibardine

Harvey's of Edinburgh
Speyside

Kilchoman Distillery Co.
Kilchoman

Cuthbert family
Daftmill

Mark Tayburn
Abhainn Dearg

Aurora Brewing Ltd
Wolfburn

Strathearn Distillery Ltd
Strathearn

Annandale Distillery Co.
Annandale

Adelphi Distillery Co.
Ardnamurchan

Wemyss
Kingsbarns

Ballindalloch Estate
Ballindalloch

Paul Miller
Eden Mill

Isle of Harris Distillers
Harris

The Glasgow Distillery Company
Glasgow Distillery

Closed
distilleries

The distilleries on the following pages have all been closed
and some of them even demolished. The only one that still had a chance of
producing again was Bladnoch which was transferred to this chapter
because the owning company went into receivership in 2014. No new
owner materialised until end of July 2015 when the layout for the book had
already been set. More details can be found below and on page 268.
New whiskies from some of the distilleries appear on a regularly basis but
for most of them chances are very slim of ever finding annother bottling.

Banff

Owner:	Region:	Founded:	Status:
Diageo	Speyside	1824	Demolished

Banff's tragic history of numerous fires, explosions and
bombings have contributed to its fame. The most spectacular incident was when a lone Junkers Ju-88 bombed one
of the warehouses in 1941. Hundreds of casks exploded
and several thousand litres of whisky were destroyed.
The distillery was closed in 1983 and the buildings were
destroyed in a fire in 1991. The distillery was owned for
80 years by the Simpson family but when their company
filed for bankruptcy in 1932, it was sold to Scottish Malt
Distillers which later would be a part of Diageo. When the
distillery was at its largest it produced 1 million litres per
year in three pairs of stills.

Bottlings:
There has only been one official Rare Malts bottling from
2004. A couple of independent bottlings from 1975 were
released in 2012 and a 49 year old from Gordon & MacPhail
(distilled in 1966) appeared in 2015.

Bladnoch

Owner:	Region:	Founded:	Status:
David Prior	Lowlands	1817	Closed

Owned by the McClelland family until 1911 when the Irish
whisky broker Dunville & Co. took over. Production was
intermittent during their regime and in 1937 the distillery
closed. In 1956, new owners installed four new stills and
re-started production. The distillery was bought by United
Distillers in 1985 but they only ran it for eight years until
it was sold to Ramond Armstrong who´s intentions were
to convert it into a guest house. The plans were changed
and whisky production started in 2000. In 2010, production
stopped again and in 2014 the company went into
liquidation. In July 2015, Australian business man David
Prior bought the distillery with the intention of refurbishing
it and to start production in 2016. Read more on 268.

Bottlings:
In the last two years before liquidation a number of official
bottlings were released, ranging from 12 to 22 years old.
Independent bottlings still appear from time to time.

Brora

[bro•rah]

Owner:	**Region/district:**
Diageo	Northern Highlands
Founded: **Status:**	**Capacity:**
1819 Closed	-

Although founded under the name Clynelish distillery in 1819, it is under the name Brora that the single malt has enjoyed its newfound fame during the past two decades.

The whisky has mostly appealed to peat freaks around the world but, for the first 140 years, it actually wasn't that peated. In 1967 DCL decided to build a new, modern distillery on the same site. This was given the name Clynelish and it was decided the old distillery should be closed. Shortly after, the demand for peated whisky, especially for the blend Johnnie Walker, increased and the old site re-opened but now under the name Brora and the "recipe" for the whisky was changed to a heavily peated malt. This continued from 1969 to 1973 and after that the peatiness was reduced, even if single peated batches turned up until the late seventies. Brora closed permanently in 1983 but the buildings still stand next to the new Clynelish. The two stills, the feints receiver, the spirit receiver and the brass safe remain, while the warehouses are used for storage of spirit from Clynelish.

The first distillery was built in the time referred to as the Highland Clearances. Many land-owners wished to increase the yield of their lands and consequently went into large-scale sheep farming. Thousands of families were ruthlessly forced away and the most infamous of the large land-owners was the Marquis of Stafford who founded Clynelish (Brora) in 1819.

Since 1995 Diageo has regularly released different expressions of Brora in the Rare Malts series. The latest, which also became the last, appeared in 2003. In 2002 a new range was created, called Special Releases and bottlings of Brora have appeared ever since. Although not confirmed by Diageo at the time of writing, the official registration of a label indicate that a Brora 37 years old, bottled at 50.4% could be this year's Special Release. Probably due for release in the beginning of October, this will be the 14th since 2002.

History:

1819 The Marquis of Stafford, 1st Duke of Sutherland, founds the distillery as Clynelish Distillery.

1827 The first licensed distiller, James Harper, files for bankruptcy and John Matheson takes over.

1828 James Harper is back as licensee.

1834 Andrew Ross takes over the license.

1846 George Lawson & Sons takes over.

1896 James Ainslie & Heilbron takes over and rebuilds the facilities.

1912 Distillers Company Limited (DCL) takes over together with James Risk.

1925 DCL buys out Risk.

1930 Scottish Malt Distillers takes over.

1931 The distillery is mothballed.

1939 Production restarts.

1960 The distillery becomes electrified (until now it has been using locally mined coal from Brora).

1967 A new distillery is built adjacent to the first one, it is also named Clynelish and both operate in parallel from August.

1968 'Old' Clynelish is mothballed in August.

1969 'Old' Clynelish is reopened as Brora and starts using a very peaty malt over the next couple of years

1983 Brora is closed in March.

1995 Brora 1972 (20 years) and Brora 1972 (22 years) are launched as Rare Malts.

2002 A 30 year old is the first bottling in the Special Releases.

2014 The 13th release of Brora – a 35 year old.

2015 The 14th release of Brora.

35 years old

Caperdonich

Owner:	Region:	Founded:	Status:
Chivas Bros.	Speyside	1897	Demolished

The distillery was founded by James Grant, owner of Glen Grant which was located in Rothes just a few hundred metres away. Five years after the opening, the distillery was shut down and was re-opened again in 1965 under the name Caperdonich. In 2002 it was mothballed yet again, never to be re-opened. Parts of the equipment were dismantled to be used in other distilleries within the company. In 2010 the distillery was sold to the manufacturer of copper pot stills, Forsyth´s in Rothes, and the buildings were demolished. In the old days a pipe connected Caperdonich and Glen Grant for easy transport of spirit, ready to be filled.

Bottlings:

An official cask strength was released in 2005. Recent independent bottlings are a 21 year old distilled in 1992 and a 41 year old (1972), both from Duncan Taylor. There is also Caperdonich Batch 4 from That Boutique-y Whisky Company.

Coleburn

Owner:	Region:	Founded:	Status:
Diageo	Speyside	1897	Dismantled

Like so many other distilleries, Coleburn was taken over by DCL (the predecessor of Diageo) in the 1930s. Although the single malt never became well known, Coleburn was used as an experimental workshop where new production techniques were tested. In 1985 the distillery was mothballed and never opened again. Two brothers, Dale and Mark Winchester, bought the buildings in 2004 with the intention of transforming the site into an entertainment centre - a plan that never materialised. Since 2014, the warehouses are used by Aceo Ltd, who bought independent bottler Murray McDavid in 2013, for storing their own whiskies as well as stock belonging to clients

Bottlings:

There has been one official Rare Malts bottling from 2000, while Independent bottlings are also rare. After a 36 year old was released in 2006 by Signatory it took another 7 years until the next appeared - a 1972 from Gordon & MacPhail.

Convalmore

Owner:	Region:	Founded:	Status:
Diageo	Speyside	1894	Dismantled

This distillery is still intact and can be seen in Dufftown next to Balvenie distillery. The buildings were sold to William Grant´s in 1990 and they now use it for storage. Diageo, however, still holds the rights to the brand. In the early 20th century, experimental distilling of malt whisky in continuous stills (the same method used for producing grain whisky) took place at Convalmore. The distillery closed in 1985. One of the more famous owners of this distillery was James Buchanan who used Convalmore single malt as a part of his famous blend Black & White. He later sold the distillery to DCL (later Diageo).

Bottlings:

A 28 year old was released by the owners in 2005. In autumn 2013, as part of the Special Releases, Diageo released a 36 year old distilled in 1977. The latest independent bottling was a 40 year old from 1975 released by Gordon & MacPhail in 2015.

Dallas Dhu

Owner:	Region:	Founded:	Status:
Diageo	Speyside	1898	Closed

Dallas Dhu distillery is located along the A96 between Elgin and Inverness and is still intact, equipment and all, but hasn´t produced since 1983. Three years later, Diageo sold the distillery to Historic Scotland and it became a museum which is open all year round. In spring 2013 a feasibility study was commissioned by Historic Scotland to look at the possibilities of re-starting production again. One of the founders of the distillery, Alexander Edwards, belonged to the more energetic men in the 19th century Scotch whisky business. Not only did he start Dallas Dhu but also established Aultmore, Benromach and Craigellachie and owned Benrinnes and Oban.

Bottlings:

There are two Rare Malts bottlings from Diageo, the latest in 1997. The latest from independents is a 1980 bottled in 2014 by Gordon & MacPhail.

Glen Albyn

Owner:	Region:	Founded:	Status:
Diageo	N Highlands	1844	Demolished

Glen Albyn was one of three Inverness distilleries surviving into the 1980s. Today, there is no whisky production left in the city. The first forty years were not very productive for Glen Albyn. Fire and bankruptcy prevented the success and in 1866 the buildings were transformed into a flour mill. In 1884 it was converted back to a distillery and continued producing whisky until 1983 when it was closed by the owners at the time, Diageo. Three years later the distillery was demolished.

Bottlings:
Glen Albyn has been released as a Rare Malt by the owners on one occasion. It is rarely seen from independents as well. In 2010, Signatory released a 29 year old and in 2012 a 1976 was bottled by Gordon & MacPhail.

Glenesk

Owner:	Region:	Founded:	Status:
Diageo	E Highlands	1897	Demolished

Few distilleries, if any, have operated under as many names as Glenesk; Highland Esk, North Esk, Montrose and Hillside. The distillery was one of four operating close to Montrose between Aberdeen and Dundee. Today only Glencadam remains. At one stage the distillery was re-built for grain production but reverted to malt distilling. In 1968 a large drum maltings was built adjacent to the distillery and the Glenesk maltings still operate today under the ownership of Boortmalt, the fifth largest producer of malt in the world. The distillery building was demolished in 1996.

Bottlings:
The single malt from Glen Esk has been bottled on three occasions as a Rare Malts, the latest in 1997. It is also very rare with the independent bottlers. Last time it appeared was in 2014 when Gordon & MacPhail released a 34 year old distilled in 1980.

Glenlochy

Owner:	Region:	Founded:	Status:
Diageo	W Highlands	1898	Demolished

Glenlochy was one of three distilleries in Fort William at the beginning of the 1900s. In 1908 Nevis merged with Ben Nevis distillery (which exists to this day) and in 1983 (a disastrous year for Scotch whisky industry when eight distilleries were closed), the time had come for Glenlochy to close for good. Today, all the buildings have been demolished, with the exception of the kiln with its pagoda roof and the malt barn which both have been turned into flats. For a period of time, the distillery was owned by an energetic and somewhat eccentric Canadian gentleman by the name of Joseph Hobbs who, after having sold the distillery to DCL, bought the second distillery in town, Ben Nevis.

Bottlings:
Glenlochy has occurred twice in the Rare Malts series. Recent independent bottlings from 2012 are a 32 year old from Signatory and 33 year old from Gordon & MacPhail.

Glen Mhor

Owner:	Region:	Founded:	Status:
Diageo	N Highlands	1892	Demolished

Glen Mhor is one of the last three Inverness distilleries and probably the one with the best reputation when it comes to the whisky that it produced. When the manager of nearby Glen Albyn, John Birnie, was refused to buy shares in the distillery he was mana-ging, he decided to build his own and founded Glen Mhor. Almost thirty years later he also bought Glen Albyn and both distilleries were owned by the Birnie family until 1972 when they were sold to DCL. Glen Mhor was closed in 1983 and three years later the buildings were demolished. Today there is a supermarket on the site.

Bottlings:
Glen Mhor has appeared on two ocasions as Rare Malts. In the last couple of years there have been a 30 year old (1982) from Douglas Laing and a 29 year old (1982) from Signatory.

Glenury Royal

Owner:	Region:	Founded:	Status:
Diageo	E Highlands	1825	Demolished

Glenury Royal did not have a lucky start. Already a few weeks after inception in 1825, a fire destroyed the whole kiln, the greater part of the grain lofts and the malting barn, as well as the stock of barley and malt. Just two weeks later, distillery worker James Clark, fell into the boiler and died after a few hours. The founder of Glenury was the eccentric Captain Robert Barclay Allardyce, the first to walk 1000 miles in 1000 hours in 1809 and also an excellent middle-distance runner and boxer. The distillery closed in 1983 and part of the building was demolished a decade later with the rest converted into flats.

Bottlings:

Bottled as a Rare Malt on three occasions. Even more spectacular were three Diageo bottlings released 2003-2007; two 36 year olds and a 50 year old. In early 2012 a 40 year old was released. There are few independent bottlings, the latest being a 38 year old released in 2012 by Gordon & MacPhail.

Imperial

Owner:	Region:	Founded:	Status:
Chivas Bros	Speyside	1897	Demolished

Rumours of the resurrection of this closed distillery have flourished from time to time during the last decade. Eight years ago, the owner commissioned an estate agent to sell the buildings and convert them into flats. Shortly after that, Chivas Bros withdrew it from the market. In 2012, the owners announced that a new distillery would be built on the site, ready to start producing in 2015. Demolition of the old distillery began and in spring 2015 the new Dalmunach distillery was commissioned. In over a century, Imperial distillery was out of production for 60% of the time, but when it produced it had a capacity of 1,6 million litres per year.

Bottlings:

The 15 year old official bottling is impossible to find these days but independents are more frequent. Signatory released a 19 year old in 2014 and in spring 2015, Gordon & MacPhail released a 19 year old from 1995.

Littlemill

Owner:	Region:	Founded:	Status:
Loch Lomond Co.	Lowlands	1772	Demolished

Until 1992 when production stopped, Littlemill was Scotland's oldest working distillery and could trace its roots back to 1772, possibly even back to the 1750s! Triple distillation was practised at Littlemill until 1930 and after that some new equipment was installed, for example, stills with rectifying columns. The stills were also isolated with aluminium. The goal was to create whiskies that would mature faster. Two such experimental releases were Dunglas and Dumbuck. In 1996 the distillery was dismantled and part of the buildings demolished and in 2004 much of the remaining buildings were destroyed in a fire.

Bottlings:

Official bottlings still occur – in August 2015 a 25 year old was released. Several independent bottlings were released in 2013/2014 including a 25 year old distilled in 1988 from Douglas Laing. The latest release was a 22 year old from 1992, bottled by Hunter Laing in 2014.

Lochside

Owner:	Region:	Founded:	Status:
Chivas Bros	E Highlands	1957	Demolished

Originally a brewery for two centuries, In the last 35 years of production Lochside was a whisky distillery. The Canadian, Joseph Hobbs, started distilling grain whisky and then added malt whisky production in the same way as he had done at Ben Nevis and Lochside. Most of the output was made for the blended whisky Sandy MacNab´s. In the early 1970s, the Spanish company DYC became the owner and the output was destined for Spanish blended whisky. In 1992 the distillery was mothballed and five years later all the equipment and stock were removed. All the distillery buildings were demolished in 2005.

Bottlings:

There are no recent official bottlings. In 2012, the second edition of the unusual Lochside single blend (malt and grain distilled at the same distillery) distilled in 1965 was released by Adelphi Distillery and in 2015 a Lochside single malt from 1981 was launched by Gordon & MacPhail.

Millburn

Owner: Diageo **Region:** N Highlands **Founded:** 1807 **Status:** Dismantled

The distillery is the oldest of those Inverness distilleries that made it into modern times and it is also the only one where the buildings are still standing. It is now a hotel and restaurant owned by Premier Inn. With one pair of stills, the capacity was no more than 300,000 litres. The problem with Millburn distillery was that it could never be expanded due to its location, sandwiched in between the river, a hill and the surrounding streets. It was bought by the London-based gin producer Booth´s in the 1920s and shortly after that absorbed into the giant DCL. In 1985 it was closed and three years later all the equipment was removed.

Bottlings:
Three bottlings of Millburn have appeared as Rare Malts, the latest in 2005. Other bottlings are scarce. The most recent was a 33 year old distilled in 1974, released by Blackadder in 2007.

North Port

Owner: Diageo **Region:** E Highlands **Founded:** 1820 **Status:** Demolished

The names North Port and Brechin are used interchangeably on the labels of this single malt. Brechin is the name of the city and North Port comes from a gate in the wall which surrounded the city. The distillery was run by members of the Guthrie family for more than a century until 1922 when DCL took over. Diageo then closed 21 of their 45 distilleries between 1983 and 1985 of which North Port was one. It was dismantled piece by piece and was finally demolished in 1994 to make room for a supermarket.The distillery had one pair of stills and produced 500,000 litres per year.

Bottlings:
North Port was released as a Rare Malt by Diageo twice and in 2005 also as part of the Special Releases (a 28 year old). Independent bottlings are very rare - the latest (distilled in 1981) was released by Duncan Taylor in 2008.

Parkmore

Owner: Edrington **Region:** Speyside **Founded:** 1894 **Status:** Dismantled

The distillery, located in Dufftown close to Glenfiddich, was built during the great whisky boom in the late 1890s. In 1900, it was taken over by James Watson & Co. which, in turn, was bought by John Dewar & Sons. From 1925 to the last distillation in 1931, the distillery was owned by Distiller´s Company Limited. The whisky from Parkmore never maintained a high quality, apparently because of problems with the water source. The water was taken from an area that is now used as a limestone quarry. The beautiful buildings remain today and are used by Edrington for warehousing.

Bottlings:
No bottles of Parkmore single malt have been offered for sale since 1995 when one was auctioned by Christie´s. It is not unlikely though, that the odd bottle may be hiding in private collections.

Pittyvaich

Owner: Diageo **Region:** Speyside **Founded:** 1974 **Status:** Demolished

The life span for this relatively modern distillery was short. It was built by Arthur Bell & Sons on the same ground as Dufftown distillery which also belonged to them and the four stills were exact replicas of the Dufftown stills. Bells was bought by Guinness in 1985 and the distillery was eventually absorbed into DCL (later Diageo). For a few years in the 1990s, Pittyvaich was also a back up plant for gin distillation (in the same way that Auchroisk is today) in connection with the production of Gordon´s gin having moved from Essex till Cameronbridge. The distillery was mothballed in 1993 and has now been demolished.

Bottlings:
An official 12 year old Flora & Fauna can no longer be found but in 2009 a 20 year old was released by the owners. Although not confirmed by Diageo at the time of writing, the official registration of a label indicate that a 25 year old Pittyvaich could be part of this year´s Special Releases.

Port Ellen

[port ell•en]

Owner:	**Region/district:**
Diageo	Islay
Founded: **Status:**	**Capacity:**
1825 Dismantled	-

When Port Ellen closed in 1983 it was one of three Islay distilleries owned by Diageo (then DCL). The other two were Lagavulin and Caol Ila who had been operating uninterruptedly for many years.

Port Ellen, mothballed since 1930, had only been producing for 16 years since re-opening, which made it easy for the owners to single out which Islay distillery was to close when malt whisky demand decreased. It was also the smallest of the three, with an annual output of 1,7 million litres of alcohol.

The stills were shipped abroad early in the 1990s, possibly destined for India, and the distillery buildings were destroyed shortly afterwards. The whisky from Port Ellen is so popular, however, that rumours of distilling starting up again, do flourish from time to time.

Today, the site is associated with the huge drum maltings that was built in 1973. It supplies all Islay distilleries and a few others, with a large proportion of their malt. There are seven germination drums with a capacity of handling 51 tonnes of barley each. Three kilns are used to dry the barley and for every batch, an average of 6 tonnes of peat are required which means 2,000 tonnes per year. The peat was taken from Duich Moss until 1993 when conservationists managed to obtain national nature reserve status for the area in order to protect the thousands of Barnacle Geese that make a stop-over there during their migration. Nowadays the peat is taken from nearby Castlehill.

Besides a couple of versions in the Rare Malts series, Diageo began releasing one official bottling a year in 2001 and, although not confirmed by Diageo at the time of writing, the official registration of a label indicate that Port Ellen 32 years old from 1983 (the last year the distillery was producing), bottled at 53.9% could be this year's Special Release. Probably due for release in the beginning of October, this will be the 15th since 2001.

History:

1825 Alexander Kerr Mackay assisted by Walter Campbell founds the distillery. Mackay runs into financial troubles after a few months and his three relatives John Morrison, Patrick Thomson and George Maclennan take over.

1833 John Ramsay, a cousin to John Morrison, comes from Glasgow to take over.

1836 Ramsay is granted a lease on the distillery from the Laird of Islay.

1892 Ramsay dies and the distillery is inherited by his widow, Lucy.

1906 Lucy Ramsay dies and her son Captain Iain Ramsay takes over.

1920 Iain Ramsay sells to Buchanan-Dewar who transfers the administration to the company Port Ellen Distillery Co. Ltd.

1925 Buchanan-Dewar joins Distillers Company Limited (DCL).

1930 The distillery is mothballed.

1967 In production again after reconstruction and doubling of the number of stills from two to four.

1973 A large drum maltings is installed.

1980 Queen Elisabeth visits the distillery and a commemorative special bottling is made.

1983 The distillery is mothballed.

1987 The distillery closes permanently but the maltings continue to deliver malt to all Islay distilleries.

2001 Port Ellen cask strength first edition is released.

2014 The 14th release of Port Ellen - a 35 year old from 1978.

2015 The 15th release of Port Ellen.

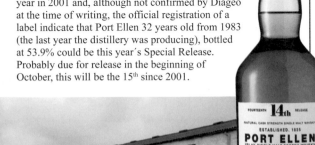

35 years old

Rosebank

Owner:	Region:	Founded:	Status:
Diageo	Lowlands	1798	Dismantled

When Rosebank in Falkirk was mothballed in 1993, there were only two working malt distilleries left in the Lowlands – Glenkinchie and Auchentoshan. The whisky from the distillery has always had a great amount of supporters and there was a glimmer of hope that a new company would start up the distillery again. At the beginning of 2010 though, most of the equipment was stolen and furthermore, Diageo has indicated that they are not interested in selling the brand. The buildings are still intact and most of them have been turned into restaurants, offices and flats. The whisky from Rosebank was triple distilled.

Bottlings:

The official 12 year old Flora & Fauna is now almost impossible to find but in 2014 a 21 year old Special Release appeared. The latest independent bottlings were released in 2013/2014 - a 21 year old from Douglas Laing and a 24 year old from Gordon & MacPhail.

St Magdalene

Owner:	Region:	Founded:	Status:
Diageo	Lowlands	1795	Dismantled

At one time, the small town of Linlithgow in East Lothian had no less than five distilleries. St Magdalene was one of them and also the last to close in 1983. The distillery came into ownership of the giant DCL quite early (1912) and was at the time a large distillery with 14 washbacks, five stills and with the possibility of producing more than 1 million litres of alcohol. Ten years after the closure the distillery was carefully re-built into flats, making it possible to still see most of the old buildings, including the pagoda roofs.

Bottlings:

These include two official bottlings in the Rare Malts series. In 2008/2009 a handful of independent releases appeared, all of them distilled in 1982 and released by Ian MacLeod, Douglas Laing, Blackadder, Signatory and Berry Brothers. The latest was a 31 year old, distilled in 1982 and released by Hart Brothers in 2013.

Ben Wyvis

Owner:	Region:	Founded:	Status:
Whyte & Mackay	N Highlands	1965	Dismantled

The large grain distillery, Invergordon, today producing 36 million litres of grain whisky per year, was established in 1959 on the Cromarty Firth, east of Alness. Six years later a small malt distillery, Ben Wyvis, was built on the same site with the purpose of producing malt whisky for Invergordon Distiller's blends. The distillery was equipped with one mash tun, six washbacks and one pair of stills. Funnily enough the stills are still in use today at Glengyle distillery. Production at Ben Wyvis stopped in 1976 and in 1977 the distillery was closed and dismantled.

Bottlings:

There have been only a few releases of Ben Wyvis. The first, a 27 year old, was released by Invergordon in 1999, followed by a 31 year old from Signatory in 2000 and finally a 37 year old from Kyndal (later Whyte & Mackay) in 2002. It is highly unlikely that there will be more Ben Wyvis single malt to bottle.

Inverleven

Owner:	Region:	Founded:	Status:
Chivas Bros	Lowlands	1938	Demolished

Dumbarton was the largest grain distillery in Scotland when it was built in 1938. It was mothballed in 2002 and finally closed in 2003 when Allied Domecq moved all their grain production to Strathclyde. On the same site, Inverleven malt distillery was built, equipped with one pair of traditional pot stills. In 1956 a Lomond still was added and this still (with the aid of Inverleven's wash still), technically became a second distillery called Lomond. Inverleven was mothballed in 1991 and finally closed. The Lomond still is now working again since 2010 at Bruichladdich.

Bottlings:

The first official bottling of Inverleven came in 2010 when Chivas Bros released a 36 year old in a range called Deoch an Doras. The latest independent was a 34 year old released by Duncan Taylor in 2012. In 2015, Wm Grant released Ghosted Reserve, a 26 year old vatting of Ladyburn and Inverleven.

Glen Flagler/Killyloch

Owner:	Region:	Founded:	Status:
InverHouse	Lowlands	1964	Closed

In 1964 Inver House Distillers was bought by the American company, Publicker Industries, and that same year they decided to expand the production side as well. Moffat Paper Mills in Airdrie was bought and rebuilt into one grain distillery (Garnheath) and two malt distilleries (Glen Flagler and Killyloch). A maltings was also built which, at the time, became the biggest in Europe. The American interest in the Scotch whisky industry faded rapidly and Killyloch was closed in 1975, while Glen Flagler continued to produce for another decade.

Bottlings:

Glen Flagler was bottled as an 8 year old by the owners in the 1970s. The next release came in the mid 1990s when Signatory released both Glen Flagler and Killyloch (23 year old) and finally in 2003 when Inver House bottled a Glen Flagler 1973 and a Killyloch 1967. A peated version of Glen Flagler, produced until 1970, was called Islebrae.

Kinclaith

Owner:	Region:	Founded:	Status:
Chivas Bros	Lowlands	1957	Demolished

This was the last malt distillery to be built in Glasgow and was constructed on the grounds of Strathclyde grain distillery by Seager Evans (later Long John International). Strathclyde still exists today and produces 40 million litres of grain spirit per year. Kinclaith distillery was equipped with one pair of stills and produced malt whisky to become a part of the Long John blend. In 1975 it was dismantled to make room for an extension of the grain distillery. It was later demolished in 1982.

Bottlings:

There are no official bottlings of Kinclaith. The latest from independents came in 2009 when Signatory released a 40 year old distilled in 1969. In 2005, Duncan Taylor and Signatory both released 35 year old bottlings.

Glenugie

Owner:	Region:	Founded:	Status:
Chivas Bros	E Highlands	1831	Demolished

Glenugie, positioned in Peterhead, was the most Eastern distillery in Scotland, producing whisky for six years before it was converted into a brewery. In 1875 whisky distillation started again, but production was very intermittent until 1937 when Seager Evans & Co took over. Eventually they expanded the distillery to four stills and the capacity was around 1 million litres per year. After several ownership changes Glenugie became part of the brewery giant, Whitbread, in 1975. The final blow came in 1983 when Glenugie, together with seven other distilleries, was closed never to open again.

Bottlings:

The first official bottling of Glenugie came as late as in 2010 when Chivas Bros (the current owners of the brand) released a 32 year old single sherry cask in a new range called Deoch an Doras. Recent independent bottlings include a 33 year old with 8 years Oloroso finish from Signatory, released in 2011.

Ladyburn

Owner:	Region:	Founded:	Status:
W Grant & Sons	Lowlands	1966	Dismantled

In 1963 William Grant & Sons built their huge grain distillery in Girvan in Ayrshire. Three years later they also decided to build a malt distillery on the site which was given the name Ladyburn. The distillery was equipped with two pairs of stills and they also tested a new type of continuous mashing. The whole idea was to produce malt whisky to become a part of Grant´s blended whisky. The distillery was closed in 1975 and finally dismantled during the 1980s. In 2008 a new malt distillery opened up at Girvan under the name Ailsa Bay.

Bottlings:

A 27 year old official bottling was released in 2001. Independent bottlings have appeared, sometimes under the name Ayrshire. The most recent was a 37 year old, released by Signatory in 2013 and Ayrshire Batch 1 from That Boutique-y Whisky Company. In 2015, Wm Grant released Ghosted Reserve, a 26 year old vatting of Ladyburn and Inverleven.

Convalmore Distillery, Dufftown

Single Malts
from Japan

In August 2015, Japanese astronaut Kimiya Yui
manoeuvred a cargo ship into a docking station on the
International Space Station. When he opened it up, he found food,
equipment and samples of Suntory whisky. Sadly for him, the samples were
sent to space to study the effects of gravity on the ageing process.
And the image of Suntory malts rocketing through the stratosphere
provided a nice metaphor for the year in Japanese whisky.

Any visitors to Japan who decided to leave
their whisky shopping until the duty-free shop
on the way home this year would have been
in for a rude shock. Not only had the selection
shrunk to a handful of blends and super high-
ends, but the prices had gained a zero or two.
Last year, US$650 dollars might have seemed
a lot for an airport bottle of Hakushu 25 years
old. This year the same bottle came in a fancier
box for an eye-watering US$2,000.

You can blame all the juries that furnished awards on
Japanese whiskies over the last 10 years. You can blame
the journalists who wrote about the rising whisky stars
in the east. You can blame Japan's national broadcaster,
which revved up domestic interest in Nikka Whisky with a
weekday drama telling the life story of founder Masataka
Taketsuru. And you can certainly blame Jim Murray, who
named a Yamazaki the world's best single malt and set off
a buying frenzy. But you can't blame the distillers, who
have found themselves with dwindling stock and rocketing
demand. Throttling the output and hiking the prices was the
only sensible move.

Nikki officially discontinued the full range of Yoichi and Miyagikyo malts, but issued a new NAS for each distillery. Beam Suntory says they're still selling the full 12, 18 and 25 year range of malts from both distilleries, but you'll be hard pressed to find any. Instead, you can try a new product from a distillery that's usually happy in the background: the Chita grain facility in central Japan. "The Chita" uses what Suntory describes as "clean, medium and heavy" grain whiskies, aged in Spanish oak and wine casks.

Meanwhile, there's a new generation of Japanese distillers stirring. Two distilleries are under construction at time of writing, and several others are being planned. There's even a rumour of a Tokyo microdistillery.

Most likely to fire up the stills first is Shizuoka Distillery, named for the city in which it's based. It's the brainchild of Taiko Nakamura, CEO of whisky importer Gaia Flow, who says he started thinking about starting a distillery three years ago but, just like in classic distillery-founding lore, it took him years to find the perfect location. When he saw the Tamagawa district of Shizuoka city, he immediately felt it was the perfect place. The water from the Abe river was of high quality and the verdant, mountainous setting was "just like Scotland". It's also a quick drive to Tokyo or Nagoya, and boasted farmers who grow barley and city officials who were willing to lease the land.

Shizuoka Distillery will reuse equipment from that most iconic of closed Japanese distilleries, Karuizawa. Nakamura purchased the entire equipment at a public auction, though it was clear that most of it was not going to be of much use. The distillery had been mothballed since 2000, so the washbacks were rotten and the mashtun was rusty. But some of the equipment was salvageable, including the newest of the pot stills. That will be put to use at Shizuoka, though with a new heating system. Nakamura will also be using Karuizawa's Porteus malt mill and a hoops press machine.

The distillery will use new pot stills from Scotland and a hybrid still from Germany. The plan is to start building in September 2015 and have the distillery running by the spring of 2016. Nakamura says he's aiming to produce single malts and blended whisky, gin, brandy and liqueurs. The production target for whisky is around 200,000 litres, which would leapfrog Chichibu, Eigashima and Miyashita on the production volume list.

The aim, says the boss, is a bright, fruity and delicate whisky whose "beautiful aromas you can lose yourself in", and if all goes to plan, we'll see the first release by September 2019, just in time for the Tokyo Olympics.

Hot on the heels of Shizuoka Distillery will be Akkeshi Distillery. Tokyo-based import-export firm Kenten plans to build Hokkaido's second distillery on the island's Pacific coast. It's an area of wild beauty with fog rolling in from the ocean over seemingly endless wetlands rich in peat. If this conjures up images of Islay, that's no accident. Distillery director Keiichi Toita is a long-time fan of Islay malts: "That smoky flavor, which struck me as an odd note the first time I tried it, at some point became something I couldn't do without." Peat will clearly be an important element in the Akkeshi malt. Toita's plan is to try out different varieties of local peat to see what flavours they impart, depending on their proximity to the mountains or the sea. Given the inspiration, it's no surprise that Ak-

keshi will adhere to traditional Scottish methods. All the equipment will be imported from Scotland. That includes two Forsyths pot stills of 5,000 and 3,600 litres, and 6 wash backs. Initially, most of the malted barley will be imported from Scotland, too.

But while his inspiration and the methods may come from Scotland, Toita wants Akkeshi to be an expression of its environment. He says he plans to source local barley to produce – at least in small batches – a "100% Akkeshi" whisky.

Another expression of terroir will be the maturation process. In addition to using bourbon and sherry casks, Toita plans to pour into mizunara wood, which grows relatively abundantly in Hokkaido. Since 2013, he's been test-aging batches of spirit purchased from two existing distilleries. He says the huge fluctuations in temperature, from –20°C in winter to 25°C in summer, accelerate the ageing process.

Kenten is hoping to have the distillery buildings completed by April 2016, the equipment installed by June, and the stills running by October. But all those targets depend on how easy it is to get the licence. In Japan, you can't even apply for one until the equipment is set up and ready to roll.

The projected output for the first year is a modest 30,000 litres but, when up to speed, the plan is to produce ten times that each year. The first Akkeshi malt is scheduled for the final months of 2019, also in time for the Olympics.

Several other new distillery projects are in the early stages of development, including a plan to start Karuizawa Distillery 2.0. After unsuccessful attempts to buy the distillery from Kirin and the subsequent purchase of the complete inventory by Number One Drinks, it seemed like the dream to rekindle the fire was well and truly dead. The removal of the equipment should have been the last nail in the coffin. But never say never. A young entrepreneur based in Tokyo is currently negotiating with Miyota city officials to buy the land and distillery buildings, with an eye on reviving the old Karuizawa spirit. Some of the key people who used to make Karuizawa at the old distillery have agreed to cooperate and help bring the spirit to life again. Although there are many hurdles to be cleared, this is proof that there's a new community of whisky producers gearing up in a country dominated for far too long by two powerhouses.

Nicholas Coldicott is the former editor-in-chief of Eat magazine, former drink columnist for The Japan Times and former contributing editor at Whisky Magazine Japan. He currently works for Japan's national broadcaster, NHK, and writes a drink column for CNNgo.com.

Stefan Van Eycken grew up in Belgium and Scotland and moved to Japan in 2000. Editor of Nonjatta, he is also the man behind the 'Ghost series' bottlings and the charity event 'Spirits for Small Change'. He's on the Japanese panel of the WWA and is currently working on a book about the history of whisky making in Japan.

Chichibu

Owner: Venture Whisky
Location: Saitama P.
Founded: 2008
Capacity: 80,000 l

Malt whisky range:
Occasional limited releases

Fuji Gotemba

Owner: Kirin Holdings
Location: Shizuoka P.
Founded: 1973
Capacity: 2,000,000 l

Malt whisky range:
Fujisanroku 18 years

It's hard to get hold of Chichibu whisky, even in Japan. Most bottles are spoken for before they're out the door which is remarkable for a distillery that's not yet a decade old.

One reason, of course, is owner Ichiro Akuto's background (see the Hanyu distillery text). But it's also that Akuto and his team have proven themselves with consistently quality releases. In 2015 they were literally building on their success, adding a blending room, a third warehouse and an off-site cooperage.

Akuto says it's getting harder and harder to find the casks he wants, so he's buying mizunara from Hokkaido and making his own. His team is chopping bourbon barrels into "chibidaru", his word for a quarter cask. Around ten percent of the barley they use is malted on-site, and that's a figure they hope will grow as they set their sights on being as local as possible. They've been using local peat, and devote the final two months before the summer break to distilling peated malt.

The Chichibu process is more hands-on than in many distilleries. The mashman uses a wooden paddle to stir, and the stillman opens the spirit safe and samples the distillate to decide when to cut. The two stills are small with steep arms for a rich spirit. The distillate is aged, without temperature controls, in Chichibu's seesawing climate. Founder Ichiro Akuto says the fluctuations speed up the aging process. He uses more than 20 kinds of cask, including Cognac, Madeira and rum.

As other whisky makers are shedding their age statements, Akuto says his mid-term focus is on releasing his first 10 year old. Expect that to arrive in 2020.

Tasting note The Peated 2015:

Bottled at a whopping 62.5%, this is not for the faint-of-heart. It's very peat-forward, but the smoke is supported by honeydew melon with prosciutto, grilled apple pie, crema catalana and a myriad other treats. On the palate, you get all of that with some citrus and spice. It's a great swimmer, too, so have some water handy.

Well, here's a funny thing. As the big boys of Japanese whisky hit the brakes on their output and leave the market parched, Kirin's master blender Jota Tanaka says we can expect to see more and more single malt and single grain releases from his team.

At the Tokyo International Bar Show in May, the Nikka and Suntory stands looked more like contractual obligations than marketing pushes, and the Kirin stand had much more interesting bottles to pour, including an 18 year old single malt and a 25 year old single grain. The grain whisky in particular showed Tanaka's taste. He says everything he does, from selecting his malt or grain all the way through to mashing, fermentation, distillation and aging, is to create clean & estery flavours.

Fuji Gotemba was originally a joint venture between Kirin, Seagram and Chivas, but Kirin took full control in 2002. Much of the output goes into domestic blends at the cheaper end of the market, under names with international connotations, such as Boston Club, Ocean Lucky Gold and Robert Brown. The only single malt from Fuji Gotemba is an 18 year old called Fujisanroku (literally: "at the foot of Mount Fuji"), bottled at 43%. It used to be a challenge to find a bottle in Japan, but the shrinking Suntory and Nikka presence means there's a better chance of picking up the bottles that were once squeezed out.

Fuji Gotemba also produces a blend called Fujisanroku 50, bottled, as the name suggests, at 50%. The distillery has an unusually diverse range of equipment, with two sets of pot stills, as well as kettles and various columns. They distill grain in columns and kettles. And the Four Roses connection means the warehouses are dominated by bourbon barrels.

Hakushu

Owner:	Location:	Founded:	Capacity:
Beam Suntory	Yamanashi P.	1973	4,000,000 l

Malt whisky range:
NAS, 12, 18, 25 year old plus occasional limited releases

If you visit the Hakushu complex in Japan's Southern Alps, be careful which tour group you follow. One will take you through the whisky distillery that was once the biggest in the world. The other will take you through a water bottling plant.

Japan's top-selling mineral water is the same stuff that goes into Hakushu whisky and Suntory bottles them in the same gorgeously verdant setting. If you've joined the right line, you'll be taken past 16 pot stills in the vast array of shapes and angles that characterise major Japanese distilleries. You'll also pass through a very modern warehouse, so packed with casks that it will be hard to imagine there's a whisky shortage. Only when you reach the tasting room, and then the gift shop, and find none of the rare treats that were once on offer to visitors, will you lament the surging demand for this whisky.

Back in the 1980s, when Hakushu was the world's number one, it had 36 stills turning out an estimated 30 million litres of spirit a year. If the enthusiasm for Japanese whisky continues at this pace, perhaps we'll see similar numbers again. Suntory distills non-peated, lightly peated and heavily peated malt here. With the variety of stills and casks, the blenders have a lot of styles to play with. In years past, the company would show off the variety via limited releases. It's a fair guess we won't be getting any more of those for a few years.

There's also a grain facility on site with continuous stills. It produces only a tenth the output of Suntory's main grain distillery near Nagoya, but the focus here is experimentation. They produce small-run test batches of different kinds of grain – another sign that the people at Suntory are never ready to rest on their laurels.

Tasting note Hakushu NAS:

Green in all senses of the word. Cucumber and mint on the nose. On the palate, mossy twigs and citrus, with faint smoke emerging later on. The finish is long and woody.

Hanyu

Owner:	Location:	Founded:	Status:
Toa Shuzo	Saitama P.	1941	Dismantled

Malt whisky range:
Limited single cask and other releases.

In 2000, the folks at Toa Shuzo decided to pull the plug on their Hanyu whisky operation after 17 years. The market had dried up.

Fast forward 15 years, it's all demand and no supply. The Hanyu stock was snapped up by Ichiro Akuto, grandson of the distillery's founder, and current boss of the Chichibu distillery. In a stroke of marketing genius, he created a series of single-cask releases bearing playing card labels. The format made single-cask whisky approachable, the whisky was great, and it proved to be catnip for collectors.

The last of the 52-card series came out in early 2013, two jokers appeared in 2014, and when the releases pop up on auction sites now, they can fetch up to 20 times the original price. There is still one way to buy Hanyu product: pick up a bottle of Double Distilleries, a vatting of malts from Hanyu and Chichibu.

Karuizawa

Owner:	Location:	Founded:	Status:
Kirin Holdings	Nagano P.	1956	Dismantled

Malt whisky range:
Single cask and Asama vatted malt.

U.K.-based distributor Number One Drinks can take plenty of the credit (or is it the blame?) for bringing Japanese whisky so collectable.

When they started bottling whiskies from the mothballed Karuizawa distillery, interest was so low that they were giving the stuff away. Kirin had acquired the distillery when it bought fellow drink company Mercian in 2007, but showed no interest in bringing it back to life.

Number One kept plugging away, a cask at a time, and by the time they bought the entire remaining stock in 2011, interest in the heavily sherried malts far outweighed the supply. A bottle of single cask Karuizawa from 1960 was sold at a Hong Kong auction in mid-2015 for an eye-watering $118,500, the most ever paid for a Japanese whisky.

Miyagikyo

Owner:	Location:	Founded:	Capacity:
Nikka Whisky	Miyagi P.	1969	3,000,000 l

Malt whisky range:
NAS, 10, 12, 15, 20 years old, single casks, Coffey malt.

Miyashita Shuzo

Owner:	Location:	Founded:	Capacity:
Miyashita Shuzo	Okayama P.	1922	T.b.d.

Malt whisky range:
To be determined

It's said that Masataka Taketsuru took three years to find the perfect site for his second distillery.

He settled on the valley that brings the Hirosegawa and Nikkagawa (no relation to the company name) rivers together because of the quality of the very soft water, the humidity and the crisp air. Originally known as 'Sendai', the distillery was renamed 'Miyagikyo' when Asahi took control of Nikka in 2001.

Miyagikyo is equipped with 8 huge pot stills of the boil ball type with upward lyne arms. It's designed to encourage reflux for a lighter, cleaner spirit. The site also houses two enormous Coffey stills imported from Scotland. They're used to produce grain whisky (Coffey Grain) and, occasionally, to distil malted barley (Coffey Malt).

In June 2015, Nikka announced it would discontinue the entire Miyagikyo range – which included a no-age-statement expression, a 10yo, 12yo and 15yo – because of stock shortages. In September they released the replacement: a single NAS expression. For the foreseeable future – the next five or six years, people within Nikka are whispering – this will be the only permanently available expression of Miyagkyo. A limited-edition 'Miyagikyo Sherry Cask' (also NAS; 3,000 bottles only) was released in September as well but sold out instantly.

In a sense, the distillery became a victim of the double success of Japanese whisky as a category and the spotlighting of local hero Taketsuru in the national broadcaster's morning drama 'Massan'. The number of visitors to the distillery has doubled since 'Massan', so one can imagine the sort of strain this increased interest is putting on stocks that were laid down a decade ago when whisky consumption was at an all-time low in Japan.

Tasting note Miyagikyo NAS (2015 release):

Apples and pears on the nose with grassy and light floral elements; dried fruits, vanilla and anise on the palate, with a tiny bit of bitterness and some milk chocolate on the finish.

Miyashita Shuzo makes half a million litres of alcoholic drinks a year. Sake, shochu and beer account for most of the company's output, and there's a curiosity called Doppo that's made by single-distilling a very hoppy beer, but until 2012 they'd never tried to make whisky.

Company president Buichiro Miyashita says the company's 90th anniversary (in 2012) was the impetus to start a new venture, and he went on research trips to distilleries in Japan, Tennessee and Kentucky. He initially used malt shipped over from Germany and England, but has now begun buying local barley. The company is based in Okayama, an area famous for its barley.

He's been promising the first release would be a 3 year old and would appear in 2015. At time of writing, that's still the promise, though there's not much left of the year.

The first distilling was done using one of the company's shochu stills. It may not bode well for the early output that the company has already splashed out on proper whisky stills from Germany.

They've also revised their estimated annual output from 6,000 litres a year to "more than 2,000", which would make places like Chichibu or Eigashima seem prolific. The ultimate aim, says Miyashita, is to produce a soft, delicate whisky.

Shinshu

Owner: Hombo Shuzo   **Location:** Nagano P.   **Founded:** 1985   **Capacity:** 40,000 l

Malt whisky range:
Komagatake 10yo, The Revival 2011 and occasional releases

White Oak

Owner: Eigashima Shuzo   **Location:** Huogo P.   **Founded:** 1919   **Capacity:** 47,000 l

Malt whisky range:
Akashi NAS and occasional, limited releases

No whisky producer in Japan has had a rockier history than Shinshu, but like a phoenix, they keep rising from the ashes.

Parent company Hombo Shuzo got a license to distil whisky in 1949, but for the first decade simply bought and blended malt and grain. In 1960, founder Kiichiro Iwai set up a distillery in Yamanashi prefecture and began producing a heavy, smoky style of whisky. It didn't sell well and after nine years, he called it a day. In 1978, the Hombo Shuzo people were ready to give it another go, but they were using the old site for winemaking, so they needed a new one. They set up in Kagoshima, on the southern island of Kyushu, and made whisky in Iwai's style. In 1985, they moved north to Shinshu and changed to a lighter style they felt was better suited to the Japanese palate. They were forced to mothball that plant in 1992 as people were switching to shochu as their drink of choice. In 2011, with whisky booming again, they fired up the stills once more. Production is limited to winter, after which the team switches to making beer, umeshu and other spirits. In November 2014, they replaced the old pot stills with brand new ones, built following the original blueprints. They now produce four types of distillate (non-peated and peated at 3.5, 12 and 50ppm). They're exploring the influence of climate on the maturation process by sending some casks to the old site in Kagoshima, and some to Yakushima island. Stock from the pre-1992 era is dwindling fast – fewer than 40 casks remain – so the company is pushing their post-2011 production in the marketplace.

They launched 'Komagatake Sherry & American White Oak 2011' at the end of 2014, and a new blended malt (a vatting of new Mars and Scottish malt) called 'Cosmo' in the summer of 2015. They're promising a new standard single malt expression in the summer of 2016.

Tasting note Komagatake Sherry & American White Oak 2011:
Bears its youth on its sleeve. On the nose, white chocolate, rhubarb jam, apple compote with lots of lemon; on the palate, a slightly perplexing mix of stewed fruits, goya and grapefruit peel,

Eigashima (White Oak) is, without a shadow of a doubt, the least active player on the Japanese whisky field. In fact, most of the time they don't play at all.

It's technically the oldest whisky distillery in Japan – having acquired a distilling license in 1919, four years before Yamazaki, but it took them four decades to get started. And it took them another four decades to release their first single malt. That was an 8 year old in 2007, and it wasn't until 2013 that they released their first single cask bottling. That's almost a hundred years – talk about a slow start. But whisky production was never a priority for Eigashima: they make their money with sake and shochu. When they entered the single malt market, whisky accounted for less than 1% of their total sales. They've expanded production to 5 months of the year, but we're still talking low single digits in sales percentage.

The current distillery was built in 1984. After a period of using malt with varying peat levels, they switched to Crisp Malting and now use lightly peated (5ppm). All production is matured on site, near the Akashi strait.

They used up their oldest casks for a 15 year old released in 2013, so the oldest stock left in the warehouses is approaching 8 years at the time of writing – and there's not very much of that. They mostly fill ex-Wild Turkey bourbon casks, but they also have sherry butts, cognac casks, wine and shochu casks (some made from Japanese 'konara' oak, the little brother of 'mizunara'), as well as domestically-made virgin oak. The only new releases in 2015 were a blend called 'Sea Anchor' and two 5-year-old single cask bottlings.

Tasting note Akashi NAS:
The deep colours and surprisingly warm nose, tips you off that there's a sherry influence. There's a peach tea and autumn fruits when you sip, and a fairly short finish.

Yamazaki

The Japanese whisky world was a confusion of relabelled Scotch, doctored shochu, and other shenanigans until Suntory founder Shinjiro Torii decided to do things properly. He picked a humid valley between Osaka and Kyoto and opened the country's first bona fide whisky distillery in 1923. The marketing blurbs talk of the purity of the local water and the connection to the father of the tea ceremony, Sen no Rikyu, who built his most famous tea house in Yamazaki. But it can't have hurt to have been right in the middle of the country and close to several major cities.

Owner:	Location:	Founded:	Capacity:
Beam Suntory	Osaka P.	1923	6,000,000 l

Malt whisky range:
NAS, 12, 18, 25 years old and occasional limited releases.

Yamazaki was hardly an unknown in the whisky world when Jim Murray named one of its sherry cask limited editions his top pick of the year, but the headlines that generated shone a brighter spotlight than ever on Japan's oldest distillery.

And it likely drove sales for every style of Yamazaki except the one he chose, which had sold out. Bartenders and retailers in Japan say they've seen demand increase as people remember the distillery name rather than the details of the particular release. Since the only Yamazaki widely available these days is the entry-level bottle with no age statement, it seems likely that many people have been pleasantly surprised by the affordability of the world's best whisky.

The distillery underwent a billion-dollar expansion in 2013, with engineers installing 4 more pot stills, taking the total to 16. It was the first expansion in 45 years. The stills are varied in size and shape, some with boil balls, some without, some direct fired, some using steam, some with steep arms, some with straight ones, to produce a miscellany of styles.

Yamazaki is more accessible than most Japanese distilleries – just a 15-minute train ride from Kyoto – and has one of the best tours of any Japanese distillery, with English headsets.

Tasting note Yamazaki NAS:
Zesty nose, with some sawdust and strawberry. Lots of bourbon influences on the palate, with creamy vanilla, creme brulee, but also some spice. A relatively short, sharp finish. Superb for its price bracket.

Yoichi

Owner: Nikka Whisky
Location: Hokkaido
Founded: 1934
Capacity: 2,000,000 l

Malt whisky range:
NAS and occasional limited releases.

The 'Massan' drama series – broadcast on national TV in Japan from September 2014 to March 2015 – cemented Nikka founder Masataka Taketsuru's role as 'father of Japanese whisky' in the public eye.

It also drove hordes of visitors to the far-flung distillery with its archives, museum and perfectly preserved rooms from Taketsuru's era. In 2014, roughly 470,000 people visited Yoichi. In the first half of 2015 alone, the number was 520,000.

When Taketsuru started the distillery in 1934 there was a single still that doubled as spirit and wash still. Nowadays there are 6 stills, coal-heated and featuring straight heads and downward lyne arms to produce a robust spirit.

The distillery released its first single malt, a 12 year old, in 1984, when Japan was in the grip of its first domestic whisky boom. It came out under the distillery's original name: Hokkaido. Although the 'house style' is peaty and heavy, Yoichi is set up to create a wide range of distillates. Between various peating levels, yeast strains, mash bills, fermentation times, distillation methods and maturation types, it's said that Yoichi is capable of producing 3,000 different types of malt whisky.

In 2015, Nikka replaced the entire Yoichi range (which included a no-age-statement expression as well as a 10, 12, 15 and 20yo) with a single option: a new NAS, which is closer in character to the old 10yo Yoichi than to the old NAS. A limited-edition 'Heavily Peated Yoichi' (NAS; 3,000 bottles only) released at the same time flew off the shelves in no time. They say they will no longer be offering single cask bottlings for the time being.

Tasting note Yoichi NAS, (2015 release):

Barley sweetness, pencil shavings, over-ripe orchard fruits and soft smoke on the nose; oak and peat lead the dance on the palate with some candied orange peel thrown in; the finish is earthy and vegetal, with some tea on the side.

Blends and Blended Malts
from Japan

Now that Japanese single malt whisky
is something you invest in rather than something you drink,
perhaps it's a good time to explore the world of blends and blended malts.
As you might expect, some are extraordinarily good, others best forgotten.

With all the fuss over single malts in Japan, it's easy to forget that nobody was drinking them until the 1980s. It was a blended world, and the king was a little square bottle with a tortoiseshell design. It debuted in 1937 with a label saying, simply, "Suntory Whisky." Nearly 80 years on, it's still a core brand for Japan's biggest drink company. Locals call it "kakubin" (literally: square bottle), and for many Japanese drinkers, nothing else works quite so well with soda. Kakubin's abv has yo-yoed up and down over the years, but the current offerings are a standard 40% and a richer, slightly more potent 43% "premium" version.

In 1989 Suntory launched the premium blend 'Hibiki' (meaning "resonance"). The 17 year old was first out the door. Then came a 21 year old with a more pronounced mizunara influence, and an ultra-high-end, every-so-rare 30 year old. The 12 year old arrived 6 years ago, to mark Suntory's 90th anniversary, and in a sign of how far Japanese whisky had come, it was the first to debut in Europe rather than Japan. As a team, the Hibikis are the most award-winning Japanese blends by far. There is, of course, now a NAS version too, called Japanese Harmony. It arrived in spring 2015 and it's already a lot easier to find than its age-statement siblings.

Nikka has a vast range of blends and blended malts. For consumers, it can be a bit bewildering to navigate all the styles, bottle designs and flavour profiles. Or at least, it used to be. Along with the big restructuring of their single malt portfolio, Nikka is discontinuing several of their blends and blended malts. So what remains?

First and foremost, the 'Taketsuru' range, a vatting of Yoichi and Miyagikyo malt. Crowned 'best blended malt in the world' several times in the past decade, it includes a NAS (which replaces the 12 year old), a 17 year old, 21 year old and (occasionally) a 25 year old. Don't be too quick to dismiss the NAS. We've seen it beat its older siblings in blind tastings more than once. There used to be a 35 year old from time to time, but those days are over. Limited-edition variations – finished in sherry, madeira or port wood and 'non-chill filtered' – are blink-and-you-miss-them releases.

Then there's 'Nikka Pure Malt' which comes in bare-bones, apothecary-style bottles in three expressions; Black, White and Red. Nikka is ditching the white one, a heavily-peated, Islay-style blended malt, from September 2015. So selection will become a little easier: if you're a fan of the fresh and fruity Miyagikyo style, grab a bottle of the Red; if you like a bit of smoke with your fruit, go for the Yoichi-forward Black.

'Nikka All Malt' is the black sheep of the family and much of that has to do with the look of the bottle, we suspect. Complex and very lush, it can't be beat at its price level (around €35). It's also an interesting example of how whiskies can be in different categories in different areas of the world. In Japan, it's considered a 'blended malt whisky', but in Scotland it would be a 'blended whisky', as its recipe includes malted barley distilled in a column still (which is considered 'grain whisky' there).

Clear-cut Nikka blends include the classic 'Super Nikka', a mainstay of the Japanese bar trade since 1962; 'From the Barrel', which is bigger in Europe than back home, probably because of its success on the mixology scene there; 'Nikka Blended', which is only available in Europe; and 'Black Nikka', which got a make-over and is now a trilogy of 'Clear', 'Rich Blend' and 'Deep Blend' expressions. The pick of the bunch, however, is 'The Nikka 12', launched in the summer of 2014. The presence of the Coffey grain is immediately clear (with notes of roast coconut flakes and banana leaves), but it's balanced with orchard fruits (green apples and pears) supplied by the Miyagikyo malt and a slight whiff of peat from the Yoichi malt.

Blended whiskies are the bread and butter of Kirin's output, and there are plenty to choose from, but the one to look for is 'Fujisanroku 50'. It's got a vanilla sweetness and plenty of oak, with the kick you'd expect from an entry-level whisky with 50% alcohol. You won't want to nurse it in a tasting glass, but it makes a fine highball, and it performs very well for its price.

Whisky companies in Japan don't swap stock, so people often wonder how the smaller, independently-owned producers manage to release blends. The answer is they source their grain whisky from abroad (mostly Canada). In Scotland and other traditional whisky-producing countries that wouldn't fly, but in the absence of strict regulations that define the category of 'Japanese' whisky as such, producers are very pragmatic about the selection of ingredients for their blends. An extreme example of this is 'Togouchi', which doesn't contain a single drop of whisky actually distilled in Japan.

Most blends produced by the smaller distilleries are aimed at the Japanese market as a sort of 'craft' alternative to the ubiquitous releases of the three big players (Suntory, Nikka and Kirin). In their price bracket, bottles like Eigashima's 'Akashi' and 'Akashi Red' (a more grain-forward version of 'Akashi') and Mars' 'Iwai Tradition' and '3&7' (a blend of 3 year old grain and 7 year old malt) offer great value for money.

Chichibu has a trio of unusual blends and blended malts, produced in a solera system, which explains why they're always easy to find, pretty good value, and proving popular with tourists who can't get their hands on anything else with the magic words "Ichiro's Malt" on the label. Best of the bunch is the 'Double Distilleries', using malt from the Hanyu and Chichibu. How much of each, and what else might be in there, is not clear.

Chichibu's other blend is named in typically straightforward fashion: 'Ichiro's Malt & Grain'. There have been a few incarnations, including a 33 year old ultra premium version, but the one you'll find on shelves in Japan is a white label version described as "worldwide blended whisky". That suggests there may be some elements that weren't distilled in Japan, so purists beware. The rest of us can sit back and enjoy an affordable dram composed by one of the stars of the Japanese whisky world.

Andy Watts - Master Distiller at James Sedgwick Distillery, South Africa

Distilleries
around the globe

Scotch whisky has been having a tough year, at least in the export markets with both volumes and values being down for the first time in many years. Other categories, on the other hand, continue to increase their market shares.

Irish whiskey is expanding and the interest in bourbon is growing. So is the number of new malt distilleries around the world. The establishment rate is extremely intense right now within three areas. Ireland and Northern Ireland now have 12 distilleries which are producing (compared to just three a few years ago) and at least 15 projects are at different levels of completion. Tasmania continues to impress with nine working malt distilleries and at least as many projects in the pipeline. USA, finally, has come to a stage where it is almost impossible to keep track of new distilleries. In this edition of the Malt Whisky Yearbook, we have added another 20 distilleries in comparison to last year, which brings the total accounted for in this book to 78. If one doesn't limit oneself to the ones distilling malt whisky from barley, but include bourbon, rye, corn and other types of whisky, the number of distilleries could easily be 200.

Comparing the distilleries in these three areas, it is obvious that they differ on a few points. In Ireland, the capacity is often considerable, in a few cases well over 1 million litres per year. The new distilleries in Tasmania are typically aiming for 20-100,000 litres in a year, while the majority of the latest American distilleries produce small quantities of malt whisky. On the other hand, the range of products from these distilleries is often broad with many other types of spirits in production while Tasmania and, in particular, Ireland seems to concentrate mainly on the production of whisky.

With the growing number of distilleries, some countries are now eager to regulate how whisky should be produced and what ingredients should be allowed. In Tasmania for instance, the Tasmanian Whisky Producers Association are aiming for a law that should determine what should and

shouldn't be called Tasmanian whisky. This is similar to the appellation system that wine producers use in France. The majority of Swedish distillers organized themselves a couple of years ago and are working on the same idea. In France, they have already taken it one step further, at least for one region, when Institut National des Appellations d'Origine, which is primarily responsible for the classification of agricultural products, set up the rules that define Whisky Breton (whisky from Brittany) in 2015. In Ireland, the Irish Whiskey Association was formed in 2014 with the majority of established and proposed distilleries as members. The production method and technical standards for Irish whiskey are already regulated in the Irish Whiskey Act of 1980 and their activities are, instead, aimed at promoting Irish whiskey in Ireland and abroad and to support new role players entering the market.

The main part of the distilleries in this chapter, produce small quantities of malt whisky and will not influence the global whisky market, at least not when it comes to volumes. Their interest in new production methods and alternative raw material, on the other hand, has already triggered a discussion, even if not by the ordinary whisky consumer then at least amongst whisky aficionados and other producers. However, there are a number of distilleries with financial clout, as well as having large capacity, which undoubtedly can be counted on in the international perspective.

Midleton, Tullamore Dew and Cooley in Ireland and Bushmill's in Northern Ireland are obvious examples. We also have Penderyn in Wales, St George in England, Amrut, John Distilleries and McDowell's in India, Murree in Pakistan, Kavalan and Nantou in Taiwan, James Sedgwick in South Africa, DYC in Spain, Town Branch, Balcones, Westland and High West in USA, as well as Glenora in Canada and Mackmyra in Sweden. Some of the whiskies which are released from these distilleries belong to the absolute upper echelon and have earned respect from whisky devotees around the globe.

Europe

Austria

Distillery: Whiskydistillery J. Haider, Roggenreith
Founded: 1995
Owner: Johann & Monika Haider
roggenhof.at

In the small village of Roggenreith in northern Austria, Johann and Monika Haider have been distilling whisky since 1995 and three years later, the first Austrian whisky was released. In 2005, they opened up a Whisky Experience World with guided tours, a video show, whisky tasting and exhibitions. Roggenhof was the first whisky distillery in Austria and, over the years, production has steadily increased to 35,000 litres. The capacity currently stands at 100,000 litres per annum. The wash is allowed to ferment for 72 hours before it reaches either of the two 450 litre Christian Carl copper stills. The desired strength is reached in one single distillation, thanks to the attached column.

The main part (70%) of the production are rye whiskies – Original Rye Whisky J.H., Pure Rye Malt J.H., Special Rye Malt J.H. and (since 2012) Special Rye Malt Peated J.H. The last three are made from 100% rye. A 10 year old Original Rye Selection was released recently. The current range of single malts made from barley is Single Malt J.H, Special Single Malt J.H. and Special Single Malt Peated J.H. In addition to that, some of these whiskies are also available as 9 year olds and bottled at a higher strength (46% instead of 41%). All whiskies, except the peated ones, are matured in heavily toasted Sessile oak while the peated varieties have been filled into used sweet wine barrels. From time to time releases are also made in the Rare Selection range where the whisky has matured in sweet wine barrels or Winesky barrels, Winesky being a grape brandy which is also produced at the distillery.

Distillery: Reisetbauer, Kirchberg-Thening
Founded: 1994 (whisky since 1995)
Owner: Julia & Hans Reisetbauer
reisetbauer.at

This is a family-owned farm distillery near Linz in northern Austria specialising in brandies and fruit schnapps. Since 1995, a range of malt whiskies are also produced. The distillery is equipped with five 350 litre stills. All stills are heated, using hot water rather than steam, which, according to Hans Reisetbauer, allows for a more delicate and gentle distillation. The 70 hour-long fermentation takes place in stainless steel washbacks. Approximately 20,000 litres of pure alcohol destined for whisky making are produced annually, using local barley to make the unpeated malt. Casks are sourced locally from the best Austrian wine producers. The current range of whiskies includes a 7 year old single malt which consists of a vatting of whiskies aged in casks that have previously contained Chardonnay and Trockenbeerenauslese, a 12 year old which has undergone maturation in Trockenbeerenauslese barrels and a Vintage 1998.

Other distilleries in Austria

Broger Privatbrennerei

Klaus, founded in 1976 (whisky since 2008)

www.broger.info

The production of whisky is supplementing the distilla-tion and production of eau de vie from apples and pears. For their whisky, Broger buys peated malt in the UK and unpeated malt from Germany but also floor malted barley from Bohemia. The distillery is equipped with a 150 litre Christian Carl still. The total volume of whisky produced in a year is 2,500 litres. The current range of whiskies consists of five expressions; Triple Cask which is a blend of whiskies matured in bourbon, sherry and madeira casks, Medium Smoked which has been smoked using beech wood, Burn Out, a heavily peated whisky, Riebelmais, a corn whisky and, new since spring 2014, the limited Distiller's Edition which has been maturing in madeira casks and is bottled at cask strength (58,7%).

Destillerie Rogner

Rappottenstein, founded in 1997

www.destillerie-rogner.at

This distillery in Waldviertel in the northeast part of Austria has produced spirits from fruits and berries for more than a decade. Recently, Hermann Rogner has also added whisky to the range. Two of them are called Rogner Waldviertel Whisky 3/3 with the last figures referring to barley, wheat and rye being used for one of the expressions and three different kinds of malted barley for the other. There is also a whisky from 100% rye called Rye Whisky No. 13.

Hermann Rogner, owner of Destillerie Rogner and the still house of Broger Privatbrennerei

Destillerie Weutz

St. Nikolai im Sausal, founded in 2002

www.weutz.at

A family distillery with a history of producing schnapps and liqueur from fruits and berries. In 2004 Michael Weutz started cooperation with the brewer Michael Löscher and since then Weutz has added whisky to its produce based on the wash from the brewery. Since 2004, 14 different malt whiskies have been produced. Some of them are produced in the traditional Scottish style: Hot Stone, St. Nikolaus and the peated Black Peat. Others are more unorthodox, for example Green Panther, in which 5% pumpkin seeds are added to the mash, and Franziska based on elderflower. Annual production is currently at approximately 15,000 litres and for maturation casks made of French Limousin and Alliere oak are used.

Old Raven

Neustift, founded in 2004

www.oldraven.at

In 2004, a distillery was added to the Rabenbräu brewery by Andreas Schmidt. More than 250,000 litres of beer are produced yearly and the wash from the brewery is used for distillation of the 2,000 litres of single malt whisky every year. Old Raven, which is triple distilled, comes in three expressions – Old Raven, Old Raven Smoky and Old Raven R1 Smoky. The last one was filled into a PX sherry cask which had been used to mature Islay whisky.

Wolfram Ortner Destillerie

Bad Kleinkirchheim, founded in 1990

www.wob.at

Fruit brandies of all kinds make up the bulk of Wolfram Ortner´s produce, as well as cigars, coffee and other luxuries. For the last years he has also been producing malt whisky. New oak of different kinds (Limousin, Alolier, Nevers, Vosges and American) is used for the maturation process. His first single malt, WOB DÖ MALT Vergin, began selling in 2001 and an additional product line, in which Ortner mixes his whisky with other distillates such as orange/moscatel, is called WOB Mariage.

Waldviertler Granit Destillerie

Waidhofen/Thaya, founded in 1995

www.granitdestillerie.at

The distillery has from 1995 established a comprehensive product portfolio of liquers and schnapps from all kinds of delectable berries and fruit. Whisky production started in 2006 and the owner, Günther Mayer, has not only released two different smoked single malts, but is also working with rye and dinkel.

Keckeis Destillerie

Rankweil, founded in 2003

www.destillerie-keckeis.at

Like so many other Austrian distilleries, it started with schnapps and eau de vie from fruit, in Keckeis´ case, mostly pears and apples. Whisky production started in 2008 and today one expression, Keckeis Single Malt is for sale as well as the new make Keckeis Baby Malt. Part of the barley has been smoked with beech and maturation takes place in small ex-sherry casks made of French Limousin oak.

Destillerie Hermann Pfanner

Lauterach, founded in 1854

www.pfanner-weine.com

Founded as an inn and brewery more than 150 years ago, the production soon turned to distillation of eau de vie and schnapps. In 2005, the current owner, Walter Pfanner, started whisky

production and today 10,000 litres per year are filled into casks previously used for maturing sherry and sweet wines. The core expression is Pfanner Single Malt Classic and in 2013 two limited versions were released – Single Malt Red Wood with a maturation in red wine casks and, in cooperation with fellow distiller Johann Zauser, the peated Whisky Brigantium.

Dachstein Destillerie

Radstadt, founded in 2007

www.mandlberggut.com

In 2007, Doris and Bernhard Warter added a distillery to their farm Mandlberggut. Apart from production of various spirits from berries, malt whisky is also produced. Maturation takes place in a mix of different casks – new Austrian oak, ex-sherry casks and red wine casks. Their only release so far is the five year old Rock-Whisky which is distilled 2,5 times.

Brennerei Ebner

Absam, founded in 1930

www.brennereiebner.at

A fourth generation brewer and distiller, Arno Pauli, began to make whisky at his combination of a guesthouse, brewery and distillery in 2005. The whisky production is just a small component of the business but, besides a single malt from barley, Pauli has also released whiskies made from maize, dinkel and wheat.

Belgium

Distillery:	Het Anker Distillery, Blasfeld
Founded:	1471 (whisky since 2003)
Owner:	Charles Leclef
	hetanker.be

Charles Leclef started out as a brewer and currently maintains this role at Brouwerij Het Anker. He also experimented with distillation of his own beer into whisky with some assistance from a nearby genever distiller. The first bottles under the name Gouden Carolus Singe Malt, appeared on the market in 2008.

In 2010, he started a distillery of his own at the Leclef family estate, Molenberg, at Blaasfeld. The stills have been made by Forsyth´s in Scotland with a wash still capacity of 3,000 litres and a spirit still of 2,000 litres. The wash for the distillation is made at their own brewery in Mechelen and it is basically a Gouden Carolus Tripel beer without hops and spices and with a fermentation time of four to five days. Around 100,000 litres of alcohol are produced per year. The core expression is Gouden Carolus Single Malt wheer the whisky has matured for 30 months in ex-bourbon casks and the final six months in Anker casks which are rejuvenated and re-charred casks made of European oak. In November 2014 the limited Gold Fusion was released which is a vatting of the three year old Gouden Carolus and different 6 year old distillates made before their own distillery was built. End of 2014, a visitor centre was opened where you can buy a special version of their single malt called Pure Taste.

Distillery:	The Owl Distillery, Grâce Hollogne
Founded:	1997
Owner:	Etienne Bouillon, Christian Polis, Pierre Roberti
	belgianwhisky.com

In October 2007, Belgium's first single malt, 'The Belgian Owl', was released. The next bottling came in 2008 but was exclusively reserved for private customers. The first commercial bottling was introduced in November 2008.

A limited cask strength expression, 44 months old, was released in 2009 and more cask strength versions have followed. In 2014 and 2015, there were several releases, 3 to 4 years old with most of them bottled at 46%, but also a couple of cask strength expressions. The distillery is equipped with a mash tun holding 2,1 tonnes per mash, four washbacks with a fermentation time of 60-100 hours and two stills from the 19th century (11,000 and 8,000 litres respectively) that had previously been used at the now demolished Caperdonich distillery in Rothes, Speyside. All the barley used for production comes from six farms close to the distillery. For 2015, they intend to produce 50,000 litres of pure alcohol.

Czech Republic

Distillery: Gold Cock Distillery
Founded: 1877
Owner: Rudolf Jelinek a.s.
rjelinek.cz

The distilling of Gold Cock whisky started already in 1877. Today it is produced in three versions – a 3 year old blended whisky, a 12 year old single malt and the 22 year old Small Batch 1992 single malt. Production was stopped for a while but after the brand and distillery were acquired by R. Jelinek a.s., the leading Czech producer of plum brandy, the whisky began life anew. The malt whisky is double distilled in 500 litre traditional pot stills. The new owner has created a small whisky museum which is also home to the club Friends of Gold Cock Whisky with private vaults, where any enthusiast can store his bottlings of Gold Cock.

Denmark

Distillery: Braunstein, Köge
Founded: 2005 (whisky since 2007)
Owner: Michael & Claus Braunstein
braunstein.dk

Denmark's first micro-distillery was built in an already existing brewery in Køge, just south of Copenhagen. The wash comes from the own brewery. A Holstein type of still, with four plates in the rectification column, is used for distillation and the spirit is distilled once. Peated malt is bought from Port Ellen, unpeated from Simpsons, but as much as 40% is from ecologically grown Danish barley. The lion's share of the whisky is stored on ex-bourbon (peated version) and first fill Oloroso casks (unpeated) from 190 up to 500 litres. The Braunstein brothers filled their first casks in 2007 and have since produced 50,000 litres annually. Their first release and the first release of a malt whisky produced in Denmark was in 2010 – a 3 year old single Oloroso sherry cask called Edition No. 1 which was followed the same year by Library Collection 10:1, bottled at 46%. The most recent releases are Library Collection 15:1 (sherry matured with a Ch d´Yquem finish) in spring 2015 and Edition No: 6 (peated and bourbon matured) in December 2014. Library Collection 15:2 will be released in autumn 2015 and Edition 7 a few months later. For duty free, there is a special version called Danica. Braunstein has also made an impact on the Chinese market where 5-6,000 bottles are sold yearly.

Distillery: Stauning Whisky, Stauning
Founded: 2006
Owner: Stauning Whisky A/S
stauningwhisky.dk

The first Danish purpose-built malt whisky distillery entered a more adolescent phase in 2009, after having experimented with two small pilot stills bought from Spain. More stills were installed in 2012 and they now have two wash stills and two spirit stills and a yearly production of 15,000 litres. A new warehouse was built in summer 2014.

The aim has always been to be self-sustaining and Danish barley is bought and turned into malt on an own malting floor. The germinating barley usually has to be turned 6-8 times a day, but Stauning has constructed an automatic "grain turner" to do the job. Two core expressions were decided on – Peated Reserve and Traditional Reserve – and the peat for the first one is acquired from one of few remaining peat bogs in Denmark. In June 2012 the first edition of the two versions were released with slightly more than 700 bottles of each. The distillery has now established a core range including Traditional and Peated single malts as well as a Young Rye. Limited versions are also released; PX and oloroso finishes as well as the single malt finished in their own rye casks. In 2014, Kaos was launched, which is a vatting of all three whisky styles.

The Caperdonich stills at Owl Distillery

Other distilleries in Denmark

Fary Lochan Destilleri

Give, founded in 2009

www.farylochan.dk

This distillery, owned by Jens Erik Jørgensen, is situated in Jutland and the first cask was filled in December 2009. Jens Erik Jørgensen imports most of the malted barley from the UK, but he also malts some Danish barley by himself. A part of his own malted barley is dried using nettles instead of peat to create a special flavour. After mashing, it is fermented for five days in a 600 litre stainless steel washback. Distillation is performed in two traditional copper pot stills from Forsyth´s in Scotland – a 300 litre wash still and a 200 litre spirit still and a third still (1,500 litres) will be installed in 2016. The spirit is matured in ex-bourbon barrels, some of which have been remade into quarter casks. The first whisky, lightly smoked, was released in September 2013 and was followed by a new edition in March 2014 where nettles had been used. A further two bottlings have been released since then. A major expansion of the the distillery was made in summer 2015 which increased both production as well as storage capacity.

Trolden Distillery

Kolding, founded in 2011

www.trolden.com

The distillery is a part of the Trolden Brewery which started in 2005. Michael Svendsen uses the wash from the brewery and ferments it for 4-5 days before a double distillation in a 325 litre alembic pot still. The spirit is filled in bourbon casks and production is quite small as brewing beer is the main task. The first release of the whisky, called Nimbus, came in November 2014. Only 80 bottles were made but more will be released in 2015 and 2016.

Ørbæk Bryggeri/Nyborg Destilleri

Ørbæk, founded in 1997 (whisky since 2007)

www.oerbaek-bryggeri.nu

Niels Rømer and his son, Nicolai, have since 1997 run Ørbæk Brewery on the Danish island of Fyn. It is now one of many combinations of a micro-brewery and a micro-distillery where the wash from the brewery is used to produce whisky. In 2009 the first barrels of Isle of Fionia single malt were filled and the first release was made exactly three years later. Two different expressions are produced – Isle of Fionia and the peated Fionia Smoked Whisky. It is matured in ex-bourbon barrels from Jack Daniels and ex-sherry casks. The spirit production (a mix of whisky and rom) is very small and in 2013, only twenty casks were filled. The owners have bought an old railway workshop in Nyborg which will be converted into a new and bigger distillery sometime during 2015.

Braenderiet Limfjorden

Sillerslev Havn, founded in 2013

www.braenderiet.dk

The latest distillery to come on stream in Denmark. The owner, Ole Mark, started production in June 2013 and the plan is to to do 1,000 to 1,500 litres the first year. Mashing and fermentation is carried out at a local brewery while distillation takes place in alambic type stills. The plan is to do both peated and unpeated single malt as well as rye. The first whisky will be released in 2016.

England ✚

Distillery:	St. George´s Distillery, Roudham, Norfolk
Founded:	2006
Owner:	The English Whisky Co.
	englishwhisky.co.uk

St. George´s Distillery near Thetford in Norfolk was started by father and son, James and Andrew Nelstrop, and came on stream on 12th December 2006. This made it the first English malt whisky distillery for over a hundred years. Customers, both in the UK and abroad, have had the opportunity to follow the development of the whisky via releases of new make, as well as 18 months old spirit, both peated and unpeated. These were called Chapters 1 to 4. Finally, in December 2009, it was time for the release of the first legal whisky called Chapter 5 – unpeated and without chill-filtering or colouring. This was a limited release but, soon afterwards, Chapter 6 was released in larger quantities. The next expression (Chapter 8) was a limited release of a lightly peated 3 year old,

Andrew Nelstrop - owner of The English Whisky Company

followed in June 2010 by Chapter 9 (with the same style but more readily available). Chapter 7, a 3 year old with 6 months finish in a rum cask, was launched in autumn 2010, together with Chapter 10, which has a sherry cask finish. The next bottling, Chapter 11, appeared in July 2011. This was the heaviest peated expression so far (50ppm) aged between 3 and 4 years matured in bourbon casks. October 2012 saw the release of the sherry cask matured Chapter 12. The next one, Chapter 13 is released twice a year – in spring for St. George's Day and in autumn for Halloween. In late 2013, Chapter 14, basically an older version of Chapter 6, and Chapter 15, similar to Chapter 11 but twice the age, were released. In autumn 2014 Chapter 16, the first peated sherry maturation from the distillery, was launched and September 2015, it was time for Chapter 17. Chapter 6 is at the moment the biggest seller followed by 9 and 11.

In summer 2013 a new series was introduced called The Black Range. The rationale behind the thinking is to have a more consistent product, predominantly for export and retail chains. Two expressions are available – Classic and Peated, both bourbon matured and bottled at 43%. In between the Chapter releases there are also very limited bottlings of the so called Founder's Private Cellar. These are unique casks chosen by the founding chairman, James Nelstrop who died in September 2014. The last cask by his hand was released as The Final Signature in May 2015. All the whiskies from the distillery are un chill-filtered and without colouring.

Around 60,000 bottles were sold in 2013 and important markets are Benelux, France, Scotland, Japan, Singapore and England. In 2013 they entered three, new, important markets – USA, China and Canada. The distillery is equipped with a stainless steel semi-lauter mash tun with a copper top and three stainless steel washbacks with a fermentation time of 85 hours. There is one pair of stills, the wash still with a capacity of 2,800 litres and the spirit still of 1,800 litre capacity. First fill bourbon barrels are mainly used for maturation but the odd sherry, madeira and port casks have also been filled. Non-peated malt is bought from Crisp Malting Group and peated malt from Simpson's Malt in Berwick-upon-Tweed. Around 60% of production is unpeated and the rest is peated. The distillery capacity is 104,000 litres of pure alcohol and currently they are producing 50,000 litres per year. Recently, a third warehouse was built and they now have the capacity to mature 6,000 casks on site.

Other distilleries in England

The London Distillery Company

London, founded in 2012

www.londondistillery.com

The London Distillery Company is London's first whisky distillery since Lea Valley closed its doors for the final time, more than a century ago. Founded in 2012 by whisky expert, Darren Rook, and former microbrewery owner, Nick Taylor, and located in Battersea, the distillery started distilling gin at the beginning of February 2013. The owners have one still designated for gin called Christina and a second still, Matilda, is used exclusively for whisky production. The first release from the distillery in March 2013 was Dodd's Gin. In December 2013 they finally got the licence to produce whisky and production started shortly thereafter. Both single malt and rye (100% as well as mixed with malted barley) are produced and the ratio between whisky and gin is 50:50. In September 2015, an unaged rye spirit was released under the name Spring-heeled Jack.

The owners are planning to move the gin production off site in order to focus fully on the production of whisky as from January 2015 at the Battersea plant. Darren and his colleagues experiment with a huge variety of brewer's yeast strains some of which originate from the early 1900s. The idea is to continue with these trials by mixing brewer's yeast with distiller's yeast to see how that will affect the flavour. Recently, the rare barley variety Plumage Archer was used in the production together with distiller's yeast from 1920 as well as Maris Otter in combination with a 1940 distiller's yeast and Whitbread B yeast strains.

A decision has been made to move the distillery in December 2015 to a new location that hasn't yet been confirmed. In summer 2016, the gin production will move to Battersea Power Station while the whisky production by 2018/2019 will move back to the original site in Battersea although the premises will be twice as large as they are today.

Matilda, the whisky still at The London Distillery Company

Lakes Distillery

Bassenthwaite Lake, founded in 2014

www.lakesdistillery.com

Headed by Paul Currie, who was the co-founder of Isle of Arran distillery, a consortium of private investors three years ago embarked on a plan to build the first whisky distillery in the Lake District for more than 100 years. The distillery is housed in a Victorian farm which has been converted near Bassenthwaite Lake. In June 2014, the stills arrived and, soon thereafter, all the equipment was in place. Production started in autumn 2014 and a visitor centre was opened a couple of months later. The distillery was officially opened by Princess Anne in July 2015. The £2,5m distillery is equipped with two stills for the whisky production, each with both copper and stainless steel condensers, and a third still for the distillation of gin. A combination of copper and steel condensers allow for greater permutations when it comes to the flavour of the spirit. The capacity is 240,000 litres of pure alcohol but plans for the first full year is to produce 130,000 litres. To help create a cash flow, the company launched a British Isles blended whisky called The One in autumn 2013, where whisky had been sourced from a variety of producers around the UK. Well in advance of the first launch of a whisky having been produced at the distillery (sometime in 2017), Currie has also released both a gin and a vodka.

Adnams Copper House Distillery

Southwold, founded in 2010

www.adnams.co.uk

Famous for their beer since 1872, the owners of Adnams Brewery in Suffolk installed a new brewhouse in their Sole Bay Brewery in 2008. Left with a redundant, old building, they decided to convert it into a distillery. Equipment was ordered from the famous German manufacturer, Carl GmbH; a 1,000 litre beer stripping still and an 850 litre copper pot still with a rectification column with 42 plates attached. Distillation began towards the end of December 2010 and, apart from whisky – gin, vodka and absinthe are also produced. The distillation, in part resembles American bourbon production as they are using a beer stripping still which results in low wines that are high in alcohol (85-90%). The low wines are then distilled twice in the pot still where the 16 metre high rectification column with its plates give added reflux, making the spirit exceptionally smooth and clean. The first two whiskies from the distillery were released in December 2013 – Single Malt No. 1, a 3 year old matured in new French oak and Triple Grain No. 2 from malted barley, oats and wheat and matured in new American oak. There were 4,400 bottles of each and the next batch (10,000 bottles) of the Single Malt appeared in July 2014, while the second batch of Triple Grain was due in November 2014. The owners plan to release their first 100% Rye Whisky some time in 2015.

The Cotswolds Distillery

Stourton, founded in 2014

www.cotswoldsdistillery.com

The distillery is the brainchild of Dan Szor, who left London and a career in financing and subsequently moved to Cotswolds. Dan had been a whisky lover for many years and decided to try his hand at distilling as well. He acquired the Philip's Field Estate with two old stone buildings and started to convert them into a distillery. The main part of the equipment was in place towards the end of summer 2014 and production of both whisky and gin started in September. Three stills have been installed; one wash still (2,400 litres), one spirit still (1,600 litres) and a Holstein still (500 litres) for production of gin and other spirits. The rest of the equipment includes a 0.5 ton mash tun and eight stainless steel wash backs (four of which were installed in spring 2015). With the new fermenters installed, the distillery has a capacity of producing 300,000 bottles per year. The first product for sale was their Cotswolds Dry Gin in September 2014, while the first whisky, Organic Odyssey Single Malt, will be released in 2017.

Chase Distillery

Rosemaund Farm, Hereford, founded in 2008

www.chasedistillery.co.uk

In 2002, William Chase founded Tyrrell's Crisps, a company which he later sold in 2008. During that same year, he started his new business, a distillery, on his farm in Hereford. Both businesses emanated from the same ingredient, namely potatoes. Chase's main product, and one which has already made success in more than 25 countries, is Chase Vodka which, unusually, is made of potatoes. Today more than 500,000 bottles are produced yearly and gins have also be come part of the range. By the end of 2011, the first whisky was distilled and since then around 40 casks are filled every year. No release date has been set but Chase expects the first whisky will be launched as a 5 year old. The distillery is equipped with a copper still from Carl in Germany with a five plate column and an attached rectification column with another 42 plates. At 70 feet, the column is said to be the tallest of its kind in the world

Finland

Distillery:	Teerenpeli, Lahti
Founded:	2002
Owner:	Anssi Pyysing
	teerenpeli.com

The first Teerenpeli Single Malt was sold as a 3 year old in 2005, though solely at the owner's restaurant in Lahti. Four years later, the first bottles of a 6 year old were released and in 2011, it was time for an 8 year old, which was introduced at Whisky Live in London. This was a mix of whisky from both bourbon and sherry casks. In 2012, a 100% sherry matured version called Teerenpeli Distiller's Choice Kaski was released and this was followed up a year later with Distiller's Choice Äes, an 8 year old with a rum finish. End of 2014 saw two more limited releases; Hosa (matured in ex-Islay casks) and Port Wood Finish. In October 2015, it is time for the distillery's first 10 year old and also Rasi, a moscatel finish. Two more bottlings have been announced for release later in the year - Karhi (madeira finish) and the unusual Aura which has been matured in the brewery's own porter casks.

The distillery, located in the company's restaurant in central Lahti, is equipped with one wash still (1,500 litres) and one spirit still (900 litres) and the average fermentation time in the washback is 70 hours. In 2010 a new mash tun was installed and later that month a new visitor centre was opened. A completely new distillery, with one 3,000 litre wash still and two 900 litre spirit stills, was opened in October 2015 in the same house as the brewery. The expansion means that the two units will quadruple the capacity to 160,000 litres per year. The old distillery will be used mainly for special runs such as organic or peated whiskies.

The new 10 year old from Teerenpeli

Other distilleries in Finland

Helsinki Distilling Company

Helsinki, founded in 2014

www.hdco.fi

The first privately-owned distillery in Helsinki for over a hundred years, was opened in 2014 by two Finns, Mikko Mykkänen and Kai Kilpinen, and one Irishman, Séamus Holohan. Production started in August with gin and the first whisky was distilled one month later. The distillery is equipped with one mash tun and three washbacks made of stainless steel and one 300 litre pot still with a 7 plate column. The current capacity is 12,000 litres per year but there are already plans to install a second still which would triple the output. The first gin, with lingonberry as one of the botanicals, was released in October 2014 and more gin as well as applejack has followed since. On the whisky side, the focus is on rye, either with a mash bill of 75% rye and 25% barley or 100% rye, but also single malt made from barley. The owners have plans to also open an adjacent bar during 2015.

France

Distillery:	Glann ar Mor, Pleubian, Bretagne
Founded:	1999
Owner:	Jean Donnay
	glannarmor.com

The owner of Glann ar Mor Distillery in Brittany, Jean Donnay, already started his first trials back in 1999. He then made some changes to the distillery and the process and regular production commenced in 2005. The distillery is very much about celebrating the traditional way of distilling malt whisky. The two small stills are directly fired and Donnay uses worm tubs for condensing the spirit. He practises a long fermentation in wooden washbacks and the distillation is very slow. For maturation, first fill bourbon barrels and ex-Sauternes casks are used and when the whisky is bottled, there is neither chill filtration nor caramel colouring. The full capacity is 50,000 bottles per year. Apart from France, the whisky is available in the UK, Sweden, Denmark, Germany, The Netherlands, Italy, Singapore, Taiwan and Canada.

In July 2015, an announcement was made by the owners that the distillery would be closing down as a result of the new "Whisky Breton" Geographical Indication (GI), laid down by the French INAO organization. Donnay claimed that the technical file of the GI proposal, defining what a Whisky Breton can or cannot be, was "tailor made for industrial production...and penalising for artisan distilleries" like Glann ar Mor. Four weeks after the announcement, Donnay issued a new press release saying that recent talks with the INAO, had indicated that the technical file could be revised to suit Glann ar Mor´s needs and that the distillery wouldn´t be closed after all. As of August 2015, the situation whether or not the distillery would cease production, had not been completely resolved.

There are two versions of the whisky from Glann ar Mor – the unpeated Glann ar Mor matured in bourbon barrels and the peated Kornog matured in either bourbon barrels or Sauternes casks. Apart from the Glann ar Mor venture, Jean Donnay has also specialised in double maturation Single Malts. The "Celtique Connexion" range includes whiskies distilled and matured in Scotland, then further matured at the company's seaside warehouse. The whiskies can be found at www.tregorwhisky.com.

Recently Jean Donnay has embarked on a new project which entails building the ninth distillery on Islay. The Gartbreck distillery will be situated on the shores of Loch Indaal, a few miles south of Bowmore and opposite Port Charlotte. The plan is to open the distillery for production some time in 2016.

Distillery:	Distillerie Warenghem, Lannion, Bretagne
Founded:	1900 (whisky since 1994)
Owner:	Gilles Leizour
	distillerie-warenghem.com

Leon Warenghem founded the distillery at the beginning of the 20th century and in 1967 his grandson, Paul-Henri Warenghem, together with his associate, Yves Leizour, took over the reins. They moved the distillery to its current location on the outskirts of Lannion in Brittany. Today, the distillery is owned by the Leizour family and it was Gilles Leizour, taking over in 1983, who added whisky production to the repertoire. The first whisky, a blend called

Mikko Mykkänen, Kai Kilpinen and Séamus Holohan from Helsinki Distilling Company

David Roussier and Gilles Leizour, Distillerie Warenghem

WB, was released in 1987 and in the ensuing year, the first single malt distilled in France, Armorik, was launched. The distillery is equipped with a 6,000 litres semi lautermash tun, four stainless steel washbacks and two, traditional copper pot stills (a 6,000 litres wash still and a 3,500 litres spirit still). The company currently produces 100,000 litres of pure alcohol per year and 30% thereof is grain spirit for their blends. The owners have plans to increase production from 2015 and onwards.

The single malt core range consists of Armorik Edition Originale and Armorik Sherry Finish. Both are around 4 years old, bottled at 40%, have matured in ex-bourbon casks plus a few months in sherry butts for the SherryFinish and are sold in supermarkets in France. Armorik Classic, a mix of 4 to 8 year old whiskies from bourbon and sherry casks and the 7 year old Armorik Double Maturation which has spent time in both new French oak and sherry wood are earmarked for export. Both are un chill-filtered and bottled at 46%. Armorik Millesime 2002 is a limited release and the latest expression, launched in summer 2014, was 12 years old (the oldest so far from the distillery). In October 2014, Maitre de Chai, a 6 year old with two years in a bourbon cask and a few months in first fill oloroso, was released. A new edition of that is expected in autumn 2015. Finally, in the beginning of 2015, the first French rye ever was released. Named Roof Rye, the 8 year old whisky it is a collaboration between the distillery and the famous bartender Guillaume Ferroni. The maturation, which took place in both Brittany and Marseille, was made in combination of refill sherry and virgin American oak.

There are also three blended whiskies in the range; WB Whisky Breton which is 3-4 years old with 25% malt and 75% grain whisky, Galleg, with the same age but 50% malt whisky and Breizh, which is slightly older and 50% malt whisky. A total of 250,000 bottles were sold last year.

Other distilleries in France

Distillerie Claeyssens de Wambrechies

Wambrechies, Nord-Pas de Calais, founded in 1817 (whisky since 2000)

www.wambrechies.com

Claeyssens distillery is one of the oldest in France, tracing its history back to 1817 and located in a building classified as a historic monument in 1999. Located in the town of Wambrechies situated near Roubaix, the distillery was originally famous for its genever, the traditional spirit consumed in the north of France, in Belgium and in Holland. In 2000, the distillery started to produce single malt whisky launching Wambrechies 3 year old in 2003, and an 8 year old version in 2009. In the spring of 2013, two 12 year old bottlings (the oldest French whiskies to date) were released: one aged in madeira casks and another in sherry casks, both bottled at 40%.

Domaine Mavela

Corsica, founded in 1991 (whisky since 2001)

www.domaine-mavela.com

Since 2001 whisky is produced in Corsica, the Mediterranean island off the Côte d'Azur. The creators of P&M are the brewer Dominique Sialleli, also responsible for the creation of Pietra beer in 1996, and Jean-Claude Venturini who set up the Mavela distillery in 1991. They got some help from the Alsacian Jean-Claude Meyer from the distillery of the same name. Distilled in a Holstein still and aged in ex-Corsican muscat casks, the P&M single malt was sold for the first time in 2004. Its unique taste of the Corsican maquis surprised many whisky amateurs before becoming a success. Two blends, P&M Whisky Corse and P&M Supérieur, were later released. In 2015 the line-up was modified, doing away with the blends and focusing on two single malts, Vintage and a 7 year old.

Distillerie Meyer

Hohwarth, Alsace, founded in 1958 (whisky since 2007)

www.distilleriemeyer.fr

The Meyer distillery was founded in 1958 by Fridolin Meyer, a fruit wholesaler who thought it would be an excellent way to make money from his unsold fruit. Together with his son Jean-Claude, who joined the company in 1975, Meyer soon became one of the most awarded distillers in France. At the beginning of the 2000's, Jean-Claude Meyer helped the Venturini family in Corsica to start producing whisky under the P&M brand, and this convinced him to produce his own whisky. With the help of his two sons, Arnaud and Lionel, they launched two no-age statement whiskies in 2007, just one year before the sudden death of Jean-Claude – one blend and one single malt. More than 500 casks are currently maturing in the new warehouse built in 2012. Meyer's, mainly sold in supermarkets in the Alsace area, is probably one of the best selling French whiskies right now. There are currently three different versions: Meyer's Pur Malt (a single malt), Meyer's Blend Supérieur and Oncle Meyer Blend Supérieur.

Distillerie Gilbert Holl

Ribeauvillé, Alsace, founded in 1979 (whisky since 2000)

www.gilbertholl.com

In 1979, Gilbert Holl came across distillation by chance with a friend and began to distill occasionally in the back of his wine and spirits shop. He bought his first still in 1982 in order to produce eaux-de-vie with his own fruit: cherry plums, raspberries and cherries. And at the beginning of 2000, he finally started distilling whisky in a very small still (150 litres). His first bottling, Lac'Holl, was put on sale in 2004 and was followed by Lac'Holl Junior in 2007 and Lac'Holl Vieil Or in 2009. Production of this light bodied whisky remains very limited.

Distillerie Bertrand

Uberach, Alsace, founded in 1874 (whisky since 2002)

www.distillerie-bertrand.com

The town of Uberach has only 1,000 inhabitants but two whisky distilleries, Uberach and Hepp. Distillerie Bertrand is an independent affiliate of Wolfberger, the large wine and eau-de-vie producer. The manager, Jean Metzger, gets the malt from a local brewer and then distils it in Holstein type stills. Two different types of whisky are produced. One is a single malt at 42,2%, non-chill filtered and with maturation in both new barrels and barrels which have previously contained the fortified wine Banyuls. The other is a single cask at 43,8% matured only in Banyuls barrels. The first bottles, aged 4 years, were released in late 2006. Being a great wine connoisseur, Jean Metzger started to experiment with maturation and finishing whisky in a lot of other different ex-wine casks, including Arbois and Pupillin. This new range of whisky, Cask Jaune, is bottled in 50 cl beer type bottles. In September 2013, a very limited number of bottles of Uberach 10 year old were released for the Whisky Live Paris 10th anniversary. R8, launched in 2015 was aged in ex-Rasteau casks.

Distillerie Hepp

Uberach, Alsace, founded in 1972 (whisky since 2005)

www.distillerie-hepp.com

A family-owned distillery, Hepp started producing single malt whisky in 2005, on the initiative of the then owner's son, Yannick. Besides the regular expression, a no-age statement bottled at 42% under the brand Tharcis Hepp, two limited editions have been released, the first one aged in ex-plum cask, the second one under the name Johnny Hepp. As well as producing their own whisky, Hepp also supplies the independant bottler Denis Hanns with liquid for his Authentic Whisky Alsace. AWA is a range of single cask whiskies finished in Alsacian wine (riesling, gewurtztraminer, pinot gris) and brandy (kirsch and cherry plum) casks.

Distillerie Grallet Dupic

Rozelieures, Lorraine, founded in 1860 (whisky since 2007)

www.whiskyrozelieures.com

Hubert Grallet had been distilling cherry plums for many years when his daughter married a barley farmer, Christophe Dupic: following a dare issued during a well-lubricated dinner, they decided to try their hand at whisky production, and launched the Glen Rozelieures brand in 2007, changing it to G. Rozelieures after being menaced by the SWA. Four versions are currently available: the first two are aged in ex-fino sherry casks, the third is lightly peated and aged in Sauternes casks and the fourth is peated. Towards the end of 2015, four single casks will be released. Rozelieures is also bottled under the brand name Lughnasadh ("August" in Gaelic) for the Clair de Lorraine chain-store. The company also blends and bottles a blended malt, Whisky de France (30% Alsatian whisky, 70% Lorrain whisky).

Brûlerie du Revermont

Nevy sur Seille, Franche-Comté, founded in 1991 (whisky since 2003)

www.marielouisetissot-levin.com

For many years, the Tissot family were travelling distillers offering their services to the many wine producers in the Franche-Comté area. Relying upon a very unique distillation set-up, a Blavier still with three pots, designed and built in the early 1930's for the perfume industry, they have been producing single malt whisky since 2003. Whilst at the beginning all the whisky was distilled to order by Bruno Mangin, of the Rouget de Lisle Brewery, Pascal and Joseph Tissot launched their own whisky brand Prohibition in 2011. Aged in "feuillettes" (114 litres half-casks coopered specially for macvin and vin de paille french wines), the whisky is reduced to 41% or 42% and bottled uncoloured cask by cask.

Rouget de Lisle

Bletterans, Franche-Comté, founded in 1994 (whisky since 2006),

www.brasserie-rouget-lisle.com

Rouget de Lisle (named after the man who wrote the lyrics to the French national anthem La Marseillaise) is a micro-brewery created by Bruno Mangin and his wife. Having made a first unsuccessful attempt to distill in 1998, they tried again In 2006, commissioning the Brûlerie du Revermont to do it for them. This proved to be a good

idea and the first Rouget De Lisle single malt whisky was released in 2009. In 2012, Bruno Mangin bought his own still, a Sofac Armagnac still with a capacity of 1,500 litres. Current bottlings are from the numerous casks he filled during his association with the Tissot family and which lie maturing in his own warehouse. The very first 100% Rouget de Lisle whisky won't be available until some time towards the end of 2015.

Distillerie Artisanale Leisen

Malling, Lorraine, founded in 1898 (whisky since 1998)

www.distillerie-leisen-petite-hettange.fr

Ever since 1898 the Leisen family has been distilling fruit eaux-de-vie not far from Malling in the Moselle department. Leisen is equipped with two cognac-style bain-marie pot-stills (250 litres and 350 litres), made by Firma Carl in Stuttgart. Jean-Marie Leisen set out on his whisky adventure in 1998 but had to wait until 2005 to sell his first bottles. Today he has two versions, a 7 year old reduced to 42% and a 10 year old bottled at 47%. The distillate is aged in casks made either of new wood or having contained wine from the French Moselle, but in both cases from oaks grown on the property.

Distillerie de Northmaen

La Chapelle Saint-Ouen, Haute-Normandie, founded in 1997 (whisky since 2005)

www.northmaen.com

Northmaen is a craft brewery founded in 1997 by Dominique Camus and his wife. Every year since 2005, they have bottled and sold Thor Boyo, a 3 year old single malt, distilled in a small, mobile pot still. In 2009, a 5 year old cask strength (59%) version was launched and in 2013, the oldest version to date, an 8 year old bottled at 44% under the new brand name Sleipnir was released. For the first time a peated whisky, Fafnir, will be available at the end of 2015.

Distillerie Lehmann

Obernai, Alsace, founded in 1850 (whisky since 2001)

www.distillerielehmann.com

The story of Lehmann distillery starts in 1850 when the family of the actual owner set up a still in Bischoffsheim. Yves Lehmann inherited the facility in 1982 but decided to move all the equipment to a new distillery in 1993. In 2001, he bought two Rump stills (1 200L and 600L) to produce single malt whisky which he ages exclusively in French white wine casks. The first regular bottling,

Christophe Dupic inspecting the casks at Distillerie Grallet Dupic

aged for seven years in Bordeaux casks and bottled at 40%, was launched in 2008 under the brand Elsass Whisky ("Alsace" in Alsacian dialect). A second bottling, aged for 8 years in Sauternes casks and bottled at 50%, followed soon after. The Coteaux-du-Layon casks have yet to be bottled.

Distillerie Brunet

Cognac, Poitou-Charentes, founded in 1920 (whisky since 2006)

www.drinkbrenne.com

The Cognac area needs no introduction. With more than 5 000 stills in operation, the Poitou-Charentes region boasts the highest concentration of distillers in the world. But apart from cognac there is whisky produced here as well. The Brunet distillery, owned by Stéphane Brunet, was founded in 1920 but started to distill whisky in 2006. His whisky, Tradition Malt was launched in 2009 and has already found its way in the USA where it is sold under the brand Brenne, courtesy of Allison Patel. This young woman, whisky connoisseur and enthusiast, launched this as her own brand, surfing on the single malt wave and taking advantage of the fame of cognac. Each version, bottled at 80° proof (40%) comes from a single cask. In September 2015 a 10 year old version was be released in small quantities in the USA.

Distillerie de Paris

Paris, Ile de France, founded in 2014

www.distilleriedeparis.com

The micro-distillation boom has come to Paris. Sébastien and Nicolas Julhès – two brothers in charge of one of Paris's best groceries – have set up a still in the heart of the French capital. The Distillerie de Paris is equipped with a 400 litre Holstein still configured to produce the equivalent of 50 litres of distillate at 65% per batch (double distillation). Distillation of gin and vodka started in January 2015, soon followed by brandy, rhum and grain spirit. The first single malt was distilled in June 2015.

Brasserie Michard

Limoges, Limousin, founded in 1987 (whisky since 2011)

www.bieres-michard.com

The Limousin area is very well known for its forests of oak trees cherished by the cognac industry, but not really for its eaux-de-vie, nor for its whisky. However, that could change thanks to Michard, a brewery founded in 1987 by Jean Michard. On the initiative of his daughter, Julie, he bottled his first single malt in 2011. Using their own unique yeast, the first batch of their whisky is highly original and very fruity. Available in an 800 bottle limited edition, it has been followed by a second batch in late 2013. Unfortunately, due to problems with the French customs, Jean Michard has had to stop distilling for the time being. He is currently considering building a brand new distillery.

Domaine des Hautes-Glaces

Saint Jean d´Hérans, Rhône-Alpes, founded in 2009 (whisky since 2013)

www.hautesglaces.com

At an altitude of 900 matres in the middle of the French Alps, Jérémy Bricka and Frédéric Revol have decided to produce whisky from the barley to the bottle! Apart from growing their own barley, all the parts of whisky production take place at the distillery – malting, brewing, distillation, maturation and bottling. Not only have they set up a still to create the first French single malt whisky, they are doing it organically: all of their cereal (mainly barley, but also rye) is harvested, malted, distilled and aged field by field and without any chemicals, in order to remain as faithful as possible to the expression of their unique terroir. Principium, the first whisky made at the distillery has been available since June 2014. Bottled at 50%, it was aged in a new cask for five months and then in a French vin jaune cask. The Domaine des Hautes Glaces also sell a small selection of limited edition single malt spirits, made from both barley and rye.

Distillerie Kaerilis

Belle-Île, Bretagne, founded in 2011

www.kaeriliswhisky.com

Like many others, Fabien Mueller started his activity as an independent bottler. Since 2006, Kaerilis has released a range of whiskies distilled in Scotland with an additional maturation taking place on Belle-Île, a small island a few kilometres off the south coast of Brittany. Since 2011, the company also operates a small distillery located in the back of a shop in the main city of the island, Le Palais: a single bain-marie type Müller still bought by Fabien Mueller, and some Bourbon barrels. The first bottles of An Toiseach Ar Bell'Isle are available since last June. Fabien Mueller now dreams of growing his own barley and one day releasing a 100% Kaerilis whisky (Kaer-ilis in Breton means Belle-Île).

Nicolas Julhès, owner of Distillerie de Paris

Dreumont

Neuville-en-Avesnois, Nord-Pas-de-Calais, founded in 2005 (whisky since 2011)

www.ladreum.com

Passionate about beer, which he has been brewing for the past twenty years, Jérôme Dreumont founded his distillery in 2005. Today he produces and sells some 100 hectolitres per year. In 2011 he built his own 300 litre still and ran it for the first time in November of the same year. Since then, he has been filling only one cask of double-distilled spirit per year, but intends to increase production soon. His first whisky, distilled from a mix of peated and non-peated barley, was put on sale in March 2015.

La Bercloise

Bercloux, Poitou-Charentes, founded in 2000 (whisky since 2014)

www.bercloise.fr

After many trials of brewing whilst still a student, and a course in Ireland to hone his skills, Philippe Laclie opened his own brewery in May 2000. The success he met with encouraged him to increase production from 150 to 1,500 hectolitres in less than two years. In 2007 he decided to diversify by buying in some Scotch whisky and finishing it for a few months in Pineau des Charentes barrels. He currently produces some 10,000 litres of Bercloux whisky. At the beginning of 2014, Philippe decided to take the next step and invest around 100,000 euros by purchasing a small 800 litre Stupfler type column still. After the first trial runs in June 2014, regular whisky production started in September 2014.

Distillerie du Castor

Troisfontaines, Lorraine, founded in 1985 (whisky since 2011)

www.distillerie-du-castor.com

Founded thirty years ago by Patrick Bertin, the Distillerie du Castor produces both fruit and pomace brandies. The distillery is equipped with two small stills, a 150 litres Holstein and a 250 litres Carl, which have been used since 2011 by Patrick's son to distil single malt whisky. The malt is bought from Malteurop in Metz and brewed to 8% by a local brewery. The first distillation is carried out in the two stills whereas the low wines, mixed together, are re-distilled in the Carl still only. The distillate is aged in ex-white wine (sauvignon blanc) casks and then finished in ex-sherry casks. A first bottling of 1,100 bottles (bottled at 42%) was put on sale in June 2015 under the name St Patrick (complete with green and clovers). A strange idea for a whisky made 100% in the Lorraine region, with no link whatsoever to Ireland.

Distillerie Ninkasi

Tarare, Rhône-Alpes, founded in 1998 (whisky since 2015)

www.ninkasi.fr

The Ninkasi brewery in Lyon had wanted to produce whisky ever since 2009 but had to wait until 2015 to actually be able to do so: it took them five years to move production from Tarare, where everything was housed in a former dyer, and to collect the 250,000 euros necessary to fund the project. The future whisky is distilled in a 2,500 litres still made by Prulho Chalvignac, using wash produced on site, and then aged in different types of cask. Production began in September 2015 and should reach cruising speed some time next year, producing around 10,000 litres of pure alcohol. In the long-term, Ninkasi is hoping to be able to use the on-site water source and to develop their own strain of yeasts.

Sainte-Colombe

Bercloux, Poitou-Charentes, founded in 1996 (whisky since 2010)

Gonny Keizer installed a micro-brewery in the Roche aux Fées county, next to the town of Sainte-Colombe en Ille-et-Vilaine in 1996. She is the first female master-brewer in France. In 2010,

Gonny and her husband Henry bought themselves a portable automatic batch still made in 1950 in Nantes by Coyac. With a capacity of 400 litres, the still is wood-heated and is equipped with a worm-tub condenser. The first spirit was put into cask in 2010. In 2014 a distillery was built to shelter the still in a permanent manner. Today some twenty odd casks (former wine casks from the Nantes region) await bottling some time around the end of 2015 or early 2016.

Germany

Distillery:	Whisky-Destillerie Blaue Maus, Eggolsheim-Neuses
Founded:	1980
Owner:	Robert Fleischmann
	fleischman-whisky.de

This is the oldest single malt whisky distillery in Germany and it celebrated its 30th anniversary in 2013. This was also a celebration of the first German single malt whisky which was distilled in 1983. It took, however, 15 years before the first whisky, Glen Mouse 1986, appeared. Fleischmann uses unpeated malt and the whisky matures for approximately eight years in casks of fresh German Oak. A completely new distillery became operational in April 2013. All whisky from Blaue Maus are single casks and there are currently around ten single malts in the range, for example Blaue Maus, Spinnaker, Krottentaler, Schwarzer Pirat, Grüner Hund, Mary Read, Austrasier and Old Fahr. Some of them are released at cask strength while others are reduced to 40%. To celebrate the 30th anniversary, some limited releases were made in June 2013 – for example, Blaue Maus and Spinnaker, both in 20 and 25 year old versions.

Distillery:	Slyrs Destillerie, Schliersee
Founded:	1928 (whisky since 1999)
Owner:	Stetter family
	slyrs.de

Lantenhammer Destillerie in Schliersee, Bavaria was founded in 1928 and was producing mainly brandy until 1999 when whisky took preference, and in 2003 Slyrs Destillerie was founded. The malt, smoked with beech, comes from locally grown grain and the spirit is distilled in 1,500 litre stills. Maturation takes place in charred 225-litre casks of new American White Oak from Missouri. Investments in three new fermentation tanks (washbacks) and a malt silo during 2009/2010 increased the capacity and they are currently producing 150,000 bottles per year.

The non chill-filtered whisky is called Slyrs after the original name of the surrounding area, Schliers. Around 40,000 bottles are sold annually. The core expressions are a 3 year old bottled at 43% and a cask strength version. In 2013, three limited versions were released – a PX sherry cask finish, an Oloroso finish and a Port finish. New editions of all three have since been released. In May 2015, 1,000 bottles of the distillery's first 12 year old whisky were released.

Distillery:	Hammerschmiede, Zorge
Founded:	1984 (whisky since 2002)
Owner:	Alexander Buchholz
	hammerschmiede.de

In keeping with many other small whisky producers on mainland Europe, Hammerschmiede's main products are liqueurs, bitters and spirits from fruit, berries and herbs. But whisky distilling was only embarked on in 2002 and whisky production has now increased remarkably to 25,000 bottles per year. In 2014 the owners started the construction of a new still house with additional stills making it a total of five once completed. The first bottles were released in 2006 under the name Glan Iarran. Today, all whisky produced has changed name to Glen Els. The core range consists of four expressions; Glen Els Journey with a blend of different maturations (sherry, port, madeira, marsala, malaga and bordeaux), Ember, which is woodsmoked, Unique Distillery Edition which is always from a sherry cask and Wayfare with the same maturation as Journey but bottled at cask strength. These whiskies are complemented by the Woodsmoked Malts, a range of single cask bottlings where the whisky is always smoked but has matured in a variety of casks. The 2014 releases include port, sherry, marsala and sherry/mead. A new range in 2014 was Four Seasons with six non-woodsmoked maturations (ruby port, madeira, bordeaux, marsala and sherry). Finally, there is the Alrik by Glen Els which, according to the owner, Alexander Buchholz, is the ultimate woodsmoked malt, usually matured in a PX sherry cask with an additional finish and released a couple of times a year. The latest were 1914 Sancta Brigitta, matured in a ruby port cask and 1915 Ostara, madeira cask matured and with a tokaji finish.

Other distilleries in Germany

Spreewälder Feinbrand- & Likörfabrik

Schlepzig, founded in 2004 (whisky production)

www.spreewaldbrennerei.de

The product range consists of different kinds of beers, eau-de-vie and rum, and since 2004 malt whisky is also included. The distillery is equipped with three stills with fractionating column and is fired by using gas. The annual production of whisky and rum is 15,000 litres per year. French Oak casks, that have previously contained wine made of Sylvaner and Riesling grapes, are used for maturation, as well as new Spessart oak casks. Torsten Römer released his first whisky; Sloupisti, as a 3 year old in 2007. Recent bottlings have been older (up tp a 10 year old) and there is also a cask strength version.

Obsthof am Berg

Kriftel, founded in 1983 (whisky since 2009)

www.obsthof-am-berg.de, www.gllors.de

Holger and Ralf Heinrich are the third generation running this distillery and their focal point is to produce spirits from fruits and berries. In 2009 the two brothers started whisky production and the first release of their 3 year old single malt Gilors was in 2012. The whisky is non chillfiltered and the majority of the production is unpeated. For maturation they use 100-250 litre ex-sherry and ex-port casks with a yearly production of 1,200 litres. The two core expressions, Gilors fino sherry matured and Gilors port matured, are

both 3 years old. A 4 year old Gilors Islay cask finish (1 year in new oak and 3 years in an Islay cask) was released in 2014 and spring 2015 saw the release of a Gilors Oloroso finish.

Bayerwald-Bärwurzerei und Spezialitäten-Brennerei Liebl

Kötzting, founded in 1970 (whisky since 2006)

www.coillmor.com

In 1970 Gerhard Liebl started spirit distillation from fruits and berries and in 2006 his son, Gerhard Liebl Jr., built a brand new whisky distillery. Maturation takes place in first or second fill ex-bourbon barrels, except for whisky which is destined to be bottled as single casks. Sherry, Port, Bordeaux and Cognac casks are used here. About 20,000 litres of whisky are produced annually and in 2009 the first 1,500 bottles bearing the name Coillmór were released. The current range is American oak (4 years), Port Single cask (6 years), Albanach Peat (4 years), Sauternes Cask Chateau d´Yquem (5 years), Bordeaux (5 years) and PX Sherry (6 years). Over the next few years the entire business will gradually move to new premises. The first step was a new warehouse which was built in July 2014.

Brennerei Höhler

Aarbergen, founded in 1895 (whisky since 2001)

www.brennerei-hoehler.de

The main produce from this distillery in Hessen consists of different distillates from fruit and berries. The first whisky, a bourbon variety, was distilled in 2001 and released in 2004. Since then, Karl-Holger Höhler has experimented with different types of grain (rye, barley, spelt and oat). A couple of the more recent releases of his Whesskey (so called since it is from the province Hessen) are a Cara-Aroma Single Malt and single malts made from smoked (and not peated) barley.

Stickum Brennerei (Uerige)

Düsseldorf, founded in 2007

www.stickum.de

Uerige Brewery, founded in 1862, was completed with a distillery in 2007. The wash comes from their own brewery and the distillation takes place in a 250 litre column still. For the maturation they not only use new oak but also bourbon, sherry and port casks. The distillery produces 700 bottles of their whisky BAAS per year and the first bottling (a 3 year old) was released in December 2010. In 2014 the owners released their first 5 year old whiskies – one matured in a sherry cask and the other in new oak.

Left; Alexander Buchholz, owner of Hammerschmiede. Right; Holger and Ralf Heinrich from Obsthof am Berg

Preussische Whiskydestillerie

Mark Landin, founded in 2009

www.preussischerwhisky.de

Cornelia Bohn purchased a closed-down distillery in 2009 in the Uckermark region, one hour's drive from Berlin. The distillery had been operational for 100 years up until WWII, when Russian soldiers took it apart and the last stills disappeared in the 1950s. Bohn installed a 550 litre copper still with a 4-plate rectification column attached. The spirit is distilled very slowly five to six times and is then matured in casks made of new, heavily toasted American white oak, German fine oak or German Spessart oak. Since 2013 only organic barley is used for the distillation. The first whisky was launched as a 3 year old in December 2012 and in 2014, three releases were made (March, June and September). From March 2015, all the whiskies from the distillery will be at least 5 years old. The whisky is always un chill-filtered without colouring and bottled at cask strength.

Kleinbrennerei Fitzke

Herbolzheim-Broggingen, founded in 1874
(whisky since 2004)

www.kleinbrennerei-fitzke.de

The main commerce for the distillery is the production of eau de viex and vodka, but they also distill whisky from different grains. Mashing, fermentation and distillation all take place at the distillery and for maturation they use 30 litre oak casks. For the first six months they use virgin oak and, thereafter, the spirit is filled into used barrels for another two and a half years. The first release of the Derrina single malt was in 2007 and new batches have been launched ever since. The different varieties of Derrina are either made from malted grains (barley, rye, wheat, oats etc.) or unmalted (barley, oats, buckwheat, rice, triticale, sorghum or maize).

Rieger & Hofmeister

Fellbach, founded in 1994 (whisky since 2006)

www.rieger-hofmeister.de

Marcus Hofmeister's stepfather, Albrecht Rieger, started the distillery and when Marcus entered the business in 2006 he expanded it to also include whisky production. The first release of this Schwäbischer Whisky was in 2009 and currently there are two expressions in the range – a Single Malt matured in Pinot Noir casks and a Malt & Grain (50% wheat, 40% barley and 10% smoked barley) from Chardonnay casks. By the end of 2012 a malted rye was distilled and at the beginning of 2013 a lightly peated malt from barley.

Kinzigbrennerei

Biberach, founded in 1937 (whisky since 2004)

www.kinzigbrennerei.de

Martin Brosamer is the third generation in the family and he is also the one who expanded the production in 2004 to include whisky. The total production of whisky is 2,000 litres annually. In the beginning, Martin filled small casks (50 litres) made of new oak but has progressively moved to larger casks. The first release in 2008 was Badische Whisky, a blend made from wheat and barley. Two years later came the 4 year old Biberacher Whisky, the first single malt and in 2012, the range was expanded with Schwarzwälder Rye Whisky and the smoky single malt Kinzigtäler Whisky.

Destillerie Kammer-Kirsch

Karlsruhe, founded in1961 (whisky since 2006)

www.kammer-kirsch.de

Like so many distilleries, production of spirits from various fruits and berries is the main focus for Kammer-Kirsch and they are especially known for their Kirschwasser from cherries. In 2006 they started a cooperation with the brewery, Landesbrauerei Rothaus, where the brewery delivers a fermented wash to the distillery and

they continue distilling a whisky called Rothaus Black Forest Single Malt Whisky. The whisky was launched for the first time in 2009 and, every year in March, a new batch is released. Bourbon casks are used for maturation and besides the "original" version, there is a special edition with a wood finish released every autumn. In 2014 it was a peated finish where the character came from a BenRiach peated cask. Around 6,000 bottles are produced every year.

Alt Enderle Brennerei

Rosenberg/Sindolsheim, founded in 1988 (whisky since 2000)

www.alt-enderle-brennerei.de

While concentrating on the production of schnapps, gin, rum and absinthe, Joachim Alt and Michael Enderle produced their first malt whisky in 2000. They have two expressions for sale – an 8 year old Neccarus and a recently released 12 year old. The whisky is matured in a combination of bourbon and sherry casks. In 2013 Neccarus was awarded as the best German whisky.

Brennerei Ziegler

Freudenberg, founded in 1865

www.brennerei-ziegler.de

Like so many other distilleries in Germany, Ziegler has distillation of spirits from fruits and berries as their main business, but has also added a small whisky production. One characteristic that distinguishes itself from most other distilleries is that the maturation takes place not only in oak casks, but also in casks made of chestnut! Their current bottling is a 5 year old called Aureum 1865 Single Malt and there is also a cask strength version.

AV Brennerei

Wincheringen, founded in 1824 (whisky since 2006)

www.avadisdistillery.de

For generations, the Vallendar family have been making schnapps and edelbrände on a farm in an area where France, Luxembourg and Germany's borders meet. Since 2006 the brothers Andreas and Carlo Vallendar, also produce malt whisky. Around 2,000 bottles per annum are available for purchase and the oak casks from France have previously been used for maturing white Mosel wine. Threeland Whisky is a 3 year old and the range also consists of two finishes – Oloroso and Port.

Birkenhof-Brennerei

Nistertal, founded in 1848 (whisky since 2002)

www.birkenhof-brennerei.de

The traditional production of edelbrände made from a delightful variety of fruits and berries was complemented with whisky production in 2002. The first release was a 5 year old rye whisky under the name Fading Hill in 2008. This was followed a year later by a single malt. The most recent single malt bottling is from April 2013 – a single ex-bourbon cask distilled in 2008.

Brennerei Faber

Ferschweiler, founded in 1949

www.faber-eifelbrand.de

Established as a producer of eau-de vie from fruits and berries, Ludwig Faber – the third generation of the owners – has included whisky production during the last few years. The only whisky so far is a single malt that has matured for 6 years in barrels made of American white oak.

Steinhauser Destillerie

Kressbronn, founded in 1828 (whisky since 2008)

www.weinkellerei-steinhauser.de

The distillery is situated in the very south of Germany, near Lake Constance, close to Austria. The main products are spirits which are derived from fruits, but whisky also has its own niche. The first release was the single malt Brigantia which was released in November 2011. It was triple distilled and only 111 bottles were released. That was followed by 212 bottles in December 2012. The ultimate goal is to release a 12 year old in 2020 under the name of Constantia.

Weingut Simons

Alzenau-Michelbach, founded in 1879 (whisky since 1998)

www.feinbrenner.eu

The owner, Severin Simon, produces wine from his own vineyards, spirits from fruit as well as gin, vodka, rum and whisky. Until recently all the whisky was produced in a 150 litre still but a new still from Arnold Holstein was installed in June 2013, raising the whisky production from 300 litres per year to 3-5,000 litres. A pure pot still whisky has since been released and the first whisky from 100% malted barley was distilled in January 2013 and is due for release in 2016.

Nordpfälzer Edelobst & Whiskydestille

Winnweiler, founded in 2008

www.nordpfalz-brennerei.de

This distillery owned by Bernhard Höning is based on the production of spirits from fruits but also distilling whisky. The first release was in 2011, a 3 year old single malt by the name Taranis with a full maturation in a Sauternes cask and in October 2013 a 4 year old from ex-bourbon casks with an Amarone finish was launched. The most recent release was in September 2014. In September 2013 a second distillery including a tasting room was opened and Höning is now producing 12 barrels of whisky per year.

Destillerie Drexler

Arrach, whisky since 2007

www.drexlers-whisky.de

The main business for Reinhard Drexler is the production of spirits from herbs, fruits and berries. In between he also finds the time to produce malt whisky and from August 2013 also rye whisky. The first release was Bayerwoid in 2011 which was followed up by the 3 year old No. 1 Single Cask Malt Whisky in 2012. The whisky matures in fresh American oak and casks that have previously been used for sherry, bourbon and cognac.

Destillerie & Brennerei Heinrich Habbel

Sprockhövel, founded in 1878 (whisky since 1977)

www.habbel.com

Already in 1977, Michael Habbel produced his first whisky from 85% rye and 15% malted barley. After a 10 year maturation in bourbon casks, the whisky was transferred into stainless steel tanks and wasn´t released until a few years ago as Habbel´s Uralter Whisky. Meanwhile, the distillery has produced different eau de vie, schnapps, vodka and gin. In spring of 2014 a designated whisky distillery called Hillock Park Distillery was opened on the premises and the plan is to produce 50 casks per year. A sample of the product has already been released in the shape of a newmake – Hillock White Dog (78% rye and 22% malted barley).

Märkische Spezialitäten Brennerei

Hagen, whisky since 2010

www.msb-hagen.de

Under the brand name Bonum Bono, Klaus Wurm and Christian Vormann produce spirits and liqueurs from various fruits. In autumn 2010 they added whisky to the range. Around 20 casks are produced yearly. The spirit is distilled four times, matured in ex-bourbon barrels in the distillery warehouse for 12 months and then brought to a cave, with low temperature and high humidity, 20 kilometres from the distillery for further maturation. The first releases were

new make and aged spirit with the first 3 year old whisky, Tronje van Hagen, being released in 2013. More bottles followed in 2014. The whisky has now been renamed DeCavo (from the cave, in allusion to where the maturation takes place).

Dürr Edelbranntweine

Neubulach, founded in 2002

www.blackforest-whiskey.com

Third generation distillers, Nicolas and Sebastian Dürr, began producing whisky in 2002. It wasn't until 2012 when their first single malt (limited to 200 bottles) reached the market. Doinich Daal Batch 1 was a 4 year old matured in a combination of bourbon and cognac casks. Batch two (Malachit and Azurit editions) were released in November 2014.

Tecker Whisky-Destillerie

Owen, founded in 1979 (whisky since 1989)

www.tecker.eu

Founded by one of the German whisky pioneers, Christian Gruel, the distillery is run by his grandchild Immanuel Gruel, since five years back. Apart from a variety of eau de vie and other spirits, approximately 1,500 litres of whisky is produced annually. The core expression is the 5 year old Tecker Single Malt matured in sherry casks and a 5 year old Tecker Single Grain. Limited releases have included a cask strength 10 year old single malt and a cask strength 14 year old single grain.

Brennerei Feller

Dietenheim-Regglisweiler, founded in 1903 (whisky since 2008)

www.brennerei-feller.de

Roland Feller, the owner of this old schnapps distillery, took up whisky production in 2008 and four years later he released his first single malt, the 3 year old Valerie (recent releases are 4 years old) matured in bourbon casks. It was later followed by Augustus and Augustus Corado, both single wheat whiskies with a port finish.

Marder Edelbrände

Albbruck-Unteralpfen, founded in 1953 (whisky since 2009)

www.marder-edelbraende.de

Apart from a vast number of distillates from different fruits, Stefan Marder also produces whisky since 2009. The first release came in 2013 - the 3 year old Marder Single Malt matured in a combination of new American oak and sherry casks. One thousand bottles were released and a second edition was launched in July 2014.

Edelbrände Senft

Salem-Rickenbach, founded in 1988 (whisky since 2009)

www.edelbraende-senft.de

When Herbert Senft started whisky production, he experimented with a variety of different grains but, in future, he will be concentrating on whisky from 100% malted barley. The first 2,000 bottles of 3.5 year old Senft Bodensee Whisky were released in 2012 and they were later followed by a cask strength version (55%).

Sperbers Destillerie

Rentweinsdorf, founded in 1923 (whisky since 2002)

www.salmsdorf.de

Apart from distilling eau de vie and liqueurs from fruits, Helmut Sperber has also been producing whisky since 2002. At the moment, there are four different expressions in the range, all of them 7 years old – single malt matured in bourbon casks, single malt matured in a mix of sherry and bourbons casks, a sherrymatured single malt bottled at cask strength, as well as,a single grain whisky from a mix of bourbon, sherry and Spessart oak casks.

Schwarzwaldbrennerei Walter Seger

Calw-Holzbronn, founded in 1952 (whisky since 1990)

www.krabba-nescht.de

Incorporated with a restaurant, this distillery which, apart from producing eau de vie from berries and fruit, also produces whisky. The first single malt was launched in 2009 and at the moment, Walter Seger has two expressions in the range; the 4 year old Black-Wood single malt matured in amontillado sherry casks and an 8 year old wheat whisky.

Landgasthof Gemmer

Rettert, founded in 1908 (whisky since 2008)

www.landgasthof-gemmer.de

The current owner, Klaus Gemmer, is the fourth generation running this distillery, and he was also the one who introduced whisky production in 2008. Their only single malt is the 3 year old Georg IV which has matured for two years in toasted Spessart oak casks and finished for one year in casks that have contained Banyuls wine. Around 800 litres are produced per year.

Hausbrauerei Altstadthof

Nürnberg, founded in 1984

www.hausbrauerei-altstadthof.de

The brewery component of this establishment was founded in 1984 and later Reinhard Engel also added a distillery. He produced the first German organic single malt and the current range consists of the 4 year old Ayrer's Red bottled at 43% and 58%, both matured in new American oak, Ayrer's PX, bottled at 56% and finished in PX sherry casks, Ayrer's Bourbon, bottled at 51% and matured in bourbon barrels and Ayrer's Master Cut, bottled at an exceptional 68.2%. There is also Ayrer's White which is an 8 week old whisky spirit.

Destillerie Mösslein

Zeilitzheim, founded in 1984 (whisky since 1999)

www.frankenwhisky.de

Originally a winery but also producing a wide range of eau de vie from fruits, whisky production was brought on board in 1999. The first whisky was released in 2003 and, at the moment, the owner, Reiner Mösslein, can offer two types of whisky – a single malt and a grain whisky with a matured in red wine barriques, both 5 years old.

Brennerei Josef Druffel

Oelde-Stromberg, founded in 1792 (whisky since 2010)

www.brennerei-druffel.de

Jochen Druffel is the seventh generation to run this old family distillery. A variety of different spirits are distilled with schnapps and liqueur made from plums as a speciality. The first single malt, Prum, was released in 2013 and had matured in a mix of different casks (bourbon, sherry, red wine and new Spessart oak) and was finished in small casks made of plum tree!

Brauhaus am Lohberg

Wismar, whisky since 2010

www.brauhaus-wismar.de

This unique combination of a guesthouse, brewery and distillery is situated at the centre of Wismar in a building from the 15th century. Here Herbert Wenzel has been producing whisky since 2010. The first release of Baltach single malt was in December 2013. It was a 3 year old with a finish in sherry casks. The next edition of Baltach followed in June 2015.

Wild Brennerei

Gengenbach, founded in 1855 (whisky since 2002)

www.wild-brennerei.de

The distillation of whisky constitutes only a small part of the production but Franz Wild has released two 5 year old whiskies – Wild Whisky Single Malt which has matured for three years in American white oak and another two in either sherry or port casks and Wild Whisky Grain, made from unmalted barley and with a similar maturation.

Brennerei Volker Theurer

Tübingen, founded in 1995

www.schwaebischer-whisky.de

Located in a guesthouse, this distillery which is run by Volker Theurer has been producing whisky since 1995. The first release was a 7 year old in 2003 and the current range includes Sankt Johann, an 8 year old single malt and the 9 year old Tammer which has been double matured in bourbon and sherry casks. Theurer is also selling a blended whisky called Original Ammertal Whisky.

Iceland

Distillery:	Eimverk Distillery, Reykjavik
Founded:	2012
Owner:	Thorelsson family
	flokiwhisky.is

The country's first whisky distillery emanated from an idea in 2008 when the three Thorkelsson brothers discussed the possibility of producing whisky in Iceland. In 2011 a company was formed and the first distillation was made in the ensuing year. It took 163 different trial runs before the owners could decide on a recipe to go forward with and full scale production started in August 2013. Only organic barley grown in tIceland is used for the production and everything is malted on site. The barley variety is low in sugars and it takes 50% more barley than usual to produce a bottle of whisky. The distillery is equipped with an Arnold Holstein still and the annual capacity has recently been increased to 75,000 litres with a goal to reach 100,000 litres end of 2015. Fifty percent of the

Egill Thorkelsson - Master Distiller at Eimverk Distillery

capacity is reserved for gin and aquavite and the rest for whisky. The first, limited whisky release is expected to be in early 2016. Currently, the distillery is offering Flóki young malt, a spirit which has matured for one year in fresh oak. The owners also have Vor Gin and Víti Aquavit for sale. In summer 2014 the first smoky spirit was distilled, using both peat and sheep dung to dry the barley. Other experiments include using Icelandic birch for maturation. If succesful, the plan for the future is to launch a variety with 3 years maturation in oak and a further year in casks made of birch.

Republic of Ireland

Distillery:	Midleton Distillery, Midleton, Co. Cork
Founded:	1975
Owner:	Irish Distillers (Pernod Ricard)
	irishdistillers.ie

Midleton is by far the biggest distillery in Ireland and the home of Jameson´s Irish Whiskey. The distillery that we see today is barely 40 years old, but Jameson´s as a brand dates back much further. John Jameson, the founder, moved from Scotland to Ireland in 1777 and became part-owner in a distillery called Bow Street Distillery in Dublin. Some years later he became the sole owner and renamed the company John Jameson & Son.

In 1966, John Jameson & Son with their distillery in Bow Street, merged with John Power & Son,as well as Cork Distillery Company to form Irish Distillers Group. It was decided that the production of the three companies should move to Midleton Distillery in Cork. The result was that the Bow Street Distillery was closed in 1971. Four years later an ultra modern distillery was built next to the old Midleton distillery and this is what we can see today, while the old distillery has been refurbished asa visitor attraction. The production at Midleton comprises of two sections – grain whiskey and single pot still whiskey. The grain whiskey is needed for the blends, where Jameson´s is the biggest seller. Single pot still whiskey, on the other hand, is unique to Ireland. This part of the production is also used for the blends but is being bottled more and more on its own.

Midleton distillery is equipped with mash tuns both for the barley side and the grain side. Until recently, there were 14 washbacks for grain and 10 for barley, four, large copper pot stills and 5 column stills. The hugely increased demand for Irish whiskey, and for Jameson's in particular, has now forced the owners to greatly expand their capacitiy. The expansion, which was completed in autumn 2013, included a completely new brew house, another 24 washbacks, a new still house with three more pot stills (80,000 litres each) and six, new, larger columns replacing the existing ones. A new maturation facility with 40 warehouses has also been built in Dungourney, not far from Midleton. The investment for the whole expansion was €200m and the capacity of the distillery is now 64 million litres of pure alcohol. The sales increase over the past few years has also forced the owners to invest €17m to increase the bottling plant in Clondalkin. By 2017 they will be able to bottle 120 million bottles per year.

Of all the brands produced at Midleton, Jameson´s blended Irish whiskey is by far the biggest. In 2014 the brand sold 56 million bottles of which more than 40% was sold to the USA. Apart from the core expression with no age statement, there are 12 and 18 year olds, Black Barrel, Gold Reserve and a Vintage. Other blended whiskey brands include Paddy, Powers, the exclusive Midleton Very Rare and Tullamore Dew. The latter is produced for the brand owners, William Grants, who are planning to move that production to a distillery of its own by 2014. In recent years, Midleton has invested increasingly in their second category of whiskies, single pot still, and that range now includes Redbreast (12, 12 cask strength, 15, 21 year old and the new, sherrymatured Mano a Lámh), Green Spot (12 year old and the new 12 year old Leoville Barton bordeaux finish), Yellow Spot, Powers John´s Lane and Barry Crocket Legacy. In spring of 2015, the unique single pot still Midleton Dair Ghaelach was launched. It is the first ever Irish whiskey to be finished in virgin Irish oak. It has been matured for between 15 and 22 years in bourbon casks before the finish in Irish oak. Ten different version reflecting ten different trees were released.

Distillery:	Tullamore Dew Distillery, Clonminch, Co. Offaly
Founded:	2014
Owner:	Wm Grant & Sons
	tullamoredew.com

Until 1954, Tullamore Dew was distilled at Daly´s Distillery in Tullamore. When it closed, production was temporarily moved to Power´s Distillery in Dublin, and was later moved to Midleton Distillery, where it currently is still produced. William Grant & Sons acquired Tullamore Dew in 2010 and in March 2012, they announced that they were in the final stages of negotiations to

Midleton´s new still house

acquire a site at Clonminch, situated on the outskirts of Tullamore. Construction of a new distillery began in May 2013 and in autumn 2014, the distillery was ready to start production. The construction will be executed in three phases where the first is a pot still distillery with the possibility to distil 1,8 million litres per year. This will be increased in 2019 to 3,6 million litres and, finally, in 2021 a grain distillation unit will be added which will have the capacity of doing 8 million litres of grain spirit. The total cost for the new plant will be €35m.

Tullamore Dew is the second biggest selling Irish whiskey in the world after Jameson with 10 million bottles sold in 2014. The core range consists of Original (without age statement), 12 year old Special Reserve, 10 year old Single Malt and two limited releases, Old Bonded Warehouse and Phoenix (an Oloroso finish bottled at 55%). As an exclusive to duty free, the Tullamore Dew Cider Cask Finish was launched in summer 2015. The whiskey has been finished for three months in casks that have previously been used to produce apple cider.

Distillery: Cooley Distillery, Cooley, Co. Louth
Founded: 1987
Owner: Beam Suntory
kilbeggandistillingcompany.com

In 1987, the entrepreneur John Teeling bought the dis-used Ceimici Teo distillery and renamed it Cooley distillery. Two years later he installed two pot stills and in 1992 he released the first single malt from the distillery, called Locke´s Single Malt. Due to financial difficulties, the distillery was forced to close down but was re-opened in 1995. A number of brands were launched over the years and Teeling got several offers from companies wanting to buy Cooley. Finally, in December 2011 it was announced that Beam Inc. had acquired the distillery for $95m. The most important part of the takeover for Beam's was the Kilbeggan blended whiskey brand that had already reached some level of success on the American market. In spring 2014, Suntory took over Beam in a $16bn deal and the new company was renamed Beam Suntory.

Cooley distillery is equipped with one mash tun, four malt and six grain washbacks all made of stainless steel, two copper pot stills and two column stills. There is a production capacity of 650,000 litres of malt spirit and 2,6 million litres of grain spirit.

The Cooley range of whiskies consists of several brands. Connemara single malts, which are all more or less peated, consist of a no age, a 12 year old, a cask strength, sherry finish and the heavily peated Turf Mor. In autumn 2014, Connemara 22 years old matured in first fill bourbon barrels was released. The other brand is Tyrconnel bottled without age statement. Other Tyrconnel expressions which have now been discontinued included a 15 year old single cask and several wood finishes. The distillery also produces the single grain Greenore where there currently are three expressions – 6 year old (created for the Swedish market), 8 year old and 18 year old. In spring 2015, the owners re-branded Greenore and it now goes under the name Kilbeggan single grain.

Other distilleries in Ireland

Teeling Distillery

Newmarket, Dublin, Co. Dublin, founded in 2015

teelingwhiskey.com

After the Teeling family had sold Cooley and Kilbeggan distilleries to Beam in 2011, the family started a new company, Teeling Whiskey. A wide range of whiskeys have been released since then, all made from stock made at Cooley which the family kept while selling the distilleries. However, all the while, they have planned for a new distillery of their own. John Teeling, the father of the family, is working on his own project, turning Great Northern Brewery in Dundalk into a distillery. His sons, Jack and Stephen, beat him to it though, when they opened their own distillery in Newmarket, Dublin in June 2015. This was the first new distillery in Dublin in 125 years.

Planning permission was granted in early 2014 and, thereafter, things moved quickly. Construction work commenced during the summer, the stills arrived in the autumn and on 30th March 2015, the first distillation was made. The distillery is equipped with two wooden washbacks, four made of stainless steel and three stills made in Sienna, Italy; wash still (15,000 litres), intermediate still (10,000 litres) and spirit still (9,000 litres). The owners plan to distil around 250,000 litres in the first year, while the actual capacity is 500,000 litres of alcohol. Both pot still and malt whisky will be produced and the total investment is 10 million euros. With the distillery in Dublin, the Teeling family are coming back to their roots. The family´s involvement in the whiskey industry started in 1782 when Walter Teeling owned a distillery on Marrowbone Lane in Dublin. Another reason for choosing Dublin is the easy access for visitors to come and tour the distillery.

Teeling´s Distillery in Newmarket, Dublin - the latest addition to the Irish whisky scene

Kilbeggan Distillery
Kilbeggan, Co. Westmeath, founded in 1757
www.kilbegganwhiskey.com

It was the owners of Cooley distillery with John Teeling at the forefront, who decided in 2007 to bring this distillery back to life and it is now the oldest producing whiskey distillery in the world. In 2011, Cooley Distillery was taken over by Beam Inc. and in spring 2014, Suntory bought Beam, which means that the current owners are the newly formed Beam Suntory. The distillery is equipped with a wooden mash tun, four Oregon pine washbacks and two stills with one of them being 180 years old. The production at the distillery spans over a wide range of techniques and whiskey varieties including malt whiskey, pure pot still whiskey (malted and unmalted barley mixed), rye whiskey from pot stills and triple distillation. The first single malt whiskey release (the 3 year old Kilbeggan Distillery Reserve) from the new production came in June 2010 and limited batches have been released thereafter. The core blended expression of Kilbeggan is a no age statement bottling but limited releases of aged Kilbeggan blend (15 and 18 year old) have occurred. Recently, a 21 year old bottling wa also released. To confuse matters, since spring 2015 there is also a Kilbeggan single grain, produced at Cooley, which used to be called Greenore.

West Cork Distillers
Skibbereen, Co. Cork, founded in 2004
www.westcorkdistillers.com

Started by John O´Connell, Denis McCarthy and Ger McCarthy in 2003 in Union Hall in West Cork, the distillery moved in 2013 to the present site. The distillery is equipped with four stills with the two wash stills coming from Sweden and the two spirit stills having been manufactured in Germany. The distillery is producing both malt whiskey and grain whiskey (from barley and wheat) and some of the malting is down on site. Apart from a range of vodka, gin and liqueurs, a 10 year old single malt and a blended whiskey are sold under the name West Cork. The latest release from 2015 is The Pogues Irish Whiskey developed together with Barry Walsh and Springbank´s former production manager Frank McHardy and launched in collaboration with the Irish band The Pogues.

The Dingle Whiskey Distillery
Milltown, Dingle, Co. Kerry, founded in 2012
www.dingledistillery.ie

Permission to build a distillery in Dingle, County Kerry, was applied for in 2008 and was granted in 2009. The people whose brainchild it was, Oliver Hughes, founder of the Porterhouse group of pubs and Jerry O´Sullivan, managing director of Southbound Properties, planned to convert an old creamery into a distillery. In 2010, the former business partners went their separate ways and Oliver Hughes found a new location, the old Fitzgerald sawmills and set about transforming the site into a distillery. Three pot stills and a combined gin/vodka still were installed in June 2012 and the first production of gin and vodka was in October. Whisky distillation began in December and the plan is to produce 100,000 bottles per year. The first products that were launched were Dingle Original Gin and DD Vodka. The triple distilled whisky under the brand names, Dingle Green and Dingle Gold, will not be available until 2016. However, in December 2014, a collection of four bottles called An Cliabhán was released. One of the four 25 cl bottles contained new make spirit while the other three held 2 year old malt sprit matured in bourbon, oloroso sherry and port casks.

Alltech/Carlow Distillery
Bagenalstown, Co. Carlow, founded in 2012
www.carlowbrewing.com

This is a joint venture between the American company, Alltech, and the Irish Carlow Brewing Company. Alltech is an American biotech company which also has a brewing and distilling business in Kentucky. In 2012, the owner, Dr. Pearse Lyons, of Irish descent, initiated a co-operation with the Irish brewer, Carlow, where two copper stills were shipped to Ireland and set up in the Carlow Craft Brewing House. The wash still has a capacity of 1,900 litres and the spirit still 1,000 litres. Both have columns attached and distilling began in November 2012.

In spring 2014, Alltech bought an abandoned church in St. James Street in Dublin and has since applied for a planning permission from the Dublin City Council to transform the church into a boutique distillery. If everything goes according to plans, the stills from Carlow will be moved to the Dublin site in 2016 and equipment for mashing and fermentation will be added.

The Shed Distillery
Drumshanbo, Co. Leitrim, founded in 2015
www.thesheddistillery.com

One of the newest distilleries in Ireland, The Shed Distillery was founded by entrepreneur Pat Rigney who is a veteran in the drinks business, having worked with brands like Bailey´s, Gilbey´s and Grant´s. The distillery cost €2m to build and is equipped with three Holstein stills with columns attached. The focus for the owners will be single pot still whiskey but the product range also includes potato vodka, gin and liqueurs. The first whiskey distillation was in January 2015.

Italy

Distillery:	Puni Destillerie, Glurns, South Tyrol
Founded:	2012
Owner:	Ebensperger family
	puni.com

In a country which embraced whisky in the early 1950s, it seemed a bit odd that there was no domestic production of the spirit. The lack of a whisky distillery in Italy was rectified in February 2012 when the first spirit was distilled at Puni distillery, situated in South Tyrol in the north of Italy. It is owned and run by the Ebensperger family with Albrecht, the father, and one of his sons, Jonas, as the dominant figures. There are at least two things that distinguish this project from most others. One characteristic is the design of the distillery – a 13-metre tall cube made of red brick. The other is the raw material

Jonas Ebensperger with the new Puni Nova

that they are using. They are making malt whisky but malted barley is only one of three cereals in the recipe. The other two are malted rye and malted wheat. The family calls it Triple Malt and it is their intention to use this combination of cereals for their main line of whiskies. Potential single malts or even single rye whiskies will, in the future, just be released as limited expressions.

The distillery is equipped with five washbacks made of local larch and the fermentation time is 84 hours. There is one wash still (3,000 litres) and one spirit still (2,000 litres) and the capacity is 80,000 litres of alcohol per year. For maturation, mainly ex-Bourbon barrels and ex-Marsala casks are being used. Some of the casks are maturing in old bunkers from the Second World War.

In anticipation of the first whisky, new make and aged spirit has been released on several occasions since 2013. The very latest addition (September 2014) was Opus I, a vatting of spirit that had matured for two years in a Marsala cask and one year in an Islay cask. Finally, in October 2015, it was time for the first release of, not one but two, Italian single malt whisky – both with natural colour and un chill-filtered. Puni Nova was matured in American white oak for three years while Puni Alba had been matured for three years in Marsala casks and then finished in Islay casks. Initially, the whiskies will be available in Italy, Germany and France but the company already has distribution requests from more than 20 countries.

Liechtenstein

Distillery: Brennerei Telser, Triesen
Founded: 1880 (whisky since 2006)
Owner: Telser family
brennerei-telser.com

The first distillery in Liechtenstein to produce whisky is not a new distillery. It has existed since 1880 and is now run by the fourth generation of the family. Traditions are strong and Telser is probably the only distillery in Europe still using a wood fire to heat the small stills (150 and 120 litres). Like so many other distilleries on mainland Europe, Telser produces mainly spirits from fruits and berries, including grappa and vodka. For whisky, the distillery uses a mixture of three different malts (some peated) which are fermented and distilled separately. After an extremely long fermentation (lasting 10 days), the spirit is triple distilled and the three different spirits are blended and filled into pinot noir barriques and left to mature for a minimum of three years. The first bottling of Telsington was distilled in 2006 and released in 2009. Since then, another five releases have been made, the latest (the 5 year old Telsington VI) in 2013 together with the very limited Black Edition. In 2014, the name was changed from Telsington to Liechtenstein Whisky and edition VII, a 5 year old, matured in a pinot noir cask was released. This was followed in autumn 2015 by VIII, matured for 6 years in pinot noir cask and finished for one year in a cask that had previously contained Scotch whisky. In 2014, the distillery´s first rye, a 2 year old made from 100% malted rye, was released and it was followed by another with a Laphroaig cask finish.

The Netherlands

Distillery: Zuidam Distillers, Baarle Nassau
Founded: 1974 (whisky since 1998)
Owner: Zuidam family
zuidam.eu

Zuidam Distillers was started in 1974 as a traditional, family distillery producing liqueurs, genever, gin and vodka. The first release of a whisky, which goes by the name Millstone, was from the 2002 production and it was bottled in 2007 as a 5 year old. The current range is a 5 year old which comes in both peated and unpeated versions, American oak 10 years, French oak 10 years, Sherry oak 12 years and PX Cask 1999, a 14 year old bottled at 46%. Apart from single malts there is also a Millstone 100% Rye which is bottled at 50%. New releases for 2014 were a 6 year old, triple distilled three grain whisky (equal parts of corn, rye and

Zuidam Distillery and their Millstone Peated Whisky

malted barley) and a 3 year old, triple distilled five grain whisky (wheat, corn, rye, spelt and malted barley).

The distillery has been expanded at an impressive pace during the last years and the equipment now consists of one mash tun for malt whisky, one for rye and genever and 10 washbacks (three of them installed in 2015). Furthermore, there are a total of five stills with volumes ranging from 850 litres up to 5,000 litres. The total capacity is 280,000 litres of pure alcohol per year. But the expansion doesn´t stop at that. Patrick van Zuidam has plans to build a second distillery at a farm where they will be growing their own barley and rye as well. The distillery will be equipped with four pot stills (5,000 litres each) with the possibility of adding another six. If everything goes according to plan, the distillery could start producing sometime during 2016.

Other distilleries in The Netherlands

Us Heit Distillery

Bolsward, founded in 2002

www.usheitdistillery.nl

This is one of many examples where a beer brewery also contains a whisky distillery. Frysk Hynder was the first Dutch whisky and made its debut in 2005 at 3 years of age. The barley is grown in surrounding Friesland and malted at the distillery. Some 10,000 bottles are produced annually and the whisky is matured in various casks – sherry, bourbon, red wine, port and cognac. The current core expression is a 3 year old matured in red wine casks and there is also a 5 year old matured in cognac casks.

Kalkwijck Distillers

Vroomshoop, founded in 2009

www.kalkwijckdistillers.nl

In an old building, dating back to the early 19th century, Lisanne Benus and her father Bert opened their distillery in 2009. Kalkwijck is located in Vroomshoop, in the rural eastern part of the Netherlands. Lisanne, only 25, is one of the youngest whisky producers in the world and was trained by, amongst others, Patrick van Zuidam, the maker of Millstone whisky. The distillery is equipped with a 300 litre pot still still with a column attached. The main part of the production is jenever, korenwijn and liqueurs but whisky has been distilled since 2010. In spring 2015, the first single malt was released. Eastmoor is 3 years old, made from barley grown on the estate and bottled at 40%. The first release was only 200 bottles but a second edition was launched already in May.

Northern Ireland

Distillery:	Bushmill´s Distillery, Bushmills, Co. Antrim
Founded:	1784
Owner:	Casa Cuervo
	bushmills.com

Diageo, the owners of Bushmill´s since 2005, took the market by surprise when they announced in November 2014 that they were selling the distillery. This is Diageo´s enda närvaro in the vibrant, increasing Irish whiskey segment and commentators struggled to see the reason for the sale. The buyer was the tequila maker Casa Cuervo, producer of the well-known José Cuervo. Diageo already owned 50% of the company´s other, upscale tequila brand, Don Julio and with the deal, they got the remaining 50% as well as $408m.

Bushmills is the second biggest of the Irish distilleries after Midleton, with a capacity to produce 4,5 million litres of alcohol a year. In 1972 the distillery became a part of Irish Distillers Group which thereby gained control over the entire whiskey production in Ireland. Irish Distillers were later (1988) purchased by Pernod Ricard who, in turn, resold Bushmill´s to Diageo in 2005 at a price tag of €295.5 million. Since the take-over, Diageo invested heavily into the distillery and it now has ten stills with a production running seven days a week, which means 4,5 million litres a year. Two kinds of malt are used, one unpeated and one slightly peated.

Bushmills Distillery

The distillery uses triple distillation, which is the traditional Irish method.

Bushmill`s core range of single malts consists of a 10 year old, a 16 year old Triple Wood with a finish in Port pipes for 6-9 months and a 21 year old finished in Madeira casks for two years. There is also a 12 year old Distillery Reserve which is sold exclusively at the distillery and the 1608 Anniversary Edition. Black Bush and Bushmill´s Original are the two main blended whiskeys in the range. Like so many other whisky producers, Bushmill´s jumped on the bandwagon in 2012 when they introduced their first flavoured version of the brand – Bushmill´s Irish Honey.

With 9 million bottles sold in 2014, Bushmill´s is the third most sold Irish whiskey after Jameson and Tullamore Dew.

Other distilleries in Northern Ireland

Echlinville Distillery

Kircubbin, Co. Down, founded in 2013

www.echlinville.com

Shane Braniff, who launched the Feckin Irish Whiskey brand in 2005, has had plans for a distillery of his own for some time now. Feckin Irish Whiskey, which has become popular in USA, has so far been produced at Cooley Distillery but when Beam Inc. bought the distillery in 2012, the new owners decided not to supply whiskey to independent bottlers. Shane had already started the building of the new distillery located near Kircubbin on the Ards Peninsula and some of the equipment was already in place when they were granted a license by Customs and Excise in May 2013. Only a few weeks later, the first spirit was distilled. There are already plans to install two more stills later and to open a visitor´s centre in 2015. Apart from Feckin Irish Whiskey, Braniff has also recently revived the old Dunville´s brand of blended whiskey and released a 10 year old with a finish in PX sherry casks.

Norway

Distillery:	Det Norske Brenneri, Grimstad
Founded:	1952 (whisky since 2009)
Owner:	Norske Brenneri AS
	detnorskebrenneri.no

The company was founded in 1952 by Karl Gustav Puntervold and for more than 50 years it mainly produced wine from apples and other fruits. The company was taken over by Karl Gustav´s son, Ole, in 1977. In July 2005 the state monopoly in terms of production of spirits in Norway was abolished and Ole decided to take advantage of that. He started to produce aquavit, among other products, and the first products were launched during autumn 2005. Whisky production started in 2009 and a Holstein still from Germany is used for the distillation. In November 2012 the first single malt produced in Norway was released (1,750 bottles). It was a single ex-sherry butt called Audny and in spring 2013 another 1,750 bottles were released. The third release of Audny appeared in May 2015 and at the same time the company announced a September release of the new 3,5 year old, oloroso matured Eiktyrne. In January 2014, K.G. Puntervold and Det Norske Brenneri (known as Agder Brenneri at the time) were sold to the companies´ biggest customer – Norske Brenneri AS.

Other distilleries in Noway

Myken Distillery

Myken, founded in 2014

www.facebook.com/mykdest/info

The latest whisky distillery to open in Norway, was built in the most unlikely place one can imagine. Myken is a group of islands in the Atlantic ocean, 32 kilometres from mainland Norway and 25 kilometres north of the Arctic Circle. The largest island has 20 people living the year round and this is where Myken distillery was opened in 2014. Founded by a group of friends (six women and six men), the distillery produced the first spirit in December 2014.

The Northern Lights as seen from Myken Distillery in Norway

The distillery is equipped with one wash still (1,000 litres), one spirit still (700 litres) and one gin still (300 litres) – all alambic style, made in Spain and direct fired using propane gas burners. The fermentation time is 60-140 hours and the capacity is 20,000 litres per year. During 2015, the owners expect to do 5,000 litres.

So far, only un-peated malt has been used but peated production is in the pipeline. The production water is desalinated sea water, supplied by a municipal facility and the distillery is blessed with cold Arctic water the year round for the cooling. Only gin has been released so far but there are plans to start releasing newmake and partially aged spirit in late 2015 or early 2016. The distillery is open for tours by prior appointment.

Arcus

Gjelleråsen, founded in 1996 (whisky since 2009)

www.arcus.no

Arcus is the biggest supplier and producer of wine and spirits in Norway with subsidiaries in Denmark, Finland and Sweden. They are also the largest aquavit producer in the world and are involved in cognac production in France. The first whisky produced by the distillery was released in 2013. Under the name Gjoleid, two whiskies made from malted barley and malted wheat were released – one matured in ex-bourbon American oak and the other in ex-oloroso American oak. Both whiskies are 3,5 years old.

Spain

Distillery:	Distilerio Molino del Arco, Segovia
Founded:	1959
Owner:	Distilerias y Crianza del Whisky (DYC)
	dyc.es

Spain's first whisky distillery is not a small artisan distillery like so many others on these pages. Established by Nicomedes Garcia Lopez in 1959 (with whisky distilling commencing three years later), this is a distillery with capacity for producing eight million litres of grain whisky and two million litres of malt whisky per year. In addition to that, vodka and rum are produced and there are in-house maltings which safeguard malted barley for the production. The distillery is equipped with six copper pot stills and there are 250,000 casks maturing on site. The blending and bottling plant has been relocated to the Anis Castellana plant at Valverde del Majano.

The big seller when it comes to whiskies is a blend simply called DYC which is around 4 years old. It is supplemented by an 8 year old blend and, since 2007, also by DYC Pure Malt, i. e. a vatted malt consisting of malt from the distillery and from selected Scottish distilleries. A brand new expression was also launched in 2009 to commemorate the distillery's 50th anniversary – a 10 year old single malt, the first from the distillery. A recent extension of the range is DYC Red One which is a cherry infused whisky-based spirit which is bottled at 30%.

Other distilleries in Spain

Destilerias Liber

Padul, Granada, founded in 2001

www.destileriasliber.com

This distillery is quite a bit younger than its competitor in Segovia, DYC. Destilerias Liber was founded in 2001 but did not start production until late 2002. Like so many other, newly established distilleries, they started distilling rum, marc and vodka – spirits that do not require maturation and can also instantaneously generate cash to the company. For the whisky production, the spirit is double distilled after a fermentation of 48-72 hours. Maturation takes place in sherry casks. The only available whisky on the market is a 5 year old single malt called Embrujo de Granada.

Sweden

Distillery:	Mackmyra Svensk Whisky, Valbo
Founded:	1999
Owner:	Mackmyra Svensk Whisky AB
	mackmyra.se

Mackmyra's first distillery was built in 1999 and, ten years later, the company revealed plans to build a brand new facility in Gävle, a few miles from the present distillery at Mackmyra. In 2012, the distillery was ready and the first distillation took place in spring of that year. The total investment, which included a whisky village to be built within a ten year period, was expected to amount to approximately £50 million. The construction of the new distillery is quite extraordinary and with its 37 metre structure, it is perhaps one of the tallest distilleries in the world. The reason for its exceptional height is that it is a gravity fed distillery with malt and water coming in at the top of the building and then the entire production process "works itself" downwards to reach the stills at the bottom. Since April 2013, all the distillation takes place in the new gravitation distillery, while the old distillery will be used for special runs and marketing activities.

In February 2014, the company ran into financial troubles and 15 persons were laid off. Part of the explanation for the problems was the substantial investment in the new distillery, as well as slow sales in USA, Canada and Western Europe. It was therefore decided that during the interim, the company would withdraw from these markets to concentrate exclusively on Scandinavia and northern Germany. The production level for 2014 was also lowered to 100,000 bottles compared to 750,000 during the previous year. In March 2014 a rights issue brought the company a windfall of £2,2m and one more round in June 2015, generated another £900,000.

Mackmyra whisky is based on two basic recipes, one resulting in a fruity and elegant whisky, the other being more peaty. The peatiness does not stem from peat, but from juniper wood and bog moss. The whisky is matured in a variety of casks, which to some extent include Swedish oak. The first release in 2006/2007 was a series of six called Preludium. The first "real" launch was in June 2008 – 'Den Första Utgåvan' (The First Edition) and this is still

Sommartid - the latest limited release from Mackmyra

part of the core range although renamed Svensk Ek. The other two expressions in the core range are Brukswhisky, launched in 2010 and matured in first fill bourbon casks, with an addition of whisky matured in sherry casks and Swedish oak and Svensk Rök. The latter is the first peated release from the distillery and was introduced in September 2013. A range of limited editions called Moment was introduced in December 2010 and consists of exceptional casks selected by the Master Blender Angela D'Orazio. The latest was the 10 year old Vintage from February 2015, an un-peated version with a combined maturation in wine casks from Ukraine, tokaji casks and American oak. In late 2013, a limited, seasonal expression called Midvinter (mid winter) was released and this was followed in June 2014 by Midnattssol (midnight sun) and in early 2015 by Iskristall (ice crystal).

Distillery: Box Destilleri, Bjärtrå

Founded: 2010

Owner: Box Destilleri AB

boxwhisky.se

The company was founded in 2005 by Mats and Per de Vahl. Buildings from the 19th century that had previously been used both as a box factory (hence the distillery name) and a powerplant, were restructured and equipped with a four-roller Boby mill, a semilauter mash tun with a capacity of 1,5 tonnes and three stainless steel washbacks holding 8,000 litres each. The wash still (3,800 litres) and the spirit still (2,500 litres) were both ordered from Forsyth's in Scotland. The first distillation was made in November 2010 and the distillery has a capacity of 115,000 litres.

Box Destilleri is making two types of whisky – fruity/unpeated and peated. As regards the former, the malted barley comes from Sweden, whereas the peated malt is imported from Belgium where it has been dried using peat from Islay. The distillery manager, Roger Melander, wants to create a new make which is as clean as possible through a very slow distillation process with lots of copper contact in the still. The flavour of the spirit is also impacted by the effective condensation using what might be the coldest cooling water in the whisky world, 2 to 6°C, from a nearby river. A fermentation time of 72-96 hours also affects the character. A majority of the casks (80%), from 500 litres down to 40 litres, are first fill bourbon

but Oloroso casks, virgin oak and casks made from Hungarian oak have also been filled.

A variety of spirit bottlings under the name Försmak have been released and in autumn 2013 a box of five different spirits, Advanced Master Class, was released. The idea was to showcase differences in flavour owing to cask toasting levels. The first whisky, The Pioneer, was released 5th June 2014 and all 5,000 bottles were sold out in less than 7 hours! Between 3 and 4 years old, it was a vatting of unpeated and lightly peated whisky, predominantly from bourbon casks but also a small amount of ex-sherry. The whisky was the first in a range of four called Early Days Collection. The next release, Challenger (also 5,000 bottles), sold out in 50 minutes in December 2014 and was followed by Explorer (10,000 bottles) in May 2015 and Messenger six months later. Recent limited releases include Archipelago in January 2015 followed by the Box Festival bottling in September. In June 2014 a new visitor centre was opened and warehouse capacity has also been expanded during the year.

Other distilleries in Sweden

Smögen Whisky

Hunnebostrand, founded in 2010

www.smogenwhisky.se

In August 2010, Smögen Whisky on the west coast of Sweden, produced its first spirit and thus became the country's third whisky distillery, following Mackmyra and Spirit of Hven. Pär Caldenby – a lawyer, whisky enthusiast and the author of Enjoying Malt Whisky is behind it all. He has designed the facilities himself and much of the equipment is constructed locally. The distillery is equipped with three washbacks (1,600 litres each), a wash still (900 litres) and a spirit still (600 litres). The distillery has the capacity to produce 35,000 litres of alcohol a year. Pär practices a slow distillation with unusually long foreshots (45 minutes) in order not to get a newmake with too many fruity esters. The maturation takes place in casks made of new, toasted French Oak but some of them will also have held sherry. Ex-bourbon barrels made of American white oak and Sauternes cask are also used. The cask size ranges from 28 to 500 litres. Heavily peated malt is imported from Scotland and the vision is to produce an Islay-type of whisky.

The Explorer - the latest release from BOX Distillery

Pär Caldenby (left), the owner of Smögen Distillery

The first release (1,600 bottles) from the distillery was the 3 year old Primör in March 2014. It was a vatting of eight casks made of new, European oak and one cask that had previously held Bordeaux wine and it was bottled at 63,7%. Starting November 2014, a new, limited range called Sherry Project was launched. A number of casks, all filled in July 2011, will show the different flavour profiles deriving from maturation time and type of casks. The first release in late 2014 was Sherry Project 1:1 with 2 years and 10 months in heavily toasted European oak quarter casks followed by four months in a sherry butt. The next release in March 2015, 1:2, had been aged for another four months in the sherry butt and the final two versions will be launched during autumn 2015. Various single casks are continuously released, the latest, a Sauternes maturation, in August 2015. Pär Caldenby has also released his own gin, Strane, in three versions – Merchant Strength (47.4%), Navy Strength (57.1%) and Uncut Strength (75.3).

Spirit of Hven

Hven, founded in 2007

www.hven.com

The second Swedish distillery to come on stream, after Mackmyra, was Spirit of Hven, a distillery situated on the island of Hven right between Sweden and Denmark. The first distillation took place in May 2008. Henric Molin, founder and owner, is a trained chemist and very concerned about choosing the right oak for his casks. The oak is left to air dry for three to five years before the casks are loaned to, especially, wine producers in both the USA and Europe. The distillery is equipped with a 0,5 ton mash tun, six washbacks made of stainless steel and one pair of stills – wash still 2,000 litres and spirit still 1,500 litres. A long fermentation time of 90-120 hours is used in order to achieve a more full flavoured product with high citric notes and a nutty character. Part of the barley is malted on site and for part of it he uses Swedish peat, sometimes mixed with seaweed and sea-grass, for drying. Around 15,000 litres of whisky is distilled per year and other products include rum made from sugar beet, vodka, gin, aquavit and calvados. The distillery was expanded in summer 2013 with a new warehouse, an upgraded bottling line and an advanced spirits laboratory.

The first whisky to hit the market, was the lightly peated Urania which was released in 2012. The second launch, in early 2013, was the start of a new series of limited releases called The Seven Stars. The first expression was the 5 year old, lightly peated Dubhe, which was followed a year later by Merak and in 2015 by Phecda. In June 2014 the limited cask strength single cask Sankt Ibb was released and later that year saw the launch of Sankt Claus, also bottled at cask strength and matured in ex-merlot casks. All the above have been limited releases but in autumn 2015 it was time for the distillery´s first readily available bottling - Tycho´s Star, named in honour of the famous astronomer Tycho Brahe, who lived and worked on the island in the 16th century. The mash bill is a mix of pale ale malt, chocolate malt and heavily peated malt and the whisky is bottled at 41.8%.

Tycho´s Star
- the first core bottling from Spirit of Hven

Grythyttan Whisky

Lillkyrka, founded in 2010

www.grythyttanwhisky.se

The company was founded as a result of the initiative of Benny Borgh in 2007 and the distillery came on stream in October 2010. In common with most Swedish distilleries, the stills (900 litres wash still and 600 litres spirit still) were made at Forsyth´s in Scotland and the three washbacks are made of Oregon pine. The capacity of the distillery is 24,000 litres per year. The first whisky, a 3 year old unpeated from ex-sherry casks, was released in 2013.

Production stopped in November 2013 and a surprising announcement was made by the owners – the distillery would be moved to Orkney! All the current stock would remain in Sweden and the name of the new distillery would be The Longship Distillery. Neither a definite location nor a time schedule were presented and in spring 2014 it became apparent that the move would not take place. So far, the production at the Swedish distillery has not been resumed.

Norrtelje Brenneri

Norrtälje, founded in 2002 (whisky since 2009)

www.norrteljebrenneri.se

This distillery, situated 70 kilometres north of Stockholm, was founded on a farm which has belonged to the owner´s family for five generations. The production consists mainly of spirits from ecologically grown fruits and berries. Since 2009, a single malt whisky from ecologically grown barley is also produced. The whisky is double distilled in copper pot stills (400 and 150 litres respectively) from Christian Carl in Germany. Most of the production is matured in 250 litre Oloroso casks with a finish of 3-6 months in French oak casks which have previously held the distillery´s own apple spirit. The character of the whisky will be fruity and lightly peated (6ppm) and the first bottling was released in summer 2015.

Gammelstilla Whisky

Torsåker, founded in 2005

www.gammelstilla.se

Less than 30 kilometres from the better known Mackmyra lies another distillery since 2011 – Gammelstilla. The company was already founded in 2005 by three friends but today there are more than 200 shareholders. Unlike most of the other Swedish whisky distilleries, they chose to design and build their pot stills themselves. The wash still has a capacity of 600 litres and the spirit still 300 litres and the annual capacity is 20,000 litres per year. The mashing and fermentation takes place in a brewery which has moved into the same facilities. The first distillation took place in April 2012 with a plan to launch the first bottlings in late 2015 or 2016. However, there is a possibility to buy malt spirit aged for 16 months already now.

Gotland Whisky

Romakloster, founded in 2011

www.gotlandwhisky.se

The company behind this distillery was already founded in 2004 with the intention of building a distillery at a farmstead near Klinte. The plans were changed and it was decided to build the distillery at the present location instead, a decommissioned sugar works south of Visby. The distillery is equipped with a wash still (1,600 litres) and a spirit still (900 litres) – both made by Forsyth´s in Scotland. The local barley is ecologically grown and malted on site. The floor malting is made easier through the use of a malting robot of their own construction which turns the barley. The whisky is matured in a warehouse situated four metres underground. The goal is to produce two kinds of single malts, unpeated and peated and the capacity is 60,000 litres per year. The distillery came on stream in May 2012 and the plan is to release the first whisky, under the name Isle of Lime, in 2015.

Switzerland 🇨🇭

Distillery: Whisky Castle, Elfingen, Aargau
Founded: 2002
Owner: Ruedi Käser

whisky-castle.com

The first whisky from this distillery in Elfingen, founded by Ruedi Käser, reached the market in 2004. It was a single malt under the name Castle Hill. Since then the range of malt whiskies has been expanded and today include Castle Hill Doublewood (3 years old matured both in casks made of chestnut and oak), Whisky Smoke Barley (at least 3 years old matured in new oak), Fullmoon (matured in casks from Hungary) and Terroir (4 years old made from Swiss barley and matured in Swiss oak). All these are bottled at 43%. Adding to these are Cask Strength (5 years old and bottled at 58%) and Edition Käser (71% matured in new oak casks from Bordeaux).

A couple of years ago, Ruedi´s son, Raphael, was brought into the business with the express goal of releasing older whiskies and narrowing down the range of expressions. New, open top fermenters were installed in 2010 to add fresher and fruitier notes to the new-make. The owners have also cut down on the number of casks made from new oak and have added a variety of other casks to influence the spirit. The new style of whisky will be available in about two years´ time, but some trials may be released earlier. Another goal is to use only local, organically grown grain. The yearly production is between 8,000 and 10,000 litres and on the premises one can have a complete visitor's experience, which includes a restaurant as well as a shop.

Distillery: Brauerei Locher, Appenzell
Founded: 1886 (whisky since 1999)
Owner: Locher family

saentismalt.com

This old, family-owned brewery started to produce whisky on a small scale in 1999 when the Swiss government changed laws and allowed for spirit to be distilled from grain. From 2005, larger volumes have been produced. The whole production takes place in the brewery where there is a Steinecker mash tun holding 10,000 litres. The spirit ferments in stainless steel vats and, for distillation, Holstein stills are used. Brauerei Locher is unique in using old (70 to 100 years) beer casks for the maturation. The core range consists of three expressions; Säntis, bottled at 40%, Dreifaltigkeit which is slightly peated having matured in toasted casks and bottled at 52% and, finally, Sigel which has matured in very small casks and is bottled at 40%. A range of limited bottlings under the name Alpstein is also available. After a few years in a beer cask, these whiskies have received a further maturation in casks that previously held other spirits or wine. The most recent versions are Edition VIII, released in September 2014. followed by Edition IX (a port finish) in January 2015. Yet another limited range, Snow White, was introduced in winter 2013 where the whisky had received a second maturation in a cask that had been used for spirit made from apples. The second edition of Snow White (6 years old) in 2014, had been finished for one year in glühbier (a beer version of the famous mulled wine) casks. Finally, the brand has recently been launched in Taiwan with the special Edition Formosa.

Other distilleries in Switzerland

Langatun Distillery

Langenthal, Bern, founded in 2007

www.langatun.ch

The distillery was built in 2005 and under the same roof as the brewery Brau AG Langenthal (which was already established in 2001). The reason for this co-habitation was to access a wash for distillation and thereby avoiding investments in mashing equipment. The casks used for maturation are all 225 litres and Swiss oak (Chardonnay), French oak (Chardonnay and red wine) and ex sherry casks are used. The first whisky was produced in 2007 by way of the 3 year old single malt, Olde Deer. In 2011 the whisky underwent a name change to Old Deer and a peated version was launched under the name Old Bear. Both of them are now bottled as 5 year olds and are available at 40% and 64% respectively. Among the latest releases are a single cask rye, Old Eagle, and a single cask "bourbon", Old Mustang.

Bauernhofbrennerei Lüthy

Muhen, Aargau, founded in 1997 (whisky since 2005)

www.swiss-single-malt.ch

This farm distillery started in 1997 by producing distillates from fruit, as well as grappa, absinthe and schnapps. The range was expanded to include whisky in 2005 which was distilled in a mobile pot still distillery. Lüthy´s ambition is to only use grain from Switzerland in his production. Since it was impossible to obtain peated malt from Swiss barley, he decided to build his own floor maltings in 2009. The first single malt expression to be launched in 2008, was Insel-Whisky, matured in a Chardonnay cask. It was followed by Wyna-Whisky from a sherry cask and Lenzburg-Whisky, another Chardonnay maturation in 2009. Starting in 2010, the yearly bottling was given the name Herr Lüthy and the 9th release from these was in 2014, a chardonnay maturation.

Brennerei Stadelmann

Altbüron, Luzern, founded in 1932 (whisky since 2003)

www.schnapsbrennen.ch

Established in the 1930s this distillery was mobile for its first 70 years. Hans Stadelmann took over in 1972 and in 2001 decided to build a stationary distillery. The distillery was equipped with three Holstein-type stills (150-250 litres) and the first whisky distilled for a local whisky club in 2003. In 2005 the first Luzerner Hinterländer Single Malt was released, although not as a whisky since it was just 1 year old. A year later the first 3 year old was bottled for the whisky club under the name Dorfbachwasser and finally, in 2010, the first official bottling from the distillery in the shape of a 3 year old single malt whisky was released. In autumn

Säntis Malt from Brauerei Locher

2014, the sixth release was made, matured in a Bordeaux cask. The first whisky from smoked barley was distilled in 2012.

Etter Distillerie

Zug, founded in 1870 (whisky since 2007)

www.etter-distillerie.ch

This distillery was started in 1870 by Paul Etter and has been in the family ever since. Today it is the third and fourth generations who are running it. Their main produce is eau de vie from various fruits and berries. A sidetrack to the business was entered in 2007 when they decided to distil their first malt whisky. The malted barley was bought from a brewery (Brauerei Baar), distilled at Etter, filled into wine casks and left to mature in moist caves for a minimum of three years. The first release was made in 2010 under the name Johnett Single Malt Whisky. In 2015, the limited Johnett Trinidad rum finish was released.

Spezialitätenbrennerei Zürcher

Port, Bern, founded in 1954 (whisky from 2000)

www.lakeland-whisky.ch

The first in the Zürcher family to distil whisky was Heinz Zürcher in 2000, who released the first 1,000 bottles of Lakeland single malt in 2003. Daniel and Ursula Zürcher took over in 2004. They continued their uncle's work with whisky and launched a second release in 2006. The main focus of the distillery is specialising in various distillates of fruit, absinth and liqueur but a Lakeland single malt is also in the range. The oldest version so far, appeared in August 2014. It was an 8 year old matured in an oloroso cask.

Whisky Brennerei Hollen

Lauwil, Baselland, founded in 1999

www.swiss-whisky.ch, www.single-malt.ch

The first Swiss whisky was distilled at Hollen 1st July 1999. The whisky is stored on French oak casks, which have been used for white wine (Chardonnay) or red wine (Pinot Noir). Most bottlings are 4 years old and contain 42% alcohol. A 5 year old has also been released, which has had three years in Pinot Noir casks followed by two years in Chardonnay casks. Other expressions include a peated version and a cask strength Chardonnay-matured. In 2009 the first 10 year old was released and there has also been a 12 year old double wood, the oldest expression from the distillery so far.

Brennerei Hagen

Hüttwilen, Thurgau, founded in 1999

www.distillerie-hagen.ch

A triple distilled malt whisky is produced by Ueli Hagen in the small village of Hüttwilen in the northernmost part of Switzerland. The spirit is matured in bourbon barrels and the first produce was sold in 2002 as a 3 year old. Ueli Hagen produces mainly schnapps and absinth and distills around 300 bottles of malt whisky a year.

Wales

Distillery:	Penderyn Distillery, Penderyn
Founded:	2000
Owner:	Welsh Whisky Company Ltd.
	welsh-whisky.co.uk

In 1998 four private individuals started The Welsh Whisky Company and two years later, the first Welsh distillery in more than a hundred years started distilling. A new type of still, developed by David Faraday for Penderyn Distillery, differs from the Scottish and Irish procedures in that the whole process from wash to new make takes place in one single still. But that is not the sole difference. Every distillery in Scotland is required by law, to do the mashing and fermenting on site. At Penderyn, though, the wash (until summer 2014) was bought from a regional beer brewer and transported to the distillery on a weekly basis. Even though the distillery has been working 24 hours a day to keep up with the increasing demand, it became obvious in 2012 that they had to do something to increase its capacity. In September 2013, a second still (almost a replica of the first still) was commissioned and in June 2014, two traditional pot stills, as well as their own mashing equipment was installed. The expansion, worth £1m, increased the production from 90,000 litres to 300,000 litres of alcohol per annum and will also allow the company to experiment with new styles and expressions of single malts.

The first single malt was launched in 2004. The core range consists of Penderyn Madeira Finish (46%), Penderyn Peated (46%), Penderyn Legend (madeira finish bottled at 41% and formerly known as Penderyn 41), Penderyn Myth (full bourbon maturation and bottled at 41%) and Penderyn Celt (a peated finish launched in August 2015 and bottled at 41%) A new range of whiskies called Icons of Wales was introduced in 2012 and the fourth edition, That Try, was released in June 2015. The bottling is celebrating the famous rugby match between Barbarians and All Blacks in 1973. The main markets for Penderyn are UK, France and Germany and the owners hope to sell 180,000 bottles in 2015. A visitor centre opened in 2008 and attracted 35,000 visitors last year.

Penderyn That Try - the fourth edition of Icons of Wales

North America

USA

Distillery: Westland Distillery, Seattle, Washington
Founded: 2011
Owner: Lamb family
westlanddistillery.com

Unlike most of the new craft distilleries in the USA producing whiskey, Westland Distillery did not distil other spirits to finance the early stages of production. Until November 2012, Westland was a medium sized craft distillery where they brought in the wash from a nearby brewery and had the capacity of doing 60,000 litres of whiskey per year. During the summer of 2013 the owners, the Lamb family, moved to another location which is equipped with a 6,000 litre brewhouse, five 10,000 litre fermenters and two Vendome stills (7,560 and 5,670 litres respectively). The capacity is now 260,000 litres per year and currently they are making 170,000 litres. The malt for the production is sourced both locally, as well as from England, and the casks are predominantly heavy char, new American oak. Trials are also being conducted with ex-bourbon, ex-sherry and ex-port casks.

The first 5,500 bottles of their core expression, Westland American Single Malt Whiskey, were released in autumn 2013 followed by a limited one-off release called Deacon Seat. Both were mashed with a 5-malt grain bill and matured in heavily charred American oak. Since then, the owners have released a core range which now consists of American Single Malt Whiskey (October 2013), Peated Malt (October 2014) and Sherry Wood (December 2014). In addition to this, the distillery also has a single cask program with special, limited releases. In 2014, Westland partnered with Anchor Distilling Company to distribute their whiskey nationally. It is now available in all 50 states and there are also plans to launch it overseas.

Distillery: Stranahans Whiskey Distillery, Denver, Colorado
Founded: 2003
Owner: Proximo Spirits
stranahans.com

Stranahans, founded by Jess Graber and George Stranahan, was bought by New York based Proximo Spirits (makers of Hangar 1 Vodka and Kraken Rum among others) in 2010. A surprising decision was soon made to withdraw Stranahans Colorado Whiskey from all other markets, but Colorado. The owners claim that they want to build up a significant stock before delivering nationally (and also internationally) again.

Stranahans Colorado Whiskey is always made in batches aged from two to five years. Except for the core expression, a single barrel is launched once a year under the name Snowflake and the 17th edition (Longs Peak), was released in June 2015. Since spring 2015, there is also another limited version of Stranahans called Diamond Peak which is a vatting of casks that are around 4 years old. In spring 2014, a bourbon by the name Tincup American Whiskey was released by Proximo Spirit,s with a lot of references made to Stranahans. It turned out that the whiskey had been distilled in Indiana and shipped to Stranahans where it was cut with local water and bottled.

Distillery: Balcones Distillery, Waco, Texas
Founded: 2008
Owner: Balcones Distilling Co.
balconesdistilling.com

The distillery was founded by Chip Tate but since end of 2014, he is no longer with the company. His exit was the result of a bitter feud between Tate and the company board, a feud which also included court hearings and restraining orders. A settlement was later made between the partis which means Tate is not allowed to make the same products as Balcones until March 2016. He has plans to open a new distillery ibn the near future where both whiskey and brandy will be produced.

All of Balcones´ whisky is mashed, fermented and distilled on site and they were the first to use Hopi blue corn for distillation. Four different expressions of blue corn whiskey have been released so far – Baby Blue, bottled at 46%, True Blue which is a cask strength version, Brimstone Smoked Whiskey, a smoky version and True Blue 100, a 100 proof bottling of True Blue. The biggest seller, however, is the Texas Single Malt Whisky. Like the other whiskies it is un chill-filtered and without colouring. To celebrate the distillery´s 5th anniversary, four single barrel releases were made in 2013 including a straight Bourbon, Brimstone Resurrection, a Straight Single Malt and a Straight Corn Whisky. The most recent releases include Fr.Oak Texas single malt with a maturation in French oak casks and a single barrel staff selection of the single malt. Balcones whisky is typically not aged for more than two years

Westland Distillery warehouse and the current range of single malts

but there are plans to also release older versions.

The demand for Balcones whiskies has grown rapidly and in January 2014, another four, small stills were installed. The big step though, will be a completely new distillery which is being built 5 blocks from the current site and the plan is to start production in January 2016. The expansion has been made possible thanks to a new group of investors and the capacity will initially be 120,000 litres of pure alcohol using two new, 7,600 litre stills from Forsyth´s in Scotland. Eventually, the plan is to increase the number of stills to 8.

Distillery: Clear Creek Distillery, Portland, Oregon
Founded: 1985
Owner: Hood River Distillers
clearcreekdistillery.com

Steve McCarthy was one of the first to produce malt whiskey in the USA and, like many other, smaller distilleries, they started by distilling eau-de-vie from fruit, especially pears, and then expanded the product line into whiskey. They began making whiskey in 1996 and the first bottles were on the market three years later. There is only one expression at the moment, McCarthy´s Oregon Single Malt 3 years old. The whiskey is reminiscent of an Islay and, in fact, the malt is purchased directly from Islay with a phenol specification of 30-40 ppm. Steve only bottles twice a year and the next release is scheduled for October 2015. Maturation takes place in ex-sherry butts with a finish in new Oregon White Oak hogsheads. In early 2014, it was announced that Hood River Distillers was to take over the distillery with Steve McCarthy continuing as a consultant. Hood River Distillers, also based in Oregon, began in 1934 as a distillery but, since the 1960s, they have acted as a bottler and importer of spirits from other producers. One of their biggest sellers is the Canadian blended whisky Pendleton.

Distillery: Charbay Winery & Distillery, St. Helena, California
Founded: 1983
Owner: Miles and Marko Karakasevic
charbay.com

Charbay was founded by Miles Karakasevic – a legend in American craft distilling – and the distillery is now run by his son Marko, the 13th generation in a winemaking and distilling family. With a wide range of products such as wine, vodka, grappa, pastis, rum and port, the owners decided in 1999 to also enter in to whiskey making. That year they took 20,000 gallons of Pilsner and double distilled it in their Charentais pot still. From this distillation, a 4 year old called Double-Barrel Release One (comprising of two barrels) was launched in 2002. There were 840 bottles at cask strength and non-chill filtered. The whiskey was quite unique since a ready beer, hops and all, rather than wash from a brewery was used. It took six years before Release II appeared in 2008 and in 2013 it was time for Release III. At 14 years old, it had spent the first six years in charred, new American oak. The reaming eight years it was allowed to mature in stainless steel tanks. This was followed by yet another release where the whiskey had matured for a full 14 years in the same barrel. Other recent releases include Charbay R5 Whiskey matured in French oak for 21 months and S Whiskey Lot 211A which Marko created by distilling 6,000 gallons of Big Bear stout into 590 gallons of whiskey. It was then matured for 29 months in used French oak.

Distillery: Edgefield Distillery, Troutdale, Oregon
Founded: 1998
Owner: Mike and Brian McMenamin
mcmenamins.com

Mike and Brian McMenamin started their first pub in Portland, Oregon in 1983. It has now expanded to a chain of more than 60 pubs and hotels in Oregon and Washington. More than 20 of the pubs have adjoining microbreweries (the first opened in 1985) and it is now the fourth-largest chain of brewpubs in the United States.

The chain's first distillery opened in 1998 at their huge Edgefield property in Troutdale and their first whiskey, Hogshead Whiskey (46%), was bottled in 2002. Hogshead is still their number one seller and the production has increased to 80 barrels per year. Starting with the releases in 2012, the character of the whiskey will be more complex and a touch sweeter. The reason for this is that Head Distiller, James Whelan, back in 2008 started to use three different malted barley recipes for the wash. He also began aging the spirit in both heavily and lightly charred barrels. Another part of the range is the Devil´s Bit, a limited bottling released every year on St. Patrick´s Day. For 2015 it was a five year old that had been matured in former Syrah port barrels. A second distillery was opened in 2011 at the company´s Cornelius Pass Roadhouse location in Hillsboro.

Balcones´ stills and Fr. Oak Single Malt - one of the latest releases

Distillery: High West Distillery, Park City, Utah
Founded: 2007
Owner: David Perkins
highwest.com

David Perkins has made a name for himself mainly because of the releases of several rye whiskies. None of these have been distilled at High West distillery. Perkins has instead, bought casks of mature whiskies and blended them himself. The first (released in 2008) was Rendezvous Rye, a mix of two whiskies (16 and 6 years old). Since then, he has also released a 16 year old and a 21 year old rye. In 2012, Bourye and Son of Bourye were released – both created by blending bourbon and rye whiskies. That year Campfire, a blend of straight bourbon, straight rye and peated blended malt Scotch whisky, also saw the light of day. The most recent releases are American Prairie bourbon and Yippe ki-yay Whiskey, a rye that has been aged in used vermouth and syrah barrels. Their first single malt from 100% barley has yet to be released. High West Distillery is currently involved in a project of impressive magnitude. A new distillery operating at Blue Sky Ranch in Wanship, Utah, was opened in spring 2015. It started off with one 6,000 litre pot still with the capacity of doing 350,000 litres of pure alcohol per year. The plan, however, is to eventually have 18 washbacks and four pot stills (with rectification columns attached) with the possibility of producing 1,4 million litres!

Distillery: Prichard´s Distillery, Kelso, Tennessee
Founded: 1999
Owner: Phil Prichard
prichardsdistillery.com

When Phil Prichard started his business in 1999, it became the first legal distillery for 50 years in Tennessee. Today, it is the third largest in the state after giants Jack Daniel's and George Dickel. In 2012 the capacity was tripled with the installation of a new 1,500 gallon mash cooker and three additional fermenters. The plan is to increase the capacity even further by adding a 1,500 gallon wash still and turning the old 550 gallon wash still into a spirit still. In spring 2014, a second distillery equipped with a new 400-gallon

alembic copper still was opened at Fontanel in Nashville.

Prichard produces around 20,000 cases per year with different kinds of rum as the main track. The first single malt was launched in 2010 and later releases usually have been vattings from barrels of different age (some up to 10 years old). The whiskey range also includes rye, bourbon and Tennessee whiskey. Their bourbon range was expanded in late 2011 with the innovative Double Chocolate Bourbon where Prichard has infused the essence of chocolate beans into the bourbon.

Distillery: RoughStock Distillery, Bozeman, Montana
Founded: 2008
Owner: Kari and Bryan Schultz
montanawhiskey.com

RoughStock buys its 100% Montana grown and malted barley and then mill and mash it themselves. The mash is not drained off into a wash, but brought directly from the mash tun into two 1,000 gallon open top wooden fermenters for a 72 hour fermentation before distillation in two Vendome copper pot stills (500 and 250 gallons). Maturation is on a mix of quarter casks and 225 litre barrels made from new American oak.

In 2009, the first bottles of RoughStock Montana Pure Malt Whiskey were released. Since then a single barrel bottled at cask strength has been added (Black Label Montana Whiskey) and apart from whiskey made from 100% malted barley, the product range also includes Spring Wheat Whiskey, Sweet Corn Whiskey and Straight Rye Whiskey. In 2013 a Montana Bourbon Whiskey was added to the range.

Distillery: Town Branch Distillery, Lexington, Kentucky
Founded: 1999
Owner: Alltech Lexington Brewing & Distilling Co.
lyonsspirits.com

Most of the producers of malt whiskey in the USA have a background in brewing, winemaking or distilling other spirits. This also applies to Lexington Brewing & Distilling Company, as

High West´s new distillery in Wanship

whiskey production is derived from their production of Kentucky Ale. Dr Pearse Lyons' background is interesting – being the owner, the founder and a native of Ireland, he used to work for Irish Distillers in the 1970s. In 1980 he changed direction and founded Alltech Inc, a biotechnology company specializing in animal nutrition and feed supplements.

Alltech purchased Lexington Brewing Company in 1999, with the intent to produce an ale that would resemble both an Irish red ale and an English ale. In 2008, two traditional copper pot stills from Scotland were installed with the aim to produce Kentucky´s first malt whiskey. The first single malt whiskey was released in 2010 under the name Pearse Lyons Reserve and in 2011 it was time for a release of their Town Branch bourbon. It thereafter took until 2014 before their third whiskey was released, the 4 year old Town Branch Rye. In 2012 the stills were relocated from the brewery to a new stand alone distillery building right across the street with a capacity of 450,000 litres of pure alcohol per year. In April 2015, Alltech started construction of a new brewery and distillery in Pikeville. The new unit will cost $13m to build and the plan is to start production in early 2017.

Distillery:	St. George Distillery, Alameda, California
Founded:	1982
Owner:	Jörg Rupf/Lance Winters
	stgeorgespirits.com

The distillery is situated in a hangar at Alameda Point, the old naval air station at San Fransisco Bay. It was founded by Jörg Rupf, a German immigrant, who came to California in 1979 and who was to become one of the forerunners when it came to craft distilling in America. In 1996, Lance Winters joined him and today he is Distiller, as well as co-owner. In 2005, the two were joined by Dave Smith who now has the sole responsibility for the whisky production.

The main produce is based on eau-de-vie which is produced from locally grown fruit, and vodka under the brand name Hangar One. Whiskey production was picked up in 1996 and the first single malt appeared on the market in 1999. St George´s obtains its wash from Sierra Nevada Brewery and Factor Brewing. Recently, St George´s have also set up their own brew system to be able to produce whiskies entirely in house. Some of the malt used, has been dried with alder and beech but is non-peated. Maturation is in bourbon barrels, Tennessee whiskey barrels, French oak sherry and French oak port casks. St. George Single Malt used to be sold as a three year old but, nowadays, comes to the market as a blend of whiskeys aged from 4 to 13 years.

The latest release was Lot 15 (October 2015) and every lot is around 3-4,000 bottles. A new addition to the range can be expected in 2016 where the owners have vatted 3 to 6 year old whiskies from ex-bourbon barrels and French oak. The idea is to sell it as a whiskey for highballs

Distillery:	Tuthilltown Spirits, Gardiner, New York
Founded:	2003
Owner:	Ralph Erenzo and Brian Lee
	tuthilltown.com

Just 80 miles north of New York City, Ralph Erenzo and Brian Lee produce bourbon, single malt whiskey, rye whiskey, rum, vodka and gin. The first products came onto the shelves in 2006 in New York and the whiskey range now consists of Hudson Baby Bourbon, a 2-4 year old bourbon made from 100% New York corn and the company´s biggest seller by far, Four Grain Bourbon (corn, rye, wheat and malted barley), Single Malt Whiskey (aged in small, new, charred American oak casks), Manhattan Rye and New York Corn Whiskey. The spirits range also include Half Moon Orchard Gin (made from wheat and apples), Indigenous Vodka, Roggen´s Rum and Tuthilltown Cassis. Hudson whiskey currently sells in all 50 states, in Europe, Australia and Hong Kong. A cooperative venture was announced between Tuthilltown and William Grant

& Sons (Grants, Glenfiddich, Balvenie et al) in 2010, in which W Grants acquired the Hudson Whiskey brand line in order to market and distribute it around the world. Tuthilltown Spirits remains an independent company that will continue to produce the different spirits. In 2014, the distillery site was expanded with a new packaging building and, not least, a whole new R&D building where the antique charentais brandy still will be used for new experiments.

Distillery:	Corsair Artisan, Bowling Green, Kentucky and Nashville, Tennessee
Founded:	2008
Owner:	Darek Bell, Andrew Webber and Amy Lee Bell
	corsairartisan.com

The two founders of Corsair Artisan, Darek Bell and Andrew Webber, were based in Nashville when they came up with the idea in 2008 to start up a distillery. At that time Tennessee law didn´t allow this, so the first distillery was opened across the border in Bowling Green, Kentucky. Two years later, the legislation in Tennessee had changed and a second distillery and brewery were opened up in Nashville. Apart from producing around 20 different types of beer, the brewery is also where the wash for all the whisky production takes place. In Nashville, they also have a 240 gallon antique copper pot still. For those spirits in the range that require a second distillation, the low wines are taken to Bowling Green and the custom made 50 gallon still from Vendome Copper. In spring 2015, the company established a malting facility in Nashville where they can floor malt their own grain and eventually there will also be a drum malting station.

Corsair Artisan has a wide range of spirits – gin, vodka, absinthe, rum and whiskey. The number of different whiskies released is growing constantly and Corsair Artisan is most likely the distillery in the USA which experiments the most with different types of grain (including quinoa and triticale). The big seller is Triple Smoke Single Malt Whiskey (made from three different types of smoked malt with an addition of chocolate malt) and two of the latest additions to the range are Oatrage (malted oats and coffey malt) and Buck Yeah (buckwheat and malted barley). Recently, they also joined Few Spirits, Journeyman Distillery and Mississippi River Distilling Company in a venture where whiskey from all four distilleries were blended together.

Other distilleries in USA

Dry Fly Distilling
Spokane, Washington, founded in 2007
www.dryflydistilling.com

Dry Fly Distilling was the first grain distillery to open in Washington since Prohibition. The first batch of malt whisky was distilled in 2008 but the first bottling will probably not be released until early 2018. However, several other types of whisky have been released recently – Bourbon 101, Straight Cask Strength Wheat Whiskey, Port Finish Wheat Whiskey, Peated Wheat Whiskey and Straight Triticale Whiskey (triticale is a hybrid of wheat and rye). New limited bottlings for 2014 are two 4 year old Irish style whiskies – Caledonian Grain and Hibernian Style. The original equipment consisted of one still but, in autumn of 2008, another still was installed, as well as two additional fermenters, which raised its capacity to 10,000 cases per annum. A further three fermenters were also added recently.

Rebecca Creek Distillery
San Antonio, Texas, founded in 2010
rebeccacreekdistillery.com

With a background in the insurance business, Mike Cameron and Steve Ison started Rebecca Creek Distillery in 2010. Fermenters and a mash system were bought in Canada and the 3,000 litre copper pot still, together with the column, was made by the well-known company Christian Carl. The first product to be launched was Enchanted Rock Vodka which was followed up by a peach flavoured version. The third and final spirit in the range is a blended whiskey called Rebecca Creek Fine Texas Whiskey, launched in 2011. The main part of the whiskey is distilled on site, but it also includes an 8 year old bourbon from another distillery. Announced plans to release a single malt whiskey have yet to materialise.

Triple Eight Distillery
Nantucket, Massachusetts, founded in 2000

www.ciscobrewers.com

In 1995 Cisco Brewers was established and five years later it was expanded with Triple Eight Distillery. The Nantucket facility consists of a brewery, winery and distillery. Apart from whiskey, Triple Eight also produces vodka, rum and gin. Whiskey production was moved to a new distillery in May 2007. The first 888 bottles of single malt whiskey were released on 8th August 2008 as an 8 year old. To keep in line with its theme, the price of these first bottles was also $888. More releases of Notch (as in "not Scotch") have followed, the latest being a 12 year old in spring 2015.

Cedar Ridge Distillery
Swisher, Iowa, founded in 2003

www.crwine.com

Jeff Quint and his wife, Lauri,e started Cedar Ridge Vineyards in 2003 and expanded the business soon afterwards to also include a distillery. After a whil,e they moved to the present location in Swisher, between Cedar Rapids and Iowa City, where they now have two stills in the distillery part. The first whiskey, a bourbon, was released in 2010. Malt whiskey production started in 2005 and in 2013 the time had come for the launch of the first single malt. Four 15 gallon ex-bourbon barrels were bottled after having a finish in different secondary casks (port, rum, sherry and bourbon). More releases of the single malt have been made since then. In spring 2015, the first launch of the single malt with a changed, more fruity recipe was made. Two of the most recent whiskies were a port cask finished bourbon and a malted rye. A major expansion of the distillery was made in autumn 2014 when a grain silo, a malt mill, a new 20 barrel lauter mash tun, more washbacks and a 650 litre wash still were added. The new equipment increased the whiskey production from 200 barrels per year to 500.

Nashoba Valley Winery
Bolton, Massachusetts, founded in 1978
(whiskey since 2003)

www.nashobawinery.com

Nashoba Valley Winery is mainly about wines but over the last decade, the facilities have been expanded with a brewery and a distillery. The owner, Richard Pelletier, produces a wide range of spirits including vodka, brandy and grappa. Since 2003 malt whiskey is also being distilled. The malt is imported and the wash is produced at his own brewery. The whiskey is matured in a combination of ex bourbon barrels and American and French Oak casks, which have previously contained wine. In autumn 2009, Stimulus, the first single malt was released. The second release of a 5 year old came in 2010 and it is Richard´s intention to release a 5 year old once a year. The first 10 year old single malt was released in autumn 2015 together with a 5 year old rye whiskey.

Woodstone Creek Distillery
Cincinnati, Ohio, founded in 1999

www.woodstonecreek.com

Don and Linda Outterson opened a farm winery in Lebanon, Ohio in 1999 and relocated to Evanston i Cincinnati in 2003 where a distillery was added to the business. In autumn 2013 they were forced to move again and the distillery/winery is now located a bit farther north in St. Bernard. The first whiskey, a five grain bourbon (white and yellow corn, malted barley, malted rye, and malted

wheat) was released in 2008. In 2010, the Outtersons released a peated 10 year old single malt from malted barley which they call "The Murray Cask", named after Jim Murray who praised it in his Whisky Bible. In 2012 this was followed by a 12 year old unpeated single malt whiskey, Ridge Runner (a five-grain bourbon white dog) and a blended whiskey. The last one is made from 5% malted wheat, malted barley and malted rye which is mixed with 95% unmalted wheat distillate. The malted whiskey is 8 years old while the unmalted is 3 years. A 13 year old single malt matured in a sherry cask has also been recently released.

Ballast Point Brewing & Distilling
San Diego, California, founded in 1996 (whiskey since 2008)

www.ballastpoint.com

Jack White and Yuseff Cherney founded Ballast Point Brewing Company in 1996. Today, it is one of the most influential craft beer brewers in USA, with four production sites in the San Diego area. When distilling started in 2008, it became the first craft distillery in San Diego. The equipment consists of two hybrid pot/column stills – one 200 gallon for gin production and a second 500 gallon for whiskey and rum. The first product to appear on the market was Old Grove gin in 2009. Vodka and rum have been released thereafter. Two whiskeys are produced; a single malt Devil´s Share Whiskey and Devil´s Share Bourbon, both of them matured for a minimum of three years. They were released for the first time in 2013 with batch 2 of the single malt and batch 3 of the bourbon released in spring 2015.

House Spirits Distillery
Portland, Oregon, founded in 2004

www.housespirits.com

In September 2015, Christian Krogstad and Matt Mount moved their distillery a few blocks to bigger premises. The main products for House Spirits used to be Aviation Gin and Krogstad Aquavit but with their new equipment they drastically increased whiskey capacity from 150 barrels per year to 4,000 barrels. The first three whiskies were released locally in 2009 and this was followed up by more bottlings in 2010 and 2011. In November 2012 it was time for the first, widely available single malt under the name of Westward Whiskey. It was a 2 year old, double pot distilled and matured in 2-char, new American oak. Recent releases have been 3 years old and now each release is a single barrel.

New Holland Brewing Co.
Holland, Michigan, founded in 1996 (whiskey since 2005)

www.newhollandbrew.com

This company started as a beer brewery, but after a decade, it opened up a micro-distillery as well, and the wash used for the beer is now also used for distilling whiskey. Until 2011, the spirit was double distilled in a 225 litre, self-constructed pot still. At that time, the capacity increased tenfold, mainly as a result of the installation of a restored 3,000-litre still built in 1932. The first cases of New Holland Artisan Spirits were released in 2008 and among them were Zeppelin Bend, a 3 year old (minimum) straight-malt whiskey which is now their flagship brand. This was followed by the single malt Double Down Barley, Walley Rye and Beer Barrel Bourbon which was a matured bourbon that was given a finish for 90 days in beer barrels. Most recent releases include Malt House Malt Whiskey, made from 86% malted barley and 14% malted rye and Pitchfork Wheat Whiskey.

DownSlope Distilling
Centennial, Colorado, founded in 2008

www.downslopedistilling.com

The three founders were brought together by their interest and passion for craft-brewing when they started the distillery in 2008 and in 2009 they finally got their licence to start distilling. The distillery is equipped with two stills – one copper pot still made by Copper Moonshine Stills in Arkansas and a vodka still of an in-house design. The first whiskey, Double-Diamond Whiskey, was released in 2010. It was made from 65% malted barley and 35% rye. It is still the core whiskey in the range. It was followed by the Irish Immigrant All Malt Whiskey in 2013. A vatting of 2 year old Double Diamond Whiskey and 2 year old Single Malt has also been released, as well as a malt/rye whiskey aged in a French cognac barrel. The most recent product was a two year old with one year in a sherry cask and one year in an American oak, white wine cask. All the malt whiskies are made from floor malted Maris Otter barley.

Do Good Distillery

Modesto, California, founded in 2013

dogooddistillery.com

Founded in 2013 by six friends and family members, and headed by Jim Harrelson, the goal is to make whiskey and, in particular, single malt. First production was in early autumn 2014 but apart from rum, no products have been launched yet. Maturing in the warehouse is a wide range of whiskies; 100% peated, 65% peated, whisky smoked using beechwood, cherry wood and mesquite and a whisky based on chocolate and biscuit malt. Most of the malt they are using is imported from Europe. The distillery is considerably bigger than many other newly started operations and produce the equivalent of 150,000 bottles per year.

Copper Fox Distillery

Sperryville, Virginia, founded in 2000

www.copperfox.biz

Copper Fox Distillery was founded in 2000 by Rick Wasmund. In 2005 they moved to another site where they built a new distillery and began distilling in 2006. Rick Wasmund has become one of the most unorthodox producers of single malt. He does his own floor malting of barley and it is dried using smoke from selected fruitwood. After mashing, fermentation and distillation, the spirit is filled into oak barrels, together with plenty of hand chipped and toasted chips of apple and cherry trees, as well as oak wood. Adding to the flavour, Wasmund also believes that this procedure drastically speeds up the time necessary for maturation. The first bottles of Wasmund´s Single Malt were just four months old but the current batches are more around 12-16 months. Occasionally he does older bottlings of up to 42 months. Other expressions in the distillery range include Copper Fox Rye Whiskey with a mash bill of 2/3 Virginia rye and 1/3 malted barley and two unaged spirits – Rye Spirit and Single Malt Spirit. In summer 2014, Rick Wasmund bought the former Lord Paget Hotel in Williamsburg where he plans to open up a second distillery in 2015/2016.

Bull Run Distillery

Portland, Oregon, founded in 2011

www.bullrundistillery.com

Founded by former brewer, Lee Medoff, the distillery made its first distillation in autumn of 2011. The distillery is equipped with two pot stills (800 gallons each) and the main focus is on 100% Oregon single malt whiskey. The first release isn´t expected until 2016 but a blended version may be released in late 2015. Meanwhile, the company is selling bourbon under the label Temperance Trader sourced from other producers, but blended and sometimes matured for an additional period at Bull Run. They are also working on line extensions of different barrel finishes.

Rogue Ales & Spirits

Newport, Oregon, founded in 2009

www.rogue.com

The company started in 1988 as a combined pub and brewery. Over the years the business gradually expanded and now consists of one brewery, two combined brewery/pubs, two distillery pubs (Portland and Newport) and five pubs scattered over Oregon, Washington and California. The main business is still producing Rogue Ales, but apart from whiskey, rum and gin are also distilled.

Two malt whiskies have been released so far. The first, Dead Guy

Jim Harrelson and Ryan Woods from Do Good Distillery

Whiskey, was launched in 2009 and is based on five different types of barley. It is distilled twice in a 150 gallon Vendom copper pot still and the spirit is matured for one month in charred barrels made of American Oak. The second expression was released in 2010 under the name Chatoe Rogue Oregon Single Malt Whiskey (the name was recently changed to Rogue Farms Single Malt). It is made from barley grown and floormalted on Rogue's own farm in Tygh Valley.

Copperworks Distilling Company

Seattle, Washington, founded in 2013

www.copperworksdistilling.com

The founders of the distillery, Jason Parker and Micah Nutt, both come from a brewing background and that is also where their whiskey comes from. They obtain their wash from a local brewery and then ferment it on site. The distillery is equipped with two, large copper pot stills for the whiskey production, one smaller pot still for the gin and one column still. The whiskey is matured in 53-gallon charred, American oak barrels. The first distillation was in January 2014 and they expect to release their first whiskey in 2016. Currently for sale are vodka and gin, both based on malted barley.

Sons of Liberty Spirits Co.

South Kingstown, Rhode Island, founded in 2010

www.solspirits.com

Michael Reppucci started the distillery with the help of David Pickerell who was Master Distiller for Maker's Mark for 13 years. This distillery is equipped with a stainless steel mash tun, stainless steel, open top fermenters and one 950 litre combined pot and column still from Vendome. Sons of Liberty is first and foremost a whiskey distillery, but the first product launched was Loyal 9 Vodka. In 2011 the double distilled Uprising American Whiskey was launched, made from a stout beer and it was followed in early 2014 by Battle Cry made from a Belgian style ale. Battle Cry has been aged in charred American oak and toasted French oak.

Cut Spike Distillery (formerly Solas Distillery)

La Vista, Nebraska, founded in 2009

www.cutspikedistillery.com

Originally opened as Solas distillery in 2009, Brian McGee and Jason Payne later renamed it Cut Spike distillery. The first product to hit the market in 2009 was Joss Vodka while the Cuban-style Chava Rum was released in 2011. In 2010 single malt whiskey was distilled and the first 140 bottles in a batch of 2,000 were launched in August 2013. Every month 140 bottles were released until November 2014 when batch was launched. Batch 3 (3,600 bottles) was released in summer 2015. Cut Spike Single Malt Whiskey is two years old and bottled at 43%.

Green Mountain Distillers

Morristown, Vermont, founded in 2001

www.greendistillers.com

Green Mountain Distillers, a certified organic distillery, was started by Tim Danahy and Howie Faircloth. The first product to hit the shelves was Sunshine Vodka in 2004, followed in 2009 with two new versions – Organic Lemon and Organic Orange. However, a 100% organic malt whiskey has always been uppermost in their minds. The first batches were already distilled in September 2004, but unlike many other distillers in the USA, Tim and Howie decided to let it mature for quite a number of years and, at the moment, there has been no release. In summer 2014, the distillery moved to a new location in Morristown, Vermont.

Blue Ridge Distilling Co.

Bostic, North Carolina, founded in 2010

www.blueridgedistilling.com

After a career in commercial diving and salvage, Tim Ferris opted for a change. He took the crew from his company, Defiant Marine, and opened up a distillery in 2010. The equipment consists of a

Copperworks Distillery

lauter mash tun, stainless steel fermenters and a modified Kothe still. For maturation they use 24 months air dried, toasted American oak. At least for the time being, their minds are totally set on single malt whisky and for 2013 they plan to produce 10,000 cases with production doubling the following year. The first distillation was in June 2012 and already in December the first bottles of Defiant Single Malt Whisky were released.

Journeyman Distillery

Three Oaks, Michigan, founded in 2010

www.journeymandistillery.com

Before opening his own distillery, Bill Welter rented still time at Koval Distillery in Ravenswood to make sure he had an aged rye whiskey (Ravenswood Rye) available when his own distillery was opened. The range of whiskies distilled at his own premises now include Last Feather Rye (a name change was made due to a trademark dispute) which is their biggest seller, Featherbone bourbon, Silver Cross Whiskey (equal parts of rye, wheat, corn and barley), W.R. Whiskey (un-aged rye), Kissing Cousins (bourbon finished in a Cabernet Sauvignon barrel) and Federalist 12 Rye. The first, limited release of Three Oaks Single Malt Whiskey was in October 2013 and was followed by another batch in July 2015. A major expansion of the distillery was made in 2015, including a new 5,500 litre still, five times the size of the previous .

Santa Fe Spirits

Santa Fe, New Mexico, founded in 2010

www.santafespirits.com

Colin Keegan, the owner of Santa Fe Spirits, is collaborating with Santa Fe Brewing Company which supplies the un-hopped beer that is fermented and distilled in a 1,000 litre copper still from Christian Carl in Germany. The whiskey gets a hint of smokiness due to a special type of malt. The first product, Silver Coyote released in spring 2011, was an unaged malt whiskey. The first release of an aged (2 years) single malt whiskey, Colkegan, was in October 2013. By spring 2015 four batches had been released. Colin´s goal is to produce 24,000 bottles of whiskey a year.

Wood´s High Mountain Distillery

Salida, Colorado, founded in 2011

www.woodsdistillery.com

Even though the brothers PT and Lee Wood have released two different gins since they started production in 2012, it seems that whiskey will be their main product. Their first expression, Tenderfoot Whiskey released in June 2013, is something as rare as a triple quad. The mash bill is 77% malted barley (a mix of chocolate malt and cherrywood smoked malt), 13% malted rye and 10% malted wheat. In July 2015 batch #19 was released with every batch being around 350 bottles. The next releases were Alpine Rye Whiskey, with a mash bill of 70% malted rye and 30% malted barley and a 16 months old whiskey made from local Oilman Imperial Stout. After a very long fermentation (9 days) the spirit is double-distilled in a 350 gallon stripping still and a 50 gallon pot-column hybrid still and then filled into small casks (25-30 gallons).

Door County Distillery

Sturgeon Bay, Wisconsin, founded in 2011

www.doorcountydistillery.com

The Door Peninsula Winery was founded in 1974 and ten years later the current owners, the Pollman family, took over. It is a large facility including a shop and tasting rooms and attracts thousands of visitors every year. In 2011, the family decided to add a distillery to the site. The wash is brought in from a local brewery and distilled in a copper pot/column hybrid still. Gin, vodka and brandy are the main products but they also make roughly 100 gallons of single malt whiskey per year. The first Door County Single Malt was released as a one year old in 2013.

Immortal Spirits

Medford, Oregon, founded in 2008

www.immortalspirits.com

In the beginning, this distillery could be seen mainly as a labour of love by two home brewers, Jesse Gallagher and Enrico Carini, but they had their minds set on something bigger. The two stills (a 1,200 gallon pot still and an 88 gallon still for limited release runs) are designed and fabricated by themselves, and the wash used to come from a local brewery. Recently, however, a 2,000 gallon mash tun and three 2,000 gallon fermenters were installed, so not only will all of the production be handled at the distillery, but the volumes have increased substantially as well. While awaiting the first releases of the aged whiskey, the owners released the two months old Early Whiskey in spring 2015.

Batch 206 Distillery

Seattle, Washington, founded in 2011

www.batch206.com

The distillery was started by Jeff Steichen and his wife Daleen Esterhuizen in downtown Seattle in 2011 and its first distillation was in February 2012. Eventually they were joined by master distiller Rusty Figgins, known from the Ellensburg distillery, in November 2012. Accompanying him he also had his own brand of malt whisky, Gold Buckle Club, which in the future will be produced under his supervision at this new distillery. In the beginning, the team outsourced the distillation to House Spirits in Portland but are now responsible for all the production themselves. The first products from Batch 206 to hit the shelves were Batch 206 Vodka, Counter Gin and See 7 Stars Moonshine. The distillery is also producing bourbon, rye and, since summer 2013, rum. The first whiskey was the 7 year old Barrel Raider Bourbon which was produced at another distillery but blended at Batch 206. In 2015, the first bourbon produced in-house was released.

Deerhammer Distilling Company

Buena Vista, Colorado, founded in 2010

www.deerhammer.com

The location of the distillery at an altitude of 2,500 metres with drastic temperature fluctuations and virtually no humidity, have a huge impact on the maturation of the spirit. Owners Lenny and Amy Eckstein found that their first whiskey, based on five varieties of malted barley, was ready to be released after only 9 months´ maturation in December 2012. The next bottling of the Down Time single malt was in June 2013 and, so far, they have released around 12 batches. The distillery was upgraded in 2013 with more fermenters, a bigger mash tun and an additional 600 gallon pot still. The spirit is double distilled and matured in 30 gallon barrels made of new American oak.

Hillrock Estate Distillery

Ancram, New York, founded in 2011

www.hillrockdistillery.com

What makes this distillery unusual, at least in the USA, is that they are not just malting their own barley – they are floor malting it. This a technique that has been abandoned even in Scotland, except for a handful of distilleries. Jeff Baker founded the distillery in 2011 and equipped it with a 250 gallon Vendome pot still and five fermentation tanks. The first spirit was distilled in November 2011. The first release from the distillery was in 2012, the Solera Aged Bourbon and in 2013 it was time for the first Single Malt whiskey. This was followed by the Double Cask Rye. New releases in 2014 were two versions of a Peated Single Malt (8-hour and 14-hour peat smoke) and in 2015 a portfinished Double Cask Rye was launched. The distillery is now producing 1,000 cases a year but the owners are looking to increase that to 5,000.

Lost Spirits Distillery

Prunedale, California, founded in 2009

www.lostspirits.net

Bryan Davis and Joanne Haruta started up by building a highly original distillery all by themselves. The still was of a design called log-and-copper which means that it is made of both oak wood and copper. In 2013, however, the whole site (fermenters, the still and part of the maturing whisky) was hit by TCA which is caused by a reaction between mold and chlorine. It basically means that everything you produce will eventually taste like mold. In autumn 2013 they had to tear everything down and rebuild it, this time using a copper still, and in December they were producing again. From the very start, their goal was to make heavily peated (often more than 100ppm), single malt whisky and bottle it at cask strength. In 2012 the first two expressions were released – Leviathan I (peated at 110ppm) and Seascape (55ppm). Several bottlings have followed since.

Painted Stave Distilling

Smyrna, Delaware, founded in 2013

paintedstave.com

Like for so many other new distilleries, production for Painted Stave started with vodka and gin. Whiskey production started in 2014, first with bourbon and rye, then followed by whiskey from malted barley. Most of the whiskey production is centered on bourbon and rye but the owners, Ron Gomes and Mike Rasmussen, released a malt whisky made from peated malt, regular malt and corn in summer 2015. They hope to have more malt whiskies in the core range in a year or two.

Pinckney Bend Distillery

New Haven, Missouri, founded in 2011

www.pinckneybend.com

The founders of this distillery all pooled their vast experience as artisan brewers when they decided to build a distillery. The distillery has five stills, with the biggest (320 gallon) installed in

spring 2015. The main purpose is to produce malt whiskey, but they also distil gin, vodka and brandies. The first whiskey, and still the top seller, was Rested American Whiskey with a grain bill of 80% corn, 15% malted barley and 5% rye. New varieties of that have since been launched; sherry, port and zinfandel cask finished, single barrel versions and a high rye (55% rye) expression called Rye and Rested. The first Sweet Mash Missouri Malt Whiskey was released in summer 2015.

Long Island Spirits

Baiting Hollow, New York, founded in 2007

www.lispirits.com

Long Island Spirits is the first distillery on the island since the 1800s. The starting point for The Pine Barrens Whisky, the first single malt from the distillery, is a finished ale with hops and all. The beer is distilled twice in a potstill and matures for one year in a 10 gallon, new, American, white oak barrel. The whisky was released in April 2012 and was followed up in 2013 with a bourbon called Rough Rider and Bull Moose rye whiskey. New releases during 2015 include a cask strength rye called Rough Rider Big Stick and a cherry smoke version of the Pine Barrens single malt. The big seller for the owner, Richard Stabile, is not a whisky, however, but LiV Vodka, made 100% from potatoes and selling 70,000 bottles per year.

Parliament Distillery

Sumner, Washington, founded in 2011

www.parliament-distillery.com

This small craft distillery, founded by Jarrett Tomal and Flynn Huntington, has yet to release a 100% single malt whiskey. The first product was Ghost Owl which has changed recipe since last year and now consists of 25% of their own malt whiskey, 15% two year old rye from the midwest and the balance is made up of 8 year old Kentucky bourbon. A local brewery prepares the wort which is fermented for 2-3 days at the distillery and double distilled in a 570 litre alambic still and a 130 litre column still. The spirit is then matured in 20 litre American oak casks with a #4 char.

Painted Stave Distillery

Dark Corner Distillery

Greenville, South Carolina, founded in 2011

www.darkcornerdistillery.com

Joe Fenten, founder of this distillery in down-town Greenville, uses a handmade 80 gallon copper pot still for all his distillations. His main focus is on different varieties of whiskey (around 2,000 gallons are produced every year) which includes Lewis Redmond (South Carolina's first bourbon), Stumphouse (made from malted wheat) and a moonshine whiskey which they have recently started exporting to Singapore. In July 2015, the first straight version of the Lewis Redmond was released as a single barrel.

Van Brunt Stillhouse

Brooklyn, New York, founded in 2012

www.vanbruntstillhouse.com

To the Whisky Trail in Scotland and the Kentucky Bourbon Trail, you can now add The Brooklyn Spirits Trail in New York, where no less than 11 distillers have teamed up to showcase their different spirits. One of them is Van Brunt Stillhouse, owned by Daric Schlesselman and located near the Red Hook waterfront. They made their first release of Van Brunts American Whiskey in December 2012, a mix of malted barley, wheat and a hint of corn and rye. This was followed by a malt whiskey from 100% malted barley, a wheated bourbon and a rye.

Civilized Spirits

Traverse City, Michigan, founded in 2009

www.civilizedspirits.com

Jon Carlson and Greg Lobdell developed a passion for craft beer and artisan spirits whilst they were attending the University of Michigan. Years later they founded Northern United Brewing Company which is the parent company of Civilized Spirits. The spirits are produced at a distillery on Old Mission Peninsula, just outside Traverse City, and it opened in 2009. The still is a 1,000 litre pot still with a 24-plate column attached. The whiskey side of the business includes Civilized Single Malt (at least 3 years old), Civilized Whiskey (made from locally grown rye), Civilized White Dog Whiskey (an unoaked wheat whiskey) and, since summer 2013, Civilized Bourbon. Rum and gin are also produced.

Square One Brewery & Distillery

St. Louis, Missouri, founded in 2006

www.squareonebrewery.com

Steve Neukomm has been working with micro-breweries since 1999. In 2006 he opened a combined brewery and restaurant in St. Louis and two years later he was granted Missouri's first micro-distilling licence. Apart from rum, gin, vodka and absinthe, Steve also produces J.J. Neukomm Whiskey, a malt whiskey made from toasted malt and cherry wood smoked malt.

Great Lakes Distillery

Milwaukee, Wisconsin, founded in 2004

www.greatlakesdistillery.com

The first distillery to open in Wisconsin since Prohibition, released its first product, Rehorst Vodka, in 2006. The distillery is focused on the production of absinthe, vodka, gin, brandy and rum, but there has also been the occasional distillation of whiskey. Most of the whiskies now in stock are at least 5 years old. The only whiskey in the range at the moment is an American blended whiskey called Kinnickinnic. It is a mix of sourced bourbon, blended with malt and rye whiskey which is produced in the distillery. A one-off bottling of a straight rye, Menomonee Valley, was released in autumn 2013. Originally, the equipment consisted of a 250 litre Holstein still, three fermenters and a 1,000 litre cooker, but a second still installed in March 2015 has increased the capacity.

One Eight Distilling

Washington, DC, founded in 2014

oneeightdistilling.com

The plan for the owners, Sandy Wood and Alex Laufer, is that whiskey in the future will constitute the main part of the business, but at the moment both gin and vodka are produced as well. The range of whiskies will be called Rock Creek and the first release will probably be a rye sometime in 2016, followed by a high-rye bourbon towards the end of 2016. Production of whiskey from malted barley has not yet started.

Hamilton Distillers

Tucson, Arizona, founded in 2011

www.hamiltondistillers.com

Since 1986 Stephen Paul has been working as a manufacturer of furniture from local mesquite wood. Scraps from the shop were often brought home to fire the barbecue and he and his wife, being avid Scotch whisky drinkers, came up with the idea of drying barley over mesquite instead of peat. They started their small distillery using a 40 gallon copper alembic still, but since November 2014, a new, 500 gallon still is in place. In spring 2015, new malting equipment was installed which made it possible to malt the barley in 5,000 lbs batches instead the previous 70 lbs batches! Rather unusually, everything is done on site from malting to maturation. The first, limited releases were made in November 2013 and they now have three expressions of Whiskey del Bac in the range – aged Mesquite smoked (Dorado), aged unsmoked (Classic) and unaged Mesquite smoked (Clear).

Cornelius Pass Roadhouse Distillery

Hillsboro, Oregon, founded in 2011

www.mcmenamins.com

The distillery is owned by the McMenamin brothers who also have a chain of more than 60 pubs and hotels, as well as the Edgefield Distillery in Troutdale. The Hillsboro distillery is equipped with a 19th century Charentais alambic still. The first release was an un-aged whiskey called The White Owl (72% malted wheat and 28% malted barley) in 2012 and it was followed by a gin in 2013. In September 2014, an aged version (3 years) of the White Owl was released under the name Billy Whiskey. In addition to whiskey, the owners also produce and sell gin and brandy.

Maine Craft Distilling

Portland, Maine, founded in 2013

www.mainecraftdistilling.com

This distillery started production in January 2013 and the founder, Luke Davidson, built most of the equipment himself. Currently they are offering vodka, gin, rum and Chesuncook, which is a botanical spirit using barley and carrot distillates! Since March 2014 there is also a single malt whiskey for sale – Fifty Stone. This is currently sold out and it may take a while before there is more in stock. Luke Davidson is floor malting all the barley himself and due to a combustion of the kiln in November they had to postpone the whisky production until April 2014.

3 Howls Distillery

Seattle, Washington, founded in 2013

www.3howls.com

Inspired by whisky production in Scotland, Will Maschmeier and Craig Phalen started distillation in 2013. The Scottish connection is reflected in that all the malted barley is imported from Scotland including a small amount of peated malt. For the distillation they use a 300 gallon hybrid still with a stainless steel belly and a copper column. Their first whiskies were released at the end of 2013, a single malt and a hopped rye and these were followed in April 2014 by a rye whiskey.

Montgomery Distillery

Missoula, Montana, founded in 2012

www.montgomerydistillery.com

In 2012, Ryan and Jenny Montgomery renovated the 19th century Pipestone Mountaineering building in Missoula and opened up a distillery. The barley and the rye is milled to a fine flour on site, using a hammer mill and the wash is then fermented on the grain. Distillation takes place in a 450 litre Christian Carl pot still with a 21 plate column attached. At the moment, they have around 50 barrels of single malt and 60 of rye maturing. The first whiskey was a limited release of the rye in summer 2015 with a regular release of both malt and rye in 2016.

Ranger Creek Brewing & Distilling

San Antonio, Texas, founded in 2010

www.drinkrangercreek.com

The owners of Ranger Creek (TJ Miller, Mark McDavid and Dennis Rylander) focus on beer brewing and whiskey production. They have their own brewhouse where they mash and ferment all their beers, as well as the beer going for distillation. The still is a 1,200 litre, 6-plate column still from Arnold Holstein. The first release was Ranger Creek .36 Texas Bourbon in November 2011. It was young but had been matured in small barrels, unlike their yet to be released straight bourbon, which has been filled into larger barrels. Their first single malt, Rimfire, was launched early in 2013 and they released batch 4 in May 2014. Other expressions include Ranger Creek .44 Rye and the white dog, Ranger Creek .36 White.

Two James Spirits

Detroit, Michigan, founded in 2013

www.twojames.com

David Landrum and Peter Bailey named their distillery after their respective fathers, both named James. Equipped with a 500 gallon pot still with a rectification column attached, the distillery started production in September 2013. Vodka, gin and bourbon have already been released and they expect to launch their first single malt some time in 2016 as a 3 year old. Aged in ex-sherry casks the whiskey has been made from peated Scottish barley.

Brickway Brewery & Distillery (former Borgata)

Omaha, Nebraska, founded in 2013

www.drinkbrickway.com

Zac Triemert, who owns the distillery, together with Holly Mulkins, was involved in founding Solas distillery (later re-named Cut Spike distillery) in 2009, and left a couple of years ago to start Borgata. All the wash for the distillation comes from their own brewery and distillation takes place in a 550 gallon Canadain wash still, while the 400 gallon spirit still comes from Forsyth's in Scotland. The owners are focused on single malt whiskey but they will also produce smaller amounts of bourbon and rye. The first whisky distillation was in April 2014 and they plan to produce around 220 gallons every month. Their first whisky, Borgata American Single Malt White Whisky, was released in May 2014. End of 2014 the distillery's name was changed to Brickway following a dispute with Borgata Casino.

Seven Stills Distillery

San Francisco, California, founded in 2013

www.sevenstillsofsf.com

Founded by Tim Obert and Clint Potter, the Seven Stills distillery is actually a distillery without stills! This would be the case at least until November 2015 when they will have installed a 300 gallon copper pot still from Artisan Still Design as well as brewing equipment. Until then,they are doing all their distillations using a still at Stillwater distillery in Petaluma. Their whiskies are made from beers from different breweries, for instance E.J. Phair Brewery

in Pittsburg, California. The first release, Chocasmoke, which will also become their flagship whiskey, was made from an oatmeal stout based on peat smoked malt, chocolate malt and crystal malt. Their next two releases were Whipnose made from a double IPA and Fluxuate, made from a coffee porter. Other products in the range are vodka and bitters.

Tualatin Valley Distilling

Hillsboro, Oregon, founded in 2013

www.tvdistilling.com

Originally started as Willamette Valley Distilling, the founders (Jason O'Donnell and Corey Bowers) eventually changed the name to Tualatin Valley. The distillery is equipped with a 26 gallon, 4-plate column still for the brandies and a 100 litre pot still for whiskey. Maturation takes place in small (5 gallons) charred American oak barrels but experiments have also been made with Hungarian oak. The owners concentrate on whiskey production and the first distillation was in December 2013. In July the following year, the first two products were launched – Oregon Single Malt New Make Whiskey and 50/50 American New Make Whiskey (50% rye and 50% malted barley). Aged versions of the two styles were launched in October. The distillery only produces 1,000 bottles of whiskey per year but the owners have plans to upgrade the capacity during 2016.

Vikre Distillery

Duluth, Minnesota, founded in 2012

www.vikredistillery.com

Joel and Emily Vikre fired up the still for their first distillation in November 2013. Gin and aquavit have already found their way to shops and bars in Minnesota and they also have three different kinds of whisky maturing on site – Iron Range American Single Malt, Gunflint Bourbon and Temperance River Rye. The first releases of these are still a few years away though.

Rennaisance Artisan Distillers

Akron, Ohio, founded in 2013

renartisan.com

The distillery is an outgrowth of a homebrew supply shop run by brothers John and Jim Pastor. So far they have, apart from whiskey, produced gin, brandy, grappa and limoncello. The first whiskey release, The King's Cut single malt, was made from a grain bill including special malts such as toasted and caramel malts. It was launched in October 2014 and new batches appear every 6 months. A King's Cut made from cherry smoked barley is due for release in late 2015.

Coppercraft Distillery

Holland, Michigan, founded in 2012

coppercraftdistillery.com

Located in Holland, close to Lake Michigan, this distillery is owned and operated by Walter Catton. He uses a stainless steel mash tun, six washbacks and two stills - one stripping still with stainless steel pot and copper column and a fractioning still with both pot and column made from copper. The first three whiskies - corn, wheat and malted rye - were released in summer 2014. Walter expects to release his first single malt sometime in 2016.

John Emerald Distilling Company

Opelika, Alabama, founded in 2014

www.johnemeralddistilling.com

In October 2013, John and Jimmy Sharp (father and son) obtained the final approval to build a distillery and already in July 2014 the production started. The first distillations were made in a small, 26 gallon pilot still, but a larger pot/column hybrid still with four plates was soon installed. The mashing is done in a lauter tun and the wash is fermented on the grain in stainless steel tanks. The main product will be an Alabama Single Malt which gets its character from barley smoked with a blend of southern pecan and peach wood. The first release of the single malt was made in March 2015. Around 6,000 litres of whiskey will be produced during the first year. Other products include rum and gin.

11 Wells Distillery

St. Paul, Minnesota, founded in 2013

11wells.com

Located in the middle of St. Paul, close to the Flat Earth Brewing Company, this new distillery is run by Bob McManus and Lee Egbert. The distillery is equipped with a 650 gallon mash tun, stainless steel open-top fermentation tanks and two stills - a 250 gallon stripping still designed by themselves and a 100 gallon Kothe hybrid pot/column still. Whiskey is the main product and the first two releases, aged bourbon and rye, were released in November 2014. A wheat whiskey was released in 2015, but the owners have still to release a whiskey made entirely from malted barley. The labels on the bottles have a serial number which for instance shows oak origin, char level, yeast type, mash bill and mash type.

Blaum Bros. Distilling

Galena, Illinois, founded in 2012

blaumbros.com

Heavily influenced by Scotch whisky, the two brothers Matthew and Mike Blaum opened their distillery in 2012 and began distilling in early 2013. The equipment consists of a 2,000 litre mash tun, five 2,000 litre wash backs and a 2,000 litre Kothe hybrid still. The whiskey is matured in American oak, oloroso sherry butts, madeira casks, port barrels and rum barrels. Apart from gin and vodka, the first two releases were the sourced Knotter Bourbon and Knotter Rye. The first whiskey from their own production was a rye in 2015 which will be followed by single malt in 2016 or 2017.

Venus Spirits

Santa Cruz, California, founded in 2014

venusspirits.com

After having worked in the brewing business and organic food industry, Sean Venus decided to build his own distillery. Production, which started in May 2014, is focused on whiskey, but he has also released gin and spirits from blue agave. The first single malt was Wayward Whiskey, made from crystal malt and released in January 2015. This was followed up by a rye and later a bourbon. The distillation takes place in a hand pounded alembic still from Spain.

FEW Spirits

Evanston, Illinois, founded in 2010

fewspirits.com

Former attorney (and founder of a rock and roll band) Paul Hletko started this distillery in Evanston, a suburb in Chicago in 2010. It is equipped with three stills; a Vendome column still and two Kothe hybrid stills. Bourbon and rye have been on the market for a couple of years and the first single malt, with some of the malt being smoked with cherry wood, was released in early 2015. Apart from this, three distinctive types of gin are produced, which is unique because they have an unaged bourbon as a base instead of the usual neutral spirit. A fourth expression was launched in summer 2015. This was a gin tailored for breakfast or brunch with hints of bergamot (think Earl Grey tea).

Oak N´ Harbor Distillery

Oak Harbor, Ohio, founded in 2014

oaknharbordistillery.com

Together with his wife Andrea, Joe Helle began distilling in December 2014 and only a week later his first single malt was

on the shelves. Aptly named Six Days Seven Nights, it had been maturing for a week in small barrels made from Minnesota white oak. An aged version of the whiskey is due for release in late 2015. Other products include gin, apple brandy, rum and vodka. All spirits are distilled in a combined pot and column still from Mile Hi Distilling.

Bent Brewstillery

Roseville, Minnesota, founded in 2014

bentbrewstillery.com

After years of research and education in the arts of distilling and brewing, Bartley Blume decided to build a combined brewery and distillery. Production started in 2014 and apart from a range of beers, Blume is also producing gin and whiskey with plans for many other varieties. No whiskies have been released so far, but one called Kursed Single Malt is currently aging in a combination of charred oak and charred apple wood and yet another was distilled from an India Pale Ale. The first release, though, will probably be a bourbon towards the end of 2015.

Orange County Distillery

Goshen, New York, founded in 2013

orangecountydistillery.com

This is a true farm distillery where the owners, Bryan Ensall and John Glebocki, grow every ingredient on the farm, including sugar beet, corn, rye, barley and even the botanicals needed for their gin. They malt their own barley and even use their own peat when needed. Production started in April 2014 and the first products, gin and vodka, were selling in October of the same year. Since then they have also launched a corn whiskey, a bourbon and an unaged single malt. Next in line will be an unaged and an aged rye.

Key West Distilling

Key West, Florida, founded in 2013

kwdistilling.com

The main track for Jeffrey Louchheim is to produce rum but he is also distilling whiskey. The mash is brought in from Bone Island Brewing, fermented, distilled and filled into new barrels or used rum barrels. The first release was in December 2014 followed by another in May 2015.

Thumb Butte Distillery

Prescott, Arizona, founded in 2013

thumbbuttedistillery.com

A variety of gin, dark rum and vodka are produced by the owners, Dana Murdock, James Bacigalupi and Scott Holderness, as well as whiskey. Rodeo Rye, Bloody Basin Bourbon and three batches of Central Highlands Single Malt have all been released. Maris Otter barley is used for the malt whiskies.

Black Bear Distillery

Green Mountain Falls, Colorado, founded in 2015

blackbeardistillery.com

When people speak of craft distilleries and going back to the roots, Victor Matthews intends to take this to a new level at his distillery which opened up in July 2015. The organic barley is bought from local farmers, soaked in the creek behind the distillery and malted on the attic floor of the main building using a fire-place to dry it. It is then ground on several hundred years old millstones. Washbacks are made of cypress and distillation takes place in a hand-hammered, 400 gallon pot still. The first products will be an Irish style whiskey made from both malted and unmalted barley, followed by Throwback - a corn moonshine. Bourbon, both unaged and aged, will be next. Matthews has already released a couple of bourbons that he has distilled at Breckenridge and Distillery 291.

Seattle Distilling

Vashon, Washington, founded in 2013

seattledistilling.com

Ishan Dillon and Paco Joyce produce gin, vodka, coffee liqueur, as well as a malt whiskey. The latter, named Idle Hour, was first launched in 2013 with batch 3 released in March 2015. A small amount of honey is added during fermentation. The owners source used wine casks from a local winery and then re-cooper and char them on site.

Hewn Spirits

Pipersville, Pennsylvania, founded in 2013

hewnspirits.com

Using a 130 gallon copper still, Sean Tracy is producing a variety of spirits including rum, gin and vodka. On the whiskey side there is Red Barn Rye and the Reclamation American Single Malt Whiskey which was released for the first time in summer 2014. After maturing the malt whiskey in barrels for 1-4 months, Tracy does a second maturation in stainless steel vats where he also puts in charred staves of either chestnut or hickory wood. The second maturation lasts for two weeks.

Damnation Alley Distillery

Belmont, Massachusetts, founded in 2013

damnationalleydistillery.com

This is a very small distillery but with a wide range of whiskies. Founded in 2013 by Alison DeWolfe, Jeremy Gotsch, Jessica Gotsch, Alex Thurston and Emma Thurston, the distillery produces vodka but mainly whiskey. Maturation takes place for 4-6 months in 5 gallon barrels. Among the varieties that can be mentioned are single malt, hopped single malt, smoked single malt (smoked with fruit wood), bourbon, rye and a house whiskey from barley, corn, rye and wheat. Older, straight whiskies will be released from 2016.

Canada

Distillery:	Glenora Distillery, Glenville, Nova Scotia
Founded:	1990
Owner:	Lauchie MacLean
	glenoradistillery.com

Situated in Nova Scotia, Glenora was the first malt whisky distillery in Canada. The first launch of in-house produce came in 2000 but a whisky called Kenloch had been sold before that. This was a 5 year old vatting of some of Glenora's own malt whisky and whisky from Bowmore Distillery on Islay. The first expression, a 10 year old, came in September 2000 and was named Glen Breton and this is still the core expression under the name Glen Breton Rare. Since then several expressions have been launched, among them single casks and sometimes under the name Glenora. In 2008 a 15 year old version was available from the distillery only and in 2011 a 17 year old version was released. Recent releases have included a 14 year old and a 19 year old and for the first time these aged varities have been exported as well, in this case to Spain and France. A new expression, Glen Breton Ice (10 years old), the world's first single malt aged in an ice wine barrel, was launched in 2006. Interest was massive and more bottlings (aged 15 and 17 years) were released after that. A 15 year old version of Glen Breton single malt was released under the name Battle of the Glen in June 2010. The release commemorated the distillery's victorious outcome of the ten year-long struggle with Scotch Whisky Association.

A limited 20 year old expression was launched a few years ago for sale at the distillery only and one of the most recent releases (summer 2014) was Fiddler's Choice which was a fairly young single malt – between 4 and 9 years old. To celebrate the distillery's 25th anniversary, a limited (130 bottles) 25 year old single cask was released in summer 2015. This was complemented later in the year by yet another limited release, Jardine Specials, in honour of the founder and original owner of Glenora Distillery - Bruce Jardine.

Distillery:	Still Waters Distillery, Concord, Ontario
Founded:	2009
Owner:	Barry Bernstein and Barry Stein
	stillwatersdistillery.com

Barry Bernstein and Barry Stein started their careers in the whisky business as Canada's first independent bottlers, importing casks from Scottish distilleries and selling the whisky across Canada. The next step came in 2009 when they opened Still Waters distillery in Concord, on the northern outskirts of Toronto. The distillery is equipped with a 3,000 litre mash tun, two 3,000 litre washbacks and a Christian Carl 450 litre pot still. The still also has rectification columns for brandy and vodka production. The plan is to expand production sometime in 2016 which will increase the capacity by 2 to 3 times. The focus is on whisky but they also produce vodka, brandy and gin. Their first release was a triple distilled, single malt vodka and they have also released a Canadian whisky with distillate sourced from other producers. Their first single malt, named Stalk & Barrel Single Malt, was released in April 2013. It was a cask strength (61,3%), un chill-filtered and with no colouring. Another 30 batches have followed and in late 2014 it was time for their first rye whisky release, made from locally sourced 100% Ontario rye. In the future, rye will comprise about half of the total releases.

Distillery:	Shelter Point Distillery, Vancouver Island, British Columbia
Founded:	2009
Owner:	Patrick Evans
	shelterpointdistillery.com

In 2005, Patrick Evans and his family decided to switch from the dairy side of farming to growing crops and they bought the Shelter Point Farm just north of Comox on Vancouver Island. Eventually the idea to transform the farm into a distillery was raised and with the help of some Scottish investors headed by Andrew Currie, who co-founded Arran Distillery in the early 1990s, the construction work began. The buildings were completed in 2009 and in May 2010 all the equipment was in place. This includes a one tonne mash tun, five washbacks made of stainless steel (5,000 litres each) and one pair of stills (a 5,000 litre wash still and a 4,000 litre spirit still). Both stills and the spirit safe were made by Forsyth's in Scotland while the rest was manufactured in Canada. Evans already has plans for a major expansion of the distillery in 2017 and two 26,000 litre copper pot stills have already been acquired. The idea was to start distillation in September 2010 but federal, provincial and local licensing requirements, multiple inspections and one mechanical complication after another, delayed the start by eight months and the first casks were only filled in June 2011. To assist with the start up, Patrick Evans asked Mike Nicholson to join him and his operating manager, Jim Marinus. Mike is an experienced distiller having worked for many years at distilleries in Scotland. The barley used for the distillation is grown on the farm and being a true farm distillery also means that a lot of time is devoted to other duties, besides distillation. For now, only half of the capacity is being used which amounts to 150,000 litres of alcohol per year. Legally, a whisky could have been released in summer 2014 as a three year old, but the owners have not yet decided when to launch their first whisky. In autumn 2013, however, a Still Master Vodka was launched. It was triple distilled and new varieties and flavours have occured since.

Other distilleries in Canada

Victoria Spirits

Victoria (Vancouver Island), British Columbia, founded in 2008 (whisky since 2009)

www.victoriaspirits.com

This family-run distillery actually has its roots in a winery called Winchester Cellars, founded by Ken Winchester back in 2002. Bryan Murray, the owner of Victoria Spirits, came in as an investor, but soon started to work with Ken on the distilling part of the business. Before Ken left the business in 2008, he took part in introducing Victoria Gin, which currently is the big-selling product with 10,000 bottles a year. The Murray family left the wine part of the business in order to increase the spirits role and the next product on the list was a single malt whisky. The first batch was distilled in late 2009 by Bryan´s son, Peter Hunt, but since then, whisky production has been intermittent. The first single malt, Craigdarroch, was launched in early 2015. Only 250 bottles were released and more whisky can´t be expected for at least a couple of years. However, the owners have announced that they have plans to move from Victoria, 20 km north to Sidney and that whisky production will be increased by the end of 2015.

Pemberton Distillery

Pemberton, British Columbia, founded in 2009

www.pembertondistillery.ca

This is one of the most recently established distilleries in Canada. Distilling started in July 2009, with vodka produced from potatoes being their first product. The organically grown potatoes are sourced locally and the distillery itself is a Certified Organic processing facility. Tyler Schramm uses a copper pot still from Arnold Holstein and the first vodka, Schramm Vodka, was launched in 2009. During the ensuing year, Tyler started his first trials, distilling a single malt whisky using organic malted barley from the Okanagan Valley. His aim is to produce 6-10 200 litre casks of whisky per year. So far, they have only used ex-bourbon barrels for the maturation but they have plans to use cask made of Canadian oak that have held their own apple brandy as well. The first release was in October 2013 when 330 bottles of a 3 year old unpeated version were launched. This was followed in 2014 and 2015 by one lightly peated and one unpeated single cask. As from autumn 2015, Tyler expects to have a regular expression for sale.

Okanagan Spirits

Vernon and Kelowna, British Columbia, founded in 2004

www.okanaganspirits.com

The first distillery named Okanagan was started in 1970 by Hiram Walker to which malt whisky production was added in 1981, but it closed in 1995. The main part of the production was shipped to Japan to be used by Suntory for their whisky blending.

In 2004, forestry engineer, Frank Deiter, decided to make a career change and established Okanagan Spirits. A distillery was opened in Vernon and, later on, a second one was built in Kelowna. The Okanagan Valley is famous for its orchards and vineyards and focus has been on the production of a variety of spirits made from fruits and berries. Gin, vodka, absinthe and whisky were added to the range. Their core whisky is a blend of two 5 year old whiskies made from rye and corn with a small percentage of malted barley. In autumn 2013, a small batch of the 6 year old The Laird of Fintry was launched. This was the first release of a single malt and was followed by new editions in 2014 and in autumn 2015.

L B Distillers

Saskatoon, Saskatchewan, founded in 2012

www.lbdistillers.ca

The abbreviation for LB in the distillery name stands for Lucky Bastards and the luckiest bastard amongst the owners is Michael Goldney, who earned his nickname after winning the lottery in 2006. Joined by Cary Bowman and Lacey Crocker, he opened up the distillery in 2012. The primary goal for the trio is to produce single malt and rye whisky and recently they also tried their hand at distilling a whisky from quinoa. Until the whisky is ready for bottling in 2015, the owners have produced and released a fair amount of other spirits – vodka, gin and a variety of liqueurs.

Central City Distillery

Surrey, British Columbia, founded in 2013

www.centralcitybrewing.com

What started as a brewpub in 2003 has over the years grown into a beer brewery of quite substantial proportions. Spearheaded by Darryll Frost, founder and president of Central City Brewing, the company opened up a new brewery in 2013 with a capacity of doing two million 6-packs during 2014 and with a possibility of increasing to five million in a couple of years. The brewery was complemented with a distillery in 2013, where they began distilling single malt whisky in July. Rye followed in 2014 and during the year, the first un-aged spirits (gin and vodka) were released.

Shelter Point Distillery

Australia & New Zealand

Australia

Distillery: Lark Distillery, Hobart, Tasmania
Founded: 1992
Owner: Lark Distillery Pty Ltd.
larkdistillery.com.au

In 1992, Bill Lark was the first person for 153 years to take out a distillation licence in Tasmania. Since then he has not just established himself as a producer of malt whiskies of high quality, but has also helped to off-set several new distilleries. In honour of his great contribution to modern, Australian whisky making, he was elected to the World of Whisky Hall of Fame in London in 2015. The success of the distillery forced Bill and his wife Lyn to start thinking of how to generate future growth and in July 2013 they took on board a group of Hobart based investors as majority owners of the company. The Larks are still significant shareholders and Bill will now devote his time to being a global brand ambassador for Lark Distillery. In January 2014, they acquired Old Hobart Distillery and the Overeem brand. They will continue to work as separate entities but there has been talk of relocating Old Hobart distillery to the Mt Pleasant site.

Lark Distillery is situated on a farm at Mt Pleasant, 15 minutes' drive from Hobart. The farm grows barley for Cascade Brewery and, at the moment, that is where Lark Distillery gets its malt from. Bill´s intentions are to set up his own floor maltings but recently he became involved in a new distillery in Tasmania, namely Redlands Estate Distillery, where they have started floor malting the barley and, in the future, the capacity will be sufficient to also supply Lark distillery. The old distillery site down in Hobart at the waterfront is now a cellar door and a showcase for Lark whisky.

The core product in the whisky range is the Classic Cask Single Malt Whisky at 43%, previously released as single casks but now a marriage of several casks. There is also the Distillers Selection at 46% and a Cask Strength at 58%, both of which are also single cask. Future plans include releases of special bottlings, for example, whisky matured in rum casks or finished in the distillery´s own apple brandy casks. The whisky is double-distilled in a 1,800 litre wash still and a 600 litre spirit still and then matured in 100 litre "quarter casks". In 2015, more staff was hired and production was doubled compared to 2014. There are also plans to add another pair of stills to incresae production even further.

The demand for Lark single malt has grown significantly in recent years and the forecast indicates an annual growth of 25-30% in the years to come. Outside of Australia the whisky can also be found in the USA, Singapore, Hong Kong, Japan, UK, France, Germany, Netherlands and Spain. However, in order to meet domestic demand, export sales have been temporarily slowed down.

Distillery: Bakery Hill Distillery, North Balwyn, Victoria
Founded: 1998
Owner: David Baker
bakeryhilldistillery.com.au

Since 2008, when Bakery Hill Distillery completed the installation of a 2,000 litre brewery, David Baker has had total control of all the processes from milling the grain to bottling the matured spirit. The first spirit at Bakery Hill Distillery was produced in 2000 and the first single malt was launched in autumn 2003. Three different versions are available – Classic and Peated (both matured in ex-bourbon casks) and Double Wood (ex-bourbon and a finish in French Oak). As Classic and Peated are also available as cask strength bottlings, they can be considered two more varieties. Most of the spirit is matured in ex-bourbon American oak from Jack Daniels and the 225 litre hogsheads are rebuilt into 100 litre barrels. Recently David started doing trials with three new wood finishes but an eventual release is still a couple of years away. The whisky is double-distilled in a copper pot still. For the last couple of years David has concentrated on building stock. The interest from the home market has increased significantly and lately he´s been forced to put a temporary halt to exporting.

Distillery: Old Hobart Distillery, Blackmans Bay, Tasmania
Founded: 2005
Owner: Lark Distillery Pty Ltd.
overeemwhisky.com

Even though Casey Overeem did not start his distillery until 2007, he had spent several years experimenting with different types of distillation which were inspired by travels to Norway and Scotland. The distillery (previously known as Overeem Distillery) came on stream in 2007. The mashing is done at Lark distillery where Overeem also has his own washbacks and the wash is made to his specific requirement, among others, with his own yeast. In every

Bill Lark and part of his crew

mash, a mix of 50% unpeated barley and 50% slightly peated is used. The wash is then transported to Old Hobart Distillery where the distillation takes place in two stills (wash still of 1,800 litres and spirit still of 600 litres). The spirit is matured in casks that have previously contained either port, bourbon or sherry.

The range consists of Overeem Port Cask Matured and Overeem Sherry Cask Matured, both varieties available at 43% and 60%. In December 2013, Overeem Bourbon Cask Matured at 43% was released, followed by a 60% version in June 2014 with more bottles of both being released in September. The latest release of a bourbon cask maturation was in spring 2015 which was the last of its kind available until 2020. The whisky is sold in Australia, but export orders have also been shipped to the Netherlands and the UK (being the first Australian whisky to hit the shelves of Selfridge's!).

In January 2014, Old Hobart distillery was acquired by Lark Distillery Pty Ltd. The latter was formed during the summer of 2013 when a group of investors took a majority position in Lark distillery. Casey Overeem has now retired and his daugther, Jane, is the marketing manager for Lark and Overeem, as well as being the brand ambassador for Overeem.

Distillery:	Tasmania Distillery, Cambridge, Tasmania
Founded:	1996
Owner:	Patrick Maguire
	tasmaniadistillery.com

Three generations of whisky can trace its origin to Tasmania Distillery. The first was distilled between 1996 and 1998 and, according to the current owner, Patrick Maguire, the quality is so poor that he does not want to bottle it. The second generation was distilled from November 1999 to July 2001 and today is bottled under the name Sullivan's Cove. The third generation is the whisky distilled from 2003 until the present day under Patrick and his partners' ownership and which hasn't been bottled yet. The distillery obtains wash from Cascade Brewery in Hobart and the spirit is then double distilled, although there is only one still at the distillery. In September 2014 the distillery moved to a new building about four times the size of the current facility. Production was also ramped up and they are now distilling 7 days a week which means filling around 420 casks per year - 140 port pipes and 280 bourbon barrels.

The range comprises of Sullivan's Cove Single Cask, bottled at

47,5% and matured in either bourbon casks or French oak port casks and Sullivan's Cove Double Cask (40%) which is a marriage of port and bourbon casks. The age has increased over the years and currently they are bottling 13 and 14 year old whisky. Five years ago, 40% of the sales were made in Australia but this has now increased to 90%. The biggest export markets are USA and Canada. Sullivan's Cove is also available in the UK, Germany, France, Holland, Russia, Hong Kong and Japan.

Distillery:	Hellyers Road Distillery, Burnie, Tasmania
Founded:	1999
Owner:	Betta Milk Co-op
	hellyersroaddistillery.com.au

Hellyer's Road Distillery is the largest single malt whisky distillery in Australia with a capacity of doing 100,000 litres of pure alcohol per year. The Tasmanian barley is malted at Cascade Brewery in Hobart and peat from Scotland is used for the peated expressions. Batches of 6,5 tonnes of grist are loaded into the mash tun and then the wash is fermented for 65 hours. There is only one pair of stills but they compensate for numbers by size. The wash still has a capacity of 40,000 litres and the spirit still 20,000 litres and they practise a really slow distillation. The foreshots take around 4-5 hours and the middle cut will last for 24 hours, which is six to seven times longer compared to what is common practice in Scotland. Another interesting fact is that the pots on both stills are made of stainless steel while heads, necks and lyne arms are made of copper. Maturation takes place in ex-bourbon casks but they also use Tasmanian red wine barrels for part of it. Recently, around $1m were invested to upgrade warehouses and bottling facilities.

The first whisky was released in 2006 and there are now five varieties of Hellyers Road Single Malt Whisky in the core range: Original (with no age statement) and a peated bottled at cask strength are only available to visitors at the distillery. Original 10 year old, Slightly Peated and a Pinot Noir finish are more readily available. There is also a Hellyer's Road Roaring 40's reserved for export. In 2014, 8,000 bottles of the 12 year old Special Canister were released. End of 2014, the Henry's Legacy Range was introduced. This is a new series of cask strength, single cask bottlings with The Gorge as the first release. The second, Saint Valentine's Peak, appeared in July 2015.

Jane Overeem from Old Hobart Distillery

Distillery: Great Southern Distilling Company, Albany, Western Australia

Founded: 2004

Owner: Great Southern Distilling Company Pty Ltd./ Cameron Syme

limeburners.com.au

The distillery was built in Albany on the south-western tip of Australia in 2004 with whisky production commencing in late 2005. Throughout the initial years, production of whisky, brandy, vodka and gin took place in a set of sheds on the outskirts of Albany. A move was made in 2007 to a new, custom-built distillery with a visitor centre at Princess Royal Harbour. For the distillation, one wash still (1,800 litres) and one spirit still (580 litres) are used and a 600 litre copper pot antique gin still has also been installed. The fermentation time is unusually long – 7 to 10 days and for maturation a mix of ex-bourbon, ex-house brandy and ex-sherry barrels are used. Production was doubled in 2012 to 12,000 litres and will increase to finally reach 25,000 litres by 2016.

The first expression of the whisky, called Limeburners, was released in 2008 and this is still the core expression. In 2010, the first peated version of Limeburners was released. The whiskies are bottled either at cask strength or diluted to 43% for the unpeated and 48% for the peated version. A new addition to the range was released in 2012 – Tiger Snake, an Australian sour mash whisky based on corn, malted barley and rye. The company has plans to build a second distillery in Margaret River which will most likely be focused on producing Tiger Snake.

Distillery: Nant Distillery, Bothwell, Tasmania

Founded: 2007

Owner: Keith Batt

nant.com.au

Nant distillery, in Bothwell in the Central Highlands of Tasmania, started when Queensland businessman, Keith Batt, bought the property in 2004. He embarked on refurbishing the Historic Sandstone Water Mill on the estate that was built in 1823 and converted it into a whisky distillery. The first distillation took place in 2008. The distillery is equipped with a 1,800 litre wash still, a 600 litre spirit still and wooden washbacks for the fermentation. Quarter casks of 100 litres, which previously held port, sherry and bourbon, are used for maturation. Production has gone from 300 bottles in 2010 to 20,000 this year but there are even bigger plans installed. The idea is to purchase more stills to be able to produce 200,000 bottles from 2016. The first bottlings were released in 2010 and since 2012 the range consists of Nant Single Malt Whisky matured in either bourbon, sherry, port or pinot noir casks, bottled at 43%. There are also cask strength versions (63%) of all four. The latest additions to the range are Old Mill Reserve, matured in a combination of bourbon and sherry casks and bottled at 63% and White Oak which had been matured in virgin American oak and bottled at 43%.

In 2012, the company expanded the business to include two Nant Whisky Cellar & Bars – one in Hobart and the other in Brisbane. The bars are dedicated to all things whisky and offer single malts from Scotland as well as Tasmanian whisky. Two more bars have later been opened in Melbourne and Fortitude Valley.

Other distilleries in Australia

New World Whisky Distillery

Essendon Fields, Melbourne, Victoria, founded in 2008

www.newworldwhisky.com.au

Before David Vitale started this distillery, he worked with sales and marketing at Lark Distillery and Bill Lark has also taken part in the start-up of Victoria Valley. The distillery is fitted into an old Qantas maintenance hangar at Essendon Fields, Melbourne's original airport. The stills (an 1,800 litre wash still and a 600 litre

spirit still) were bought from Joadja Creek Distillery in Mittagong and currently the yearly production is around 20,000 cases. There are also 860 barrels of maturing whisky stored in the warehouse. David uses a variety of cask sizes for maturation (50, 100 and 200 litres) but all have previously contained sherry. The first whisky, a blend of 12 different casks and bottled at 43%, was released under the name Starward in 2013 and more batches have followed including one matured in Australian shiraz casks. A range of limited releases called New World Projects has also been launched with a double maturation (re-fill and first fill port) as the latest addition.

William McHenry and Sons Distillery

Port Arthur, Tasmania, founded in 2011

www.mchenrydistillery.com.au

William McHenry was working as an employee at a biotech company in Sydney in 2006 when he first started considering a distillery of his own. In 2011 the decision was taken and he moved to Tasmania with his family. The copper still was delivered in October of that year and in January 2012 he started distilling. The distillery is equipped with a 500 litre copper still with a surrounding water jacket to get a lighter spirit. The production averages 400 litres of newmake per month and the spirit is filled into first-fill bourbon barrels from Maker's Mark, both 100 and 200 litres. In 2013 he built a new 200 m² bond store to be able to increase production. To facilitate the cash flow, he produces a range of gin including Dry Gin, Sloe Gin, Navy Strength Gin and a Barrel Aged Gin. The first whisky release, 100 bottles made from Tasmanian Gairdner barley, aged in American oak casks from Maker's Mark and finished in French oak port pipes, is planned for December 2015. McHenry expects to make a new release every three month after that.

Timboon Railway Shed Distillery

Timboon, Victoria, founded in 2007

www.timboondistillery.com

The small town of Timboon lies 200 kilometres southwest of Melbourne. Here, Tim Marwood established his combination of a distillery and a restaurant in 2007 in a renovated railway goods shed. Using a pilsner malted barley, Marwood obtains the wash (1,000 litres) from the local Red Duck microbrewery. The wash is then distilled twice in a 600 litre pot still. For maturation, resized (20 litres) and retoasted ex-port, tokay and bourbon barrels are used. The first release of a whisky, matured in port barrels, was made in 2010 and the latest expression is a bourbon maturation. Around 4,000 litres are produced every year and Tim now has plans to create a second range of peatsmoked whiskies.

Black Gate Distillery

Mendooran, New South Wales, founded in 2012

www.blackgatedistillery.com

This boutique distillery was opened by Brian and Genise Hollingworth in January 2012. Both mashing and fermentation are done at the distillery and since autumn 2013 they have also started peatsmoking the barley on site. The first products were vodkas and liqueurs with rum following. The first release of a single malt came in the beginning of 2015 when a sherrymatured expression was launched. A few months later, another sherry maturation but this time lightly peated was released.

Redlands Estate Distillery

Plenty, Tasmania, founded in 2013

www.redlandsestate.com.au

Redlands Estate in Derwent Valley, just 35 minutes from Hobart, dates back to the early 1800s and was run mainly as a hop and grain farm. A few years ago Peter and Elizabeth Hope bought the rundown property with the goal to restore it and turn it back into a working farm. Part of the plans included a distillery. The first spirit was double distilled in March 2013 in a 900 litre copper pot still and there are plans to install yet another and bigger still (2,000 litre) in the future. The barley used is grown on the estate and it is also floor malted on site. The first whisky will be bottled during 2015 but a new make bottled at 50% is already available.

Castle Glen Distillery

The Summit, Queensland, founded in 2009

www.castleglenaustralia.com.au

Established as a vineyard in 1990 by the current owner Cedric Millar, Castle Glen moved on to open up also a brewery and a distillery in 2009. Apart from wine and beer, a wide range of spirits are produced including rum, vodka, gin, absinthe and various eau de vies. Castle Glen is also the only whiskey (they use this spelling) distillery in Queensland. Malted barley is imported from Switzerland and the first whiskey, Castle Glen Limited Edition, was released as a 2 year old in early 2012.

Joadja Distillery

Joadja, New South Wales, founded in 2014

www.joadjawhisky.com.au

This distillery was originally founded by Mark Longobardi seven years ago but never went in to production. The stills were sold to New World Whisky Distillery and the National Heritage Listed property (with no production equipment left) was bought by Valero Jimenez in March 2011. He applied for an Excise Licence and consulted with Bill Lark on how to start up the production. The first distillation was in December 2014 when the distillery was equipped with just the one still (800 litres), used for both the wash and the spirit run. Since then a 2,400 litre wash still has been installed. The owners plan to grow 30 acres of their own barley on the estate and also to malt it on site, using peat to dry it. The goal is to release the first whisky by the end of 2016.

Mackey´s/Shene Distillery

Pontville, Tasmania, founded in 2015

www.mackeysdistillery.com.au, www.shene.com.au

Shene Distillery is scheduled to open end of December 2015 but distillation actually began already in 2007 in a different location. That was the year when Damian Mackey built a minute distillery in a shed on his property in New Town outside Hobart. Over the years, Mackey has patiently been experimenting and learning the trade and in 2015 the opportunity came for him to move his production to the Shene Estate at Pontville, 30 minutes north of Hobart. The estate, dating back to 1819, has been restored and is now owned by David and Anne Kemke. No whisky has yet been released from Mackey´s production but 2018 could be the year when his first triple distilled, Irish-style malt whisky will be launched.

Mt Uncle Distillery

Walkamin, North Queensland, founded in 2001

www.mtuncle.com

When Mark Watkins founded the distillery in 2001, he started out by producing gin, rum and vodka - all of which soon became established brands on the market. After a few years, Watkins decided to add whisky production as well. Their first single malt, The Big Black Cock, released in April 2014, is produced using local Queensland barley and has been matured for five years in a combination of French and American oak.

Loch Distillery

Loch, Victoria, founded in 2014

www.lochbrewery.com.au

Situated in an old bank building, this combined brewery and distillery has been producing since summer 2014. The owner, Craig Johnson, learnt about distilling from Bill Lark (like so many others have) before ordering his stills from Portugal. Gin has already been released while whisky production didn´t start until March 2015. The wash used for the whisky production comes from their own brewery.

Fanny´s Bay Distillery

Weymouth, Tasmania, founded in 2015

One of the newest additions to the Tasmanian whisky scene, the distillery was built by Mathew and Julie Cooper in 2014. Most of the equipment was actually constructed by Mathew himself and the distillery is equipped with a 400 litre copper pot still, a 600 litre mash tun and a 300 litre washback with a 7-8 day fermentation. The whisky starts in 20 litre port barrels and is then finished in small bourbon casks.

New Zealand

Distillery:	New Zealand Malt Whisky Co., Oamary, South Island
Founded:	2000
Owner:	Extra Eight
	thenzwhisky.com, milfordwhisky.co.nz

In 2001, Warren Preston bought the entire stock of single malt and blended whisky from the decommissioned Wilsons Willowbank Distillery in Dunedin. The supplies that he acquired consisted of 400 casks of single malt whisky including production dating back to 1987. Before he bought it, the whisky was sold under the name Lammerlaw, but Preston renamed it Milford. Preston also had plans to build a distillery in Oamaru. In 2010, however, his company was evicted from its premises and later it was placed in receivership. In October 2010, rescue came in the form of a syndicate of investors led by Tasmanian-based businessman Greg Ramsay. Their capital injection revived the company and plans to build a distillery still exist. Since early 2015, the company has been distilling trial batches of whisky at Kenny Beverages in Christchurch, a small distillery owned by Doug and Anthony Lawry.

With the new ownership, the range of expressions from the old stock has increased and the following bottlings have been released (some of them are now discontinued); Milford Single Malt (10, 15, 18 and 20 years old), South Island Single Malt (18 and 21 years), Dunedin Doublewood blend 15 year old (6 years in American oak and 9 years in French oak seasoned with New Zealand red wine), single casks from 1989, 1990 and 1992, Diggers & Ditch (a blended malt with whiskies from both New Zealand and Tasmania) and Cyril´s Single Wood which was released in honour of Cyril Yates who spent over 20 years working at the Willowbank Distillery. Among the most recent bottlings is a 25 year old released in 2014.

Distillery:	Thomson Whisky Distillery, Auckland, North Island
Founded:	2014
Owner:	Thomson Whisky New Zealand Ltd.
	thomsonwhisky.com

The company started out as an independent bottler, sourcing their whiskies from the closed Willowbank Distillery in Dunedin, New Zealand. Two of the whiskies available are an 18 year old and a 21 year old single cask while the third, named Two Tone, is a vatting of whisky matured in American white oak and in European oak casks that had held New Zealand red wine. In April 2014, Rachel and Mathew Thomson opened up a small distillery (basically just a copper pot still) based at Hallertau Brewery in North West Auckland. The wash for the distillation comes from the brewery. In summer 2014, trials were made producing a whisky where the malt had been kilned using New Zealand Manuka wood. "Work-in-progress" bottles of the Manuka spirit were released in 2015 but the first whisky isn´t expected until 2017 or 2018.

Asia

India

Distillery:	Amrut Distilleries Ltd., Bangalore
Founded:	1948
Owner:	Jagdale Group
	amrutwhisky.co.uk

The family-owned distillery, based in Bangalore, south India, started to distil malt whisky in the mid-eighties. The equivalent of 50 million bottles of spirits (including rum, gin and vodka) is manufactured a year, of which 1,4 million bottles is whisky. Most of the whisky goes to blended brands, but Amrut single malt was introduced in 2004. It was first introduced in Scotland, but can now be found in more than 20 countries and has recently been introduced to the American market. Funnily enough, it took until 2010 before it was launched in India.

The distillery, with a capacity of doing 200,000 litres of pure alcohol per year, is equipped with six washbacks with a fermentation time of 140 hours and two stills, each with a capacity of 5,000 litres. The barley is sourced from the north of India, malted in Jaipur and Delhi and finally distilled in Bangalore before the whisky is bottled without chill-filtering or colouring. The owners have plans to build yet another distillery adjacent to the present and there have also been discussions about building a distillery in Himachal Pradesh in the Himalayas.

The Amrut family of single malts has grown considerably in recent years and ingenuity is great when it comes to new, limited releases. The core range consists of unpeated and peated versions bottled at 46%, a cask strength and a peated cask strength and, finally, Fusion which is based on 25% peated malt from Scotland and 75% unpeated Indian malt. Special releases include Two Continents, where maturing casks have been brought from India to Scotland for their final period of maturation: Intermediate Sherry Matured where the new spirit has matured in ex-bourbon or virgin oak, then re-racked to sherry butts and with a third maturation in ex-bourbon casks: Kadhambam which is a peated Amrut which has matured in ex Oloroso butts, ex Bangalore Blue Brandy casks and then finally in ex rum casks; Amrut Herald (released in 2011) with four years bourbon maturation in India and a final 18 months on the German

island of Helgoland and, finally, Portonova (also from 2011) with a maturation in bourbon casks, then 9 months in port pipes and back to bourbon casks for the last 8 months. Both Portonova and Intermediate Sherry are annual limited editions, whereas new expressions of Kadhambam, Herald and Two Continents are released on a more irregular basis. Lately, special single cask releases have been made exclusively for the European markets.

A big surprise for 2013 was the release of Amrut Greedy Angels, 8 years old and the oldest Amrut so far. That was an astonishing achievement in a country where the hot and humid climate cause major evaporation during maturation. In 2015 it was time for an even older expression when the 10 year old Greedy Angel's - Chairman's Reserve was launched. With a total of 320 bottles, this became the oldest and most expensive whisky from India so far. Another new release in 2015 was Amrut Naarangi, a unique whisky with orange flavour. Three year old Amrut single malt, was filled into barrels that had held wine and orange peel for three years. After another three years of maturation, the whisky had taken on distinctive orange notes.

Distillery:	McDowell's, Ponda, Goa
Founded:	1988 (malt whisky)
Owner:	United Spirits (Diageo)
	unitedspirits.in

In 1826 the Scotsman, Angus McDowell, established himself as an importer of wines, spirits and cigars in Madras (Chennai) and the firm was incorporated in 1898. In the same town another Scotsman, Thomas Leishman, founded United Breweries in 1915. Both companies were bought by Vital Mallya around 1950 and eventually, under the leadership of Vijay Mallya, United Breweries (with the spirits division United Spirits) became the second largest producer of alcohol in the world after Diageo. Since 2014, after a business settlement involving several steps, Diageo controls the majority of the shares in United Spirits.

United Spirits has more than 140 brands in their portfolio including Scotch whisky, Indian whisky, vodka, rum, brandy and wine. The absolute majority of United Spirits' whiskies are Indian whisky, made of molasses. The major brands in the group are huge sales-wise, with McDowell's No.1 whisky as the top seller (25 million cases in 2014). On the other hand, single malt sales are negligible in comparison to these figures. McDowell's Single Malt is made at the distillery in Ponda (Goa) and sells 20,000 cases each year. It has matured for 3-4 years in ex-bourbon casks.

Amrut Distillery and Greedy Angels - Chairman's Reserve 10 years old

Distillery: John Distilleries Jdl, Goa
Founded: 1992
Owner: Paul P John
pauljohnwhisky.com

Paul P John, who today is the chairman of the company, started in 1992 by making a variety of spirits including Indian whisky made from molasses. Their biggest seller today is Original Choice, a blend of extra neutral alcohol distilled from molasses and malt whisky from their own facilities. The brand, which was introduced in 1995/96 has since made an incredible journey. It is now the world's 8th most sold whisky with sales of 126 million bottles in 2014.

John Distilleries owns three distilleries and produces its brands from 18 locations in India with its head office in Bangalore. The basis for their blended whiskies is distilled in column stills with a capacity of 500 million litres of extra neutral alcohol per year. In 2007 they set up their single malt distillery which is equipped with one pair of traditional copper pot stills. The company released their first single malt in autumn 2012, a bourbon-matured single cask bottled at cask strength and around 4 years old. In the ensuing months more single cask expressions became available, some of them matured in ex-sherry casks. In May 2013 it was time for two more widely available core expressions, both made from Indian malted barley and distilled in copper pot stills in their distillery in Goa. Brilliance is unpeated and bourbon-matured while Edited, also matured in bourbon casks, has a small portion of peated barley in the recipe. The peat was brought in from Scotland but the malting process took place in India. At the beginning of 2014, two cask strength bottlings were released; Select Cask Classic (55,2%) and the quite heavily peated Select Cask Peated (55,5%). These are limited versions but will be released continuously. In 2015, finally, the third core expression was released. It was a 100% peated bottling called Bold, bottled at 46%

The whiskies from John Distilleries have no age statements but are probably quite young. It has to be remembered that the conditions in Scotland, with a cool climate and an evaporation loss (angel's share) of 1,5% per year, are quite different to those in Goa with temperatures reaching 40°C and an angel's share of 10% per year. Maturation therefore is much quicker in India. Currently the single malts are available in ten countries in Europe (including the UK, Germany and Scandinavia) as well as in Australia, Singapore, Taiwan and Malaysia.

Israel

Distillery: The Milk & Honey Distillery, Jaffa
Founded: 2013
Owner: Simon Fried, Amit Dror et al.
mh-distillery.com

This is the first whisky distillery in Israel, with a team behind the project comprising of Simon Fried, Amit Dror, Gal Kalkshtein and Nir Gilat. They have been ably assisted by the well-known whisky consultant, Jim Swan, to devise an equipment specification. When ready, the distillery will be equipped with mash tun and washbacks made of stainless steel and two copper stills (9,000 and 3,500 litres respectively). The whisky will not be matured on site near Tel Aviv. Instead, the owners want to experiment with maturation in a wide variety of climates, from dry deserts to hot and humid coastal areas. By summer 2015, the two stills had been installed and wash was brought in from an external source to start distillation of whisky but also vodka and gin. Mash tun and washbacks were expected to be ready by the end of 2015.

Pakistan

Distillery: Murree Brewery Ltd., Rawalpindi
Founded: 1860
Owner: Bhandara family
murreebrewery.com

Murree Brewery in Rawalpindi started as a beer brewery supplying the British Army. The assortment was completed with whisky, gin, rum, vodka and brandy. The core range of single malt holds two expressions – Murree's Classic 8 years old and Murree's Millenium Reserve 12 years old. In 2005 an 18 year old single malt was launched and the following year their oldest expression so far, a 21 year old, reached the market. There is also a Murree's Islay Reserve, Vintage Gold, which is a blend of Scotch whisky and Murree single malt and a number of local, blended whiskies such as Vat No. 1, Lion and Dew of Himalaya. The brewery makes its own malt (using both floor maltings and Saladin box) and produces 2,6 million litres of beer every year and approximately 440,000 litres of whisky.

John Distilleries and the new, peated Paul John Bold

The Milk & Honey Distillery

Taiwan

Distillery: Yuan Shan Distillery, Yanshan, Yilan County
Founded: 2005
Owner: King Car Food Industrial Co
kavalanwhisky.com

In a short period of time, the Taiwanese single malt, Kavalan, has swept across the world like few other brands from outside the traditional group of whisky producing countries have done before. The merits for the rapid establishment mainly lie with Master Blender, Ian Chang, and the well-known whisky consultant, Dr. Jim Swan, who have worked together since its inception in 2005.Yuan Shan distillery lies in the north-eastern part of the country, in Yilan County, just one hour from Taipei.

The distillery is equipped with a 4 ton semi-lauter stainless steel mash tun with copper top and eight closed stainless steel washbacks with a 60-72 hour fermentation time. There are two pairs of lantern-shaped copper stills with descending lye pipes. The capacity of the wash stills is 12,000 litres and of the spirit stills 7,000 litres. After 10-15 minutes of foreshots, the heart of the spirit run takes 2-3 hours. The cut points for the spirit run have been changed from 78%-72% to 65%-55% to better accommodate a more complex and richer flavour profile. The spirit vapours are cooled using tube condensers, but due to the hot climate, subcoolers are also used.

The total capacity with the above mentioned equipment is 1,5 million litres per year. However, the rapid success for the brand has now prompted the owners to increase the production. In September 2015, another 6 stills were installed which will bring up the capacity to 4.5 million litres of alcohol by the end of 2015. But it doesn't stop at that - by the end of 2016, the distillery will be equipped with a total of two mash tuns, 16 washbacks and 20 stills with a staggering capacity of 9 million litres of pure alcohol! The two warehouses are five stories high and the casks are tied together four and four due to the earthquake risk. The climate in Taiwan is hot and humid and on the top floors of the warehouses the temperature can easily reach 42°C. Hence the angel's share is quite dramatic – no less than 15% is lost every year. Predominantly American white oak is used, since the owners have found that it is the type of wood that works the best for maturation in a subtropical climate.

The brand name for the whisky produced at Yuan Shan distillery is Kavalan. The name derives from the earliest tribe that inhabited Yilan, the county where the distillery is situated. Since the first bottling was released in 2008, the range has been expanded and now holds more than ten different expressions. The best seller globally is Classic Kavalan, bottled at either 40% or 43%. In 2011, an "upgraded" version of the Classic was launched in the shape of King Car Conductor – a mix of eight different types of casks, un chill-filtered and bottled at 46%. A port finished version called Concertmaster (currently the best selling Kavalan in the USA) was released in 2009 and, later that year, two different single cask bottlings were launched under the name Solist – one ex-bourbon and one ex-Oloroso sherry. It was these two expressions that made the rest of the world aware of Taiwanese whisky. They are still part of the range with new editions being released continuously.

More expressions in the Solist series were added – Solist Fino with a fino sherry maturation and Solist Vinho Barrique, where Portuguese wine barriques had been used. By the end of 2012, two varieties of the Solist, but bottled at 46% were launched – Kavalan Bourbon Oak and Kavalan Sherry Oak – together with Kavalan Podium. The latter is a vatting of whiskies from new American oak and a selection of re-fill casks. Kavalan Distillery Reserve Peaty Cask, exclusively available at the distillery visitor centre, is a lightly peated expression that was released in 2013. The latest releases came in 2015 when a new range of shery maturations was launched - Amontillado, Manzanilla and PX. A special, limited bottling, Solist Moscatel, was also released for Whisky Luxe Taipei 2015.

The most important markets, apart from Taiwan, are Singapore, Hong Kong, Malaysia and China. The end of 2012 and the beginning of 2013 marked the start of the export of Kavalan to other markets and the whisky is now available in most European countries as well as in Australia, South Africa and USA. The company is also working to establish themselves in Russia as well.

There is an impressive visitor centre on site and it was awarded Whisky Visitor Attraction of the Year in 2011 by Whisky Magazine. No less than one million visitors come here per annum. The owning company, King Car Group, with 2,000 employees, was already founded in 1956 and runs businesses in several fields; biotechnology and aquaculture, among others. It is also famous for its ready-to-drink coffee, Mr. Brown, which is also exported to Europe.

Yuan Shan Distillery - home of Kavalan whisky with Podium and the latest release, Solist Moscatel

Africa

Other distilleries in Taiwan

Nantou Distillery

Nantou City, Nantou County, founded in 1978
(whisky since 2008)

en.ttl.com.tw

Nantou distillery is a part of the state-owned manufacturer and distributor of cigarettes and alcohol in Taiwan – Taiwan Tobacco and Liquor Corporation (TTL). Established as a government agency in the early 1900s, it was renamed Taiwan Tobacco and Wine Monopoly Bureau in 1947. Between 1947 and 1968 the Bureau exercised a monopoly over all alcohol, tobacco, and camphor products sold in Taiwan. It retained tobacco and alcohol monopolies until Taiwan's entry into the WTO in 2002.

There are seven distilleries and two breweries within the TTL group, but Nantou is the only with malt whisky production. The distillery is equipped with a full lauter Huppmann mash tun with a charge of 2.5 tonnes and eight washbacks made of stainless steel. The fermentation time is 60-72 hours and in order to regulate the fermentation, the washbacks are equipped with water-cooling jackets. There are two wash stills (9,000 and 5,000 litres) and two spirit stills (5,000 and 2,000 litres). All are equipped with shell and tube condensers, as well as aftercoolers. Malted barley is imported from Scotland and ex-sherry and ex-bourbon casks are used for maturation. Nantou Distillery also produces a variety of fruit wines and the casks that have stored lychee wine and plum wine are then used to give some whiskies an extra finish. Due to extreme temperatures during summer, distillation only takes place from October to April. The hot and humid climate also increases the angle's share which is around 6-7%. Until recently, the spirit from Nantou has been unpeated, but in 2014, trials with peated malt brought in from Scotland were made.

The main product from the distillery is a blended whisky which comprises of malt whisky from Nantou, grain whisky from Taichung distillery and imported blended Scotch. In October 2013, two cask strength single malt whiskies were launched – one from bourbon casks and the other from sherry casks. The next expressions were Omar single malt, a vatting of bourbon- and sherrymatured whisky in January 2015 and, two months later, a bourbonmatured single cask.

South Africa

Distillery: James Sedgwick Distillery, Wellington, Western Cape

Founded: 1886 (whisky since 1990)

Owner: Distell Group Ltd.

threeshipswhisky.co.za

Distell Group Ltd. was formed in 2000 by a merger between Stellenbosch Farmers' Winery and Distillers Corporation, although the James Sedgwick Distillery was already established in 1886. The company produces a huge range of wines and spirits including the popular cream liqueur, Amarula Cream. James Sedgwick Distillery has been the home to South African whisky since 1990. The distillery has undergone a major expansion recently and is now equipped with one still with two columns for production of grain whisky, two pot stills for malt whisky and one still with six columns designated for neutral spirit. There are also two mash tuns and 23 washbacks. Grain whisky is distilled for nine months of the year, malt whisky for two months (always during the winter months July/August) and one month is devoted to maintenance. Three new warehouses have been built and a total of seven warehouses now hold 180,000 casks.

In Distell's whisky portfolio, it is the Three Ships brand, introduced in 1977, that makes up for most of the sales. The range consists of Select and 5 year old Premium Select, both of which are a blend of South African and Scotch whiskies. Furthermore, there is Bourbon Cask Finish, the first 100% South African blended whisky and the 10 year old single malt. The latter was launched for the first time in 2003 and it wasn't until autumn 2010 that the next batch was released. It sold out quickly and another 8,000 bottles were launched in October 2011 and a fourth batch appeared in December 2012. The next release, 11,000 bottles and this time lightly peated, won't be until early 2016 (probably March). The reason for this is that the success of the release in 2003 took the owners by surprise and there was no stock reserved for single malt bottlings. The production all went to their blends. It wasn't until 2005 that planning again started for future, single malt releases. The happy news is that from the fifth release, there will a batch available each year. Ahead of the new bottling, the first limited single cask release (around

Nantou Distillery in Taiwan

800 bottles) delighted the fans in early autumn 2015. It had been matured in American oak for almost nine years and then finished in PX casks for another 14 months. Apart from the Three Ships range, Distell also produces two 3 year old blended whiskies, Harrier and Knight, as well as South Africa´s first single grain whisky, Bain´s Cape Mountain.

In 2013 the Distell Group acquired the Scottish whisky group, Burn Stewart Distillers, including Bunnahabhain, Tobermory and Deanston distilleries as well as the blended whisky, Scottish Leader.

.

Other distilleries in South Africa

Drayman´s Distillery

Silverton, Pretoria, founded in 2006

www.draymans.com

Being a full-time beer brewer since 1997, Moritz Kallmeyer began distilling malt whisky in July 2006. Until two years ago, production was small, but operations have now been expanded to two pot stills. The new wash still has a capacity of 1,500 litres with the old spirit still holding 800 litres. The wash is fermented for up to ten days in the washback, to allow the malolactic fermentation to transfer its character to the spirit. The whisky matures in French oak casks which have previously held red wine from the Cape area. Kallmeyer´s first whisky was released as a 4 year old in autumn 2010 under the name Drayman´s Highveld Single Malt and there has also been the release of a second batch. Other products include Mampoer, which is a local brandy, a honey liqueur and fruit schnapps. The main source of income, however, comes from production of craft beers.

James Sedgwick Distillery and the recently released Three Ships Single Cask PX

South America

Argentina

Distillery: La Alazana Distillery, Golondrinas, Patagonia
Founded: 2011
Owner: Nestor Serenelli
laalazanawhisky.com

The first whisky distillery in Argentina concentrating solely on malt whisky production was founded in 2011 and the distillation started in December of that year. Located in the Patagonian Andes to the South of Argentina, it was by Pablo Tognetti, an old time home brewer, and his son-in-law, Nestor Serenelli but end of 2014, Pablo Tognetti withdrew from the company. The distillery is equipped with a lauter mash tun, four 1,100 litre washbacks all made of stainless steel and two stills - one 500 litres and another, recently installed, 1,300 litres. The second still made it possible to double the production to 8,000 litres a year.

The owners are aiming for a light and fruity whisky but they have also filled 15 barrels with peated whisky. Maturation is mostly in ex-bourbon casks but fresh PX sherry casks and toasted Malbec casks are also used. A visitor centre has recently been built and a grand, official opening of the distillery was held in November 2014. The first, limited release was made in December 2013. A second release appeared in November 2014 and a third is planned for end of 2015. These are all limited as the owners want to keep most of the stock for older expressions.

Distillery: Justus Distillery, Bariloche, Patagonia
Founded: 2015
Owner: Pablo Tognetti
justuswhisky.com

What will become the second whisky distillery in Argentina, is now under construction. The founder, Pablo Tognetti, is no newcomer to whisky production as he was one of the owners of the Patagonian distillery La Alazana. In 2015, he left the company and brought with him some of the equipment as well as part of the maturing stock to build a new distillery in Dina Huapi, on the outskirts of the lake-side Andean ski resort Bariloche. The existing 1,300 litre copper pot still will be complemented by another still.

Brazil

Distillery: Union Distillery, Veranópolis
Founded: 1972
Owner: Union Distillery Maltwhisky Do Brasil Ltda
maltwhisky.com.br

The company was founded in 1948 as Union of Industries Ltd to produce wine. In 1972 they started to produce malt whisky and two years later the name of the company was changed to Union Distillery Maltwhisky do Brasil. In 1986 a co-operation with Morrison Bowmore Distillers was established in order to develop the technology at the Brazilian distillery. Most of the production is sold as bulk whisky to be part of different blends, but the company also has its own single malt called Union Club Whisky.

Distillery: Muraro Bebidas, Flores da Cunha
Founded: 1953
Owner: Muraro & Cia
muraro.com.br

This is a company with a wide range of products including wine, vodka, rum and cachaca and the total capacity is 10 million litres. Until recently, the blend Green Valley was the only whisky in the range. In November 2014, however, a new brand was introduced. It has the rather misleading name Blend Seven but it appears to be a malt whisky. The main market for the new whisky is The Carribean.

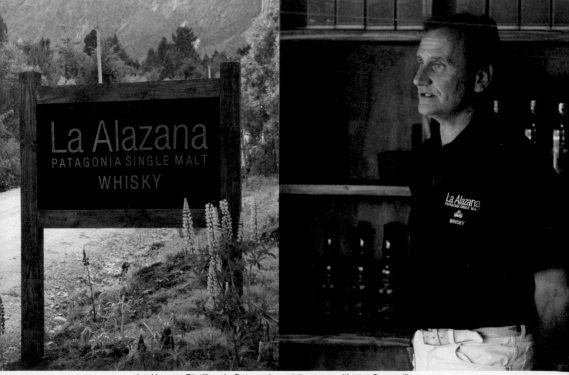

La Alazana Distillery in Patagonia and the owner, Nestor Serenelli

The Year
that was

Including the subsections:
The big players | The big brands | Changes in ownership
New distilleries | Bottling grapevine

From 2008 to 2013, the global spirits market grew by 11% and in 2014 volumes grew by a further 1.7%. The biggest spirits category, without competition, is baijiu which accounts for one quarter of the volumes and with almost 100% of the sales in China. In second place is vodka and in third place whisk(e)y, but whisky stands a fair chance of soon passing vodka, as sales of vodka continue to decline. In terms of value, whisky is already bigger than vodka.

If we look at Scotch whisky export, it does not make for happy reading. For the first time since 2005, values have decreased and by quite a lot, 7%, while volumes have slipped by 3%. There are many reasons for the decline. Firstly, bourbon has increased vigorously, not only in USA but also where export markets are concerned. Secondly, the interest in rum has continued to grow and in some of the important Asian markets, the younger consumers have turned to white spirits such as vodka and, in South Korea, soju.

It looks worse for blended Scotch, which has declined considerably, with volumes down by 8% and values even more, -15%. While all the other whisky categories have grown during 2014 (except for Canadian whisky), blended Scotch is losing grounds. It looks a lot brighter for single malt Scotch. The volume has grown to more than 100 million bottles and value has jumped by 12% to £915m. This means that single malts now represent almost a quarter of the total value of exported Scotch.

SINGLE MALT - EXPORT
Value: +12% to £915m
Volume: +11% to 105m bottles

BLENDED SCOTCH - EXPORT
Value: -15% to £2,7bn
Volume: -8% to 812m bottles

TOTAL SCOTCH - EXPORT
Value: -7% to £3,95bn
Volume: -3% to 1.19bn bottles
NB. Grain whisky, blended malt and bulk export are not included, except in the Total.

If we look at the revenue per bottle for the producers by simply dividing total values with total volumes, that figure has been growing continuously over the last decade until 2013 when the trend was broken. That year earnings were £3.47 per bottle and in 2014 it was down to £3.32.

Looking at the detailed figures for each of the nine regions, all but two (Asia and Middle East) have shown decreasing volumes, while the Middle East and European non-EU members have shown increasing value.

The European Union

The European Union (excluding the UK) is still, by far, the biggest market for Scotch whisky in the world. In terms of volume, 37% of the exports land in this region, while the value is marginally less (at 31%). During 2014 the volumes decreased by 1% while values were down by 6%.

EU — Top 3

France	volumes	+3%	values	+2%
Spain	volumes	+2%	values	-8%
Germany	volumes	-5%	values	-18%

France is not only the largest market in the EU – it is also, in terms of volume, the biggest importer of Scotch whisky in the world. Together with The Netherlands and Hungary, it was also the only country in the region where both values and volumes increased during 2014. Spain is still the second biggest market and even if sales have decreased steadily over the last 6-7 years, the pace has slightly slowed down. The most surprising numbers are the ones from Germany, which has long been a stable market for Scotch. The sharp decline during 2014 can possibly partly be ascribed to the increased sales of bourbon. Germany is the third biggest export market for the American bourbon producers. The Netherlands have proved to be a stable market during 2014 with both volumes (+4%) and values (+3%) increasing. Poland, on the other hand, went the other direction, with volumes and values dropping around 30%. A lot of the whisky that reaches Poland is exported to Russia and the trade embargo due to the conflict in Ukraine has naturally affected this trade.

North America

The second biggest region in terms of value, continues to be North America, which in 2013 eclipsed Asia into third place. The figures for 2014 though, were all but inspiring. Volumes were down 4% and values fared even worse, slipping by 8% to £914m.

North America — Top 3

USA	volumes	-7%	values	-9%
Mexico	volumes	+5%	values	-10%
Canada	volumes	-2%	values	+1%

USA is the most important market and the biggest in the world for Scotch in terms of value. The sharp decline in 2014 can, in part, be explained by a degree of caution from the importers, who wanted to use existing stock to meet the demand. Figures from US Distilled Spirits Council showed that the Scotch market in the US actually shrunk by just 1% and for the first few months in 2015, import of Scotch increased by more than 30%.

Asia

After several years as the fastest growing market for Scotch, the figures for Asia started dropping in 2012. The fall continued in 2013 and also 2014 even if the pace has subsided slightly. Values for 2014 were down 6% while volumes remained the same.

Asia — Top 3

Singapore	volumes	-41%	values	-39%
Taiwan	volumes	+23%	values	+36%
South Korea	volumes	-3%	values	+2%

Singapore is the number one market but the vast majority of the whisky is re-exported to other countries in the region, not least China. The sharp decline for the country is partially explained by a tougher legislation aimed at severely limiting consumption and gifting of expensive alcohol within government agencies, as well as public and private companies.

Taiwan, in second place, continues to be a strong market, especially for single malt. Both volumes and values increased substantially in 2014 and Taiwan is now the one country spending the most money per capita on Scotch whisky.

South Korea was one of the first emerging markets to embrace Scotch but, since 2005, sales have decreased due to a difficult economic climate, even though values showed a minor increase during 2013. Exports to China, as mentioned before, have gone down dramatically due to the austerity measures put in place by the government.

India, on the other hand, has pleased the producers of

Scotch over the last few years and 2014 was no exception. Volumes were up by 30% and values increased by 20% and the country has climbed to third place on the list of countries that import the largest volumes of Scotch. Negotiations to bring down the high import tariffs (150%) for Scotch going to India has been held during a long period of time. If these talks would be successful, India would become an even more important market in the future.

Central and South America

In fourth place in the regional list is Central and South America, an important market, particularly for blended Scotch. The region as a whole, however, was not a source of inspiration for the producers of Scotch in 2014. Volumes were down by 11% and values decreased by as much as 21%.

Central & South America — Top 3

Brazil	volumes	+-0%	values	-20%
Panama	volumes	+15%	values	-10%
Uruguay	volumes	-11%	values	+4%

The number one market is Brazil, the fifth biggest in the world in terms of volumes. It is a volatile market though with values down by 20% compared to an increase the year before by 18%. Venezuela, one of the top three markets in the region in 2013, has slipped to outside of the list for 2014 due to a dramatic decrease in both volumes (-62%)

and values (-63%). Fluctuations like these are not unusual in the region. In the same way as in Southeast Asia, the consumers quickly shift to cheaper domestic spirits in times of economic uncertainty.

Africa

Africa, exluding South Africa, is still an undeveloped market for Scotch, but most of the big producers have begun to position themselves in order to be prepared when the consumer interest starts to take off. In 2014, volumes were down by 2%, while values decreased by 11%.

Africa — Top 3

South Africa	volumes	-8%	values	-17%
Angola	volumes	+20%	values	+24%
Nigeria	volumes	-10%	values	-8%

The reason for the region´s decline is the substantial decrease noted in South Africa. The country is still the biggest market on the continent, but the last few years has seen their share of the import to Africa go from 75% in 2011 to 55% in 2014. Angola, in second place, has experienced an upward spiral during the last few years and part of the increase can be attributed to a huge amount of foreign nationals working in the country´s expanding oil industry.

After a few troublesome years, the Asian market for whisky seems to be recovering

Middle East

Middle East was the only region where both volumes (+16%) and value (+22%) increased in 2014. That having been said, the region still only accounts for 4% of global Scotch sales.

Middle East - Top 3

UAE*	volumes	+26%	values	+27%
Lebanon	volumes	+10%	values	+22%
Israel	volumes	+3%	values	+28%

* United Arab Emirates

The biggest market in the region is the United Arab Emirates but most of the whisky is not consumed within the country. Instead, UAE acts as a distribution hub for parts of Africa, Asia and India. The Top 3 markets represent 87% of the sales to the region.

European non EU-members

Export to European non EU-members have increased in recent years, perhaps not so much in terms of volume, but certainly in terms of value. Figures for 2014 show volumes remaining un-changed and value increasing by 2%.

Europe (non-EU) — Top 3

Turkey	volumes	-1%	values	+4%
Switzerland	volumes	+2%	values	+1%
Norway	volumes	-8%	values	-8%

The region is dominated by Turkey and Switzerland and The Top 3 markets represent 94% of the sales to the region.

Australasia

Australasia is the second smallest region, importing the equivalent of 30 million bottles of Scotch. Values remained un-changed in 2014 while volumes dropped by 9%.

Australasia — Top 3

Australia	volumes	-10%	values	-1%
New Zealand	volumes	+16%	values	+7%
N. Caledonia	volumes	-1%	values	+1%

Australia is dominating with more than 90% of the value, placing the country in 12th place on the global list.

Eastern Europe

Eastern Europe has long been considered to be one of the emerging markets for Scotch, but this brilliant prospect came to a halt with the sharp decline during 2014. Values were down 62% and volumes even more, minus 68%.

Eastern Europe — Top 3

Georgia	volumes	-12%	values	-21%
Ukraine	volumes	-15%	values	-21%
Russia	volumes	-93%	values	-93%

The reason for this dramatic change is, of course, the crisis in Ukraine, which has resulted in a trade embargo against Russia. Values, as well as volumes to Russia decreased by 93%, which almost obliterated the entire export to the country. The neighbouring countries, Georgia and Ukraine, were also affected but not to the same extent. The figures for the region are sometimes difficult to interpret as a result of the various volumes being imported to Russia via the Baltic states. Instead, they are reflected in the figures for Latvia and Estonia.

The big players

Diageo

We have become used to Diageo presenting good results year after year, but that soon came to an end in June 2014, when profits fell by almost 15%. Seen in the light of that, this year's figures are definitely better, but not impressive. Net sales were up by 5.4% to £10.8bn but the entire increase was due to the full and final consolidation of United Spirits, the company where Diageo now has a majority share. Without the addition of United Spirits volumes, sales were flat in organic terms. Net profits were up by 13% to £2.47bn, but operating profits rose by a mere 3.3%.

All product categories showed negative sales movement with the exception of tequila and North American whiskies (not least Bulleit bourbon and rye). Scotch whiskies were down by 5%, and are reflected in the declining figures for Johnnie Walker (down 9% to 224 million bottles), J&B, Buchanan's, Bell's, Windsor and Old Parr. The only light in the dark tunnel came from single malts which, as a category, rose by 16% with The Singleton (+20%), Talisker (+11%) and Lagavulin (+8%) being the biggest gainers.

The only one of Diageo's five sales regions that showed positive figures was Africa with sales being up by 6%. The lion's share of the volumes in that market is beer, but spirit is also increasing in several countries. The sales figures for North America (being 32% of the total Diageo sales), Western Europe (24%), Latin America (10%) and Asia-Pacific (20%) were all unchanged or down by 1-2%.

In November 2014, Diageo took the market by surprise when they announced that they were selling Bushmill's distillery. This was Diageo's only presence in the vibrant, increasing Irish whiskey segment and commentators struggled to see the reason for the sale. The buyer was the tequila maker, Casa Cuervo, producer of the well-known José Cuervo. Diageo already owned 50% of the company's other, upscale tequila brand, Don Julio and with the clinching of this deal, they acquired the remaining 50% as well as $408m.

The Diageo take-over of the Indian spirits giant, United Spirits, which started in 2012, has proven more difficult than Diageo initially expected. In addition to all the obstacles on the way, it was announced in May 2015 that the Indian government was to investigate United Spirits' books. The main reason was that there was a suspicion that the former owner, Vijay Mallya, had been using funds from United Spirits to pay off debts in some of his other companies, mainly Kingfisher Airlines. At the time of writing, Mallya still owns a minority part of United Spirits and he is also the chairman of the company.

It has been a tough year for Diageo with sales declining for top brand Johnnie Walker

In May 2015, a rumour was floating about that Diageo was the subject of a take-over by the Brazilian investment firm, 3G Capital. The investment company, controlled by Jorge Paulo Lemann, already owns one of the world´s largest beer groups, InBev, and there has been speculation that Diageo would want to get out of the beer market, which means that 3G Capital could be interested in that segment of the business at least.

Diageo, on the other hand, could be involved in a deal with Möet Hennessy. In the 1990s, Diageo acquired a 34% stake of the French drinks group, and rumour has it that LVMH (majority owner of Möet Hennessy) is now looking to buy back that stake.

Pernod Ricard

The new CEO of Pernod Ricard since February 2015, Alexandre Ricard, has set himself a goal and that is to displace Diageo as the world´s number one premium drinks company. Admittedly, he has a long way to go, but the financial statement that was released for 2014/2015 at the end of August, indicated that they are indeed on the right path. The overall figures were appreciably better than for Diageo, for the corresponding period that also ended on the 30th June.

Net sales were up by 8% to €8.56bn after last year´s decline of 14%. Operating profits had increased by 9% to €2.24bn, while net profits had fallen by 14% to €880m. The reason for the decrease in net profits could be ascribed to the performance of Absolut vodka. A considerably lower growth for the brand in the USA than was expected, forced the company to take an impairment charge of €404m.

Looking at the different regions, Asia/Rest of the World has returned to the plus side (+14%) compared to last year´s depressing results. One reason for this is the strong increase in India (+18%) but even in China, where sales last year decreased by almost 25%, the market seems to be stabilizing now. Admittedly, the annual result showed another decline, but only by 2%. Sales in the Americas increased by 11% while Europe fell by 2%.

Two brands showed particularly pleasing figures; Jameson was up by 10% while Glenlivet increased by 11%. The impressive figures for Glenlivet in recent years has led to the fact that the brand is now the biggest selling single malt in the world, surpassing Glenfiddich. Ballantines also increased while sales of Chivas Regal remained unchanged. Of the local brands of whiskies, 100 Pipers did well in India and Passport increased in Mexico and Brazil.

Just as with several of the other larger producers, Pernod Ricard also feels the changes of the drinking habits among the younger consumers in certain markets. The current trend leans towards white spirits in South Korea, while young consumers in the USA have taken a liking to bourbon. Another attempt at meeting the needs of younger Americans was this year´s launch of Barrelhound, a blended Scotch but with a flavour that imitates the sweeter profile of bourbon.

For future growth, Alexandre Ricard has outlined four markets or "battle grounds" as he calls them; USA, China, India and Africa. He also stated that the company is looking to shift focus away from being brand-centric to being more consumer-centric.

Edrington

If it hadn´t been for one particular brand, the results for Edrington last year would have been positive. The brand in question is Brugal, the rum that Edrington acquired in 2008. Sales have since been going down in key markets such as Spain and the Dominican Republic (where Brugal

Pernod Ricard´s CEO, Alexandre Ricard and the company´s new blended Scotch - Barrelhound

Company is based) and competition from other brands has been fierce. The write-down of Brugal in 2012 was followed by another £239m write-down in 2014.

The results for Edrington for the fiscal year which ended March 2015, showed net sales of £617m (-2.4%), while net profits were down by 10% to £157.6m. If Brugal was a disappointment, the other key brands of the company at least showed more pleasing figures. For Macallan single malt, volumes increased by 10% to almost 10 million bottles and the brand did particularly well in the USA and Asia. The company reports that it is now the category leader in Russia, China, Japan, South Korea and Hong Kong. The other single malt, Highland Park, increased sales volumes by 15%, which was driven mainly by growth in the USA.

The company has two major blends in the portfolio. Famous Grouse was steady on roughly the same figures as in 2014, which were 37 million bottles, while Cutty Sark increased by 4% to 9 million bottles. In 2015, the owners announced that the Famous Grouse range was due for a redevelopment into the premium whisky segment. According to CEO, Ian Curle, "It is evident that the market for mainstream blended Scotch and rum will continue to be competitive in the coming years." The company will also increase its presence on the American market with the goal to double current sales by 2018.

The Edrington Group is owned by a charitable trust, The Robertson Trust, established in 1961, which donated more than £18.2m to charitable causes in Scotland during 2013/2014.

Gruppo Campari

The CEO of Campari, Bob Kunze-Concewitz, was cautious when he commented on the results for the fiscal year ending December 2014. He said that the performance was

"in line with expectations", "organic sales performance was solid" and that "the volatility in some emerging markets... will continue also in 2015, thus limiting the visibility at this stage." The figures came in slightly better than the previous year with net sales up by 2.4% to €1560m, while net profits slipped by 14% to €128.9m. If one, however, takes one-offs into account, profits were only down by 0.7%.

Two of the most important brands for the company, Aperol and Campari, were up by 7% and 10% respectively, while Wild Turkey bourbon slipped by 3% due to a very competitive US bourbon market, as well as weak results for the flavoured version, American Honey. Scotch whiskies, on the other hand, represented by Glen Grant single malt and the blend Old Smuggler, performed much better. Sales increased by 8% and 3.3 million bottles of Glen Grant were sold.

A strong increase in Italy and a weaker performance in the USA during 2014 changed the company's sales breakdown to Americas (39%), Italy (26%), Rest of Europe (25%) and Rest of the World including duty free (10%).

The Campari group, where the Garavoglia family owns 51% of the shares, has broadened its business base over the past few years through a series of successful acquisitions. It started with SKYY vodka in 2001, continued with Glen Grant in 2006 and Wild Turkey bourbon in 2009. This trend continued with Appleton rum in 2012, as well as the Canadian whisky distiller, Forty Creek, and Fratelli Averna in 2014. In just more than a decade, Campari has invested close to €1.5bn in acquiring other companies.

Glenmorangie Company

In spite of tough trading conditions in China, Glenmorangie managed to present a good result for 2014. The turnover was up by 16% to £82m while operating profits increased by 8% to £17.7m. Both single malt brands performed well with Glenmorangie up 8% to 5.9 million bottles which makes it the fourth biggest single malt in the world. Sales of Ardbeg increased by 10% and it is now close to selling 1 million bottles in a year

Since 2003, Glenmorangie Company has also been the owners of the Scotch Malt Whisky Society with more than 20,000 members worldwide. In March 2015 it was announced that the Society had been sold to a group of private investors.

Ian Macleod Distillers

The family-owned bottler and owner of Glengoyne and Tamdhu distilleries, can reflect on a successful year. For the year ending September 2014, the turnover was admittedly down by 1.5% to £52.1m, but this was mainly due to fewer bulk trades of mature whisky. Sales of especially their own single malts soared and contributed to an increase of pre-tax profits with no less than 44% to £8.56m. According to the company's managing director, Leonard Russell, "Single malt is really what is driving our growth at the moment." The major markets for the company are Taiwan, Germany, France, The Netherlands and Belgium.

Apart from Tamdhu and Glengoyne single malts, the company also acts as an independent bottler by releasing malts from other distilleries under the name of Dun Bheagan and The Chieftain's. Isle of Skye (with four different expressions) is a blended malt and there is also Smokehead, a peated single malt from Islay.

Suntory Holdings

Suntory was founded in 1899 by Shinjiro Torii and the family still owns the major part of the business, through the company Kotobuki Fudosan. The business is run by CEO, Takeshi Niinami, who was the first leader of the company to come from outside of the firm's founding family. Until spring 2014, the company's beer and spirits division (Suntory Liquors Ltd) included such brands as Yamazaki, Hakushu, Hibiki and Kakubin. On the Scotch side, they also controlled Morrison Bowmore with Bowmore, Auchentoshan and Glen Garioch single malts. In May 2014, Suntory bought Beam Inc for $16bn and a range of spirits brands could be added to the list; Jim Beam bourbon, Teacher's blended Scotch, the two single malts Laphroaig and Ardmore, as well as Canadian Club and Courvoisier cognac.

The name of the new conglomerate, Beam Suntory, is now the third biggest spirits producer in the world. Takeshi Niinami, however, didn't seem satisfied with this when he, in February 2015, declared that Beam Suntory was aiming to "eclipse" Diageo and Pernod Ricard as the world's biggest whiskey company. The way to do that, according to Takeshi Niinami, would be to use the growing global demand for bourbon to increase sales of Jim Beam in new markets like India and Latin America. Jim Beam is already the biggest bourbon brand in the world, having sold 89 million bottles in 2014.

The 2014 financial results for Suntory Holdings was, of course, boosted by the integration of Beam Inc and sales were up by 20% to JYN2.5tn. Net profits, on the other hand, have decreased by 80% to JYN38.4bn. The explanation for the big drop lies in the exceptional increase during the previous year when Suntory Beverage & Food was listed on the Tokyo Stock Exchange, with 40% of the shares being offered to the public. If we, on the other hand, look at operating profits, it rose by 30% to JYN164.8bn.

Focusing on the Scotch whisky side of the company's business, Laphroaig was the biggest climber with an increase of 18% to 3.2 million bottles. The Islay sibling, Bowmore, however, went the other way, down 3% to 2 million bottles. Teacher's blend, which now occupies 12th place on the global sales list, increased by 2% to 24 million bottles.

The big brands

Indian whiskies continue to dominate the world's whisky market. Of the 10 most sold whiskies, eight come from India and it's only Johnnie Walker and Jack Daniels that have managed to break up the total domination. Officer's Choice is the world's most sold whisky (362 million bottles) for the third year running and it is likely that the brand will have passed Emperador brandy during next year already and thus becoming the world's second most sold liquor of all categories. On an unreachable peak position is Jinro, a soju from South Korea. With no less than 820 million bottles sold in 2014, it outsells all the top 20 Scotch whiskies put together!

If we look at the Top 30 whisky brands, the list contains 13 brands from India, 10 from Scotland, three from America, two from Canada and one each from Ireland (Jameson) and Japan (Kakubin). Among the 20 fastest growing whiskies in 2014 we find eight brands from India, six from Scotland, three from North America, two from Japan and one from Ireland.

Blended Scotch

Still unchallenged as the number one Scotch whisky in the world, Johnnie Walker got to experience something as unusual as declining sales figures during 2014. Exactly how much the decrease is, is not easily to say this year, since the figures from IWSR and Drinks International differ quite substantially when it comes to Johnnie Walker. Figures from IWSR determine there's a decline of just 2% to 225 million bottles, while Drinks International states that the decline from last year is as much as 11% to 215 million bottles. Be that as it may, Johnnie Walker has definitely lost ground, especially in the important US market. In an attempt to rejuvenate the confidence of the American consumers in the brand, Diageo announced in July 2015 that they would launch a new range of Johnnie Walker called Select Casks. The first expression has been matured for 10 years in bourbon barrels and then transferred to American rye casks for a six months finish.

Sitting at number two, a spot it has maintained since 2007, is Ballantine's with 70 million bottles sold. In third place we find Chivas Regal, reflecting a small decline last year, but still selling 55 million bottles. William Grants follows shortly thereafter with 53 million bottles, which means that the brand has lost 12% since 2010. In fifth place, J&B continues its downward journey which started

10 years ago when it was the second most sold blend in the world. During that time the brand has lost 35% of its sales with figures coming in at 43 million bottles in 2014. Famous Grouse, comes in at sixth place with 37 million bottles followed by William Lawson's with 36 million. This is Bacardi´s jewel in the crown when it comes to rapid growth. In just five years, sales of William Lawson´s has increased by 82% and in 2014, for the first time, the brand surpassed Bacardi's most well-known blend, Dewar's, which was selling 32 million bottles. The ninth and tenth place respectively are occupied by two brands produced by French companies – William Peel (Belvédère), selling 31 million bottles and Label 5 (La Martiniquaise) with 30 million.

Single Malt Scotch

If we look at the figures from IWSR, something unprecedented happened in 2014. For the first time in modern days, the category leader is no longer Glenfiddich, but Glenlivet. With an increase of 9% since last year, Chivas Brothers managed to sell 12,768,000 bottles of Glenlivet, while Glenfiddich came in at 12,576,000. Not a big difference but, if one considers the gap between the two ten years ago when Glenfiddich sold double compared to Glenlivet, one realises what a journey the latter has been through. Together, the two accounts for one quarter of all Scotch single malt sold globally.

In third place we find Macallan which has increased strongly during the last three years. For 2014, sales figures were up by 4% to 9.9 million bottles. Glenmorangie in fourth spot is also doing well. Since 2012, sales have gone up by 11% to 5.9 million bottles. The Singleton came in at fifth place with 4.9 million bottles – up by 20% since last

year. As always, the three versions (Glen Ord, Glendullan and Dufftown) are not shown separately, but it is definitely Glen Ord which accounts for the main part of sales. Aberlour, in place number 6, has now passed Glen Grant and sold 3.5 million bottles, while Glen Grant remains steadily at around 3.3 million. Laphroaig is consolidating its position as the most sold Islay malt and has taken off dramatically during the last few years, after having been steady at about two million bottles. The increase since 2013 is 17% which translates to 3.1 million bottles. Balvenie has slipped back one place, but is still showing a steady sales increase year after year. In 2014, 2.9 million bottles were sold. Finally, Talisker has, at last, made it into the Top 10 list after long term market actions by the owners Diageo. Over the past ten years, the sales have increased by 180% and 2.3 million bottles were sold in 2014 alone.

To complete the overall picture, let´s have a look at whiskies in North America, India and Ireland.

In North America, Jack Daniel´s is the undisputed leader and the sixth most sold whisky in the world with 140 million bottles, followed by the bourbon Jim Beam (89 million), the Canadian Crown Royal (64 million), the blended American whiskey Seagram´s 7 Crown (29 million) and Black Velvet from Canada (28 million).

India is the home of the big volume whiskies and changes for one brand from one year to the other are often extreme. For 2014, Officer´s Choice still reigns supreme with 362 million bottles (+27%), followed by McDowell´s No. 1 (300 million, +5%), Royal Stag (188 million, +14%), Imperial Blue (168 million, +28%) and Old Tavern (134 million, -3%).

The Irish whiskey industry is dominated by three big brands with Jameson in top spot with 56 million bottles, followed by Tullamore Dew (11 million) and Bushmill´s (9 million).

Changes in ownership

The last two years have seen a number of changes in ownership within the whisky fraternity, starting with Diageo´s gradual take-over of United Spirits which started during 2013 and ended in 2014. The acquisition must have been harder to carry through than Diageo probably anticipated, but considering India's importance as potentially being one of the world's most important markets for whisky, the future is likely to determine that the sacrifice was well worth the effort. The second major acquisition in 2013 was Distell´s take-over of Burn Stewart Distillers. Apart from adding single malts like Bunnahabhain, Tobermory and Deanston and the blend Scottish Leader to its portfolio, the South African drinks giant saw the benefits of being able to make use of Burn Stewarts strong position on the Taiwanese market for its own brand - Three Ships whisky. A third deal, albeit on a somewhat smaller scale, was reported in 2013 when BenRiach announced that they had bought Glenglassaugh distillery for an undisclosed sum.

The year 2014 started with a bang when it was announced that Beam Inc was to be sold. That had been anticipated for a couple of years but few would have thought that Suntory would be the buyer. The Japanese company already had interests in the Scottish whisky industry by virtue of its ow-

Four category leaders

nership of Morrison Bowmore, but with the acquisition of Beam, which included Laphroaig, Ardmore and Teachers, its position was significantly strengthened It also added mega-brands such as Jim Beam bourbon, Canadian Club whisky, Courvoisier cognac and Sauza tequila to its portfolio. The move made the new company, Beam Suntory, the third largest spirits producer in the world, after Diageo and Pernod Ricard. Last year continued in similar vein, with yet another Asian company showing an interest in Scotch whisky. As a direct result of Diageo's entry into United Spirits, the UK Office of Fair Trading (OFT), expressed concerns that Diageo would have too much control of the whisky market in the UK and, for reasons of fair practice and competition, it had to let go of Whyte & Mackay. Suitors had, in the meantime, been lining up since the end of 2013. One was Vivian Imerman, the South African businessman who sold Whyte & Mackay to United Spirits in the first place. Other rumoured buyers were Campari and Thai Beverages. Finally, it was announced that the Philippines brandy producer, Emperador, was the new owner. Emperador is the biggest brandy in the world and the second biggest spirits brand with almost 396 million bottles sold in 2014.

Seen in the light of these two, very turbulent years, one could almost expect the first half of 2015 to be a quiet one when it comes to mergers and acquisitions and so it was. In fact, the only distillery in Scotland that changed hands was little Bladnoch in Wigtown. Owned by Raymond Armstrong and his brother and being out of production since 2009, the company went into liquidation in 2014. Rumours of possible buyers started doing the rounds and Arran Brewery was about to close the deal, but it was never carried through. In July 2015, however, Bladnoch was sold to the Australian businessman, David Prior, who sold his successful yoghurt company five:am for $80m during the previous year. Bladnoch will need a substantial refurbishment before it can be re-opened by the end of 2016 and David Prior's intention is to increase its capacity from the current 250,000 litres to 1.5 million litres. The brand will be relaunched by the end of 2015 with a new blend, as well as single malt. Prior has managed to involve two renowned persons in his new business. They are Gavin Hewitt, former CEO of the Scotch Whisky Association who is a non-executive director in the company and the Master Blender and Distilleries Director of BurnStewart, Ian MacMillan, who will join the team as distillery manager.

New distilleries

Scotland

Without competition - the largest, new distillery to open up in Scotland in 2014, was Dalmunach which was built on the site of the demolished Imperial distillery in Carron. With a capacity of 10 million litres, it is not only one of the largest, but probably one of the most beautiful in Scotland. Commissioned in October 2014 and officially opened in July 2015, the distillery cost £25m to build.

An even bigger distillery is currently being constructed by Edrington at Macallan in Speyside. Located on the left side of the road leading up to the estate, it will be a five-part building with a visitor centre, three still houses and a mash house. Two mash tuns and a total of 12 wash stills and 24

spirit stills will give the distillery a capacity of around 16 million litres of alcohol – by far the biggest malt distillery in Scotland. The new distillery will be ready for production in spring 2017 and, at that time, the old Macallan distillery will be mothballed. The total cost of the project will be a staggering £100m!

But the buck doesn't stop there. Planning permission has been approved for a new distillery on the same premises as Glenlivet. The continued success for Glenlivet, currently the best selling single malt in the world, has necessitated the owners, Chivas Brothers, to plan for an increased capacity and the new distillery is actually two, with each unit to be equipped with one mash tun, sixteen washbacks and seven pairs of stills. When the entire expansion is to be completed, Glenlivet's capacity would have tripled to well over 30 million litres!

The idea to build a second distillery on Skye (with Talisker being the first) was presented several years ago by the late Sir Iain Noble. He had chosen a listed farm building near Torabhaig on the southeast coast as a suitable location, but the plans were never realized until Sir Iain died in 2010. A new company, Mossburn Distillers, has now taken over the plans and after a planning consent was granted , the first phase of restoration was completed in December 2014. The Torabhaig distillery will have a capacity of producing 500,000 litres of alcohol and the cost is estimated at £5m. The plan is to start whisky production by the beginning of 2017.Mossburn Distillers already produces three million bottles of wine, vodka and rum from sites in Russia, Poland and France and it is owned by the Swedish company, Haydn Holdings.

Jean Donnay, the owner of Glann ar Mor distillery in Brittany, France, initiated a project last year to build the ninth distillery on Islay. Gartbreck Distillery will be situated at Saltpan Point on the shore of Loch Indaal, just south of Bowmore distillery and overlooking Port Charlotte on the other side of the bay. An old farm will be transformed into a traditional distillery equipped with a one ton mashtun, six washbacks made of Oregon pine and one pair of directly fired stills connected to a wormtub The total capacity will be 55,000 litres of pure alcohol per year. Construction work has been delayed but will probably commence end of 2015 with a view to start production by the end of 2016 or beginning of 2017.

Duncan Taylor is one of two big independent bottlers (Douglas Laing being the other) that does not have its own distillery. In 2007, they acquired an old granary built in 1899 just outside Huntly with the express purpose of transforming it into a distillery. Following the financial crisis in 2008/2009, the plans were put on hold. In summer 2013, however, the work on the distillery began with contractors paving a new road for the diggers and trucks. The plan is to dismantle the old building brick by brick, then to install the distilling equipment and finally to reassemble the building. The latest report was that the original building had been demolished and the site cleared, ready for the foundations to go in. The owners are planning for a distillery with a capacity of 1,1 million litres.

Tim Morrison, owner of the independent bottler AD Rattray and formerly of Morrison Bowmore Distillers (owners of Bowmore, Auchentoshan and Glen Garioch, now Suntory Beam) is building a single malt distillery on

The Glenlivet Master Distiller, Alan Winchester, breaks the ground for the new distillery at Glenlivet

the banks of the river Clyde, in Glasgow. The distillery, which will be situated at the old Queens dock, will include an interactive visitor Experience and café housed in the old Pumphouse building which was built by Morrison´s great grandfather in 1877. Construction will start in November 2015 and the expectation is to have it ready for distillation by autumn 2016. The distillery will be equipped with one mash tun, eight washbacks and one pair of stills with a total capacity of 500,000 litres. After the amalgamation of Suntory and Jim Beam, The Clydeside Distillery ensures The Morrison name continues in the Scotch whisky industry as an independent family business.

Another delegation of the Morrison family represented by Tim´s brother, Brian, and his son, Jamie, are looking to build a distillery and bottling plant near the hamlet of Aberargie in Perthshire. The Morrisons, together with the Mackay family, own and operate Scottish Liqueur Centre which is situated in Bankfoot, 5 miles north of Perth. The idea with the proposed distillery is, of course, to become self-sufficient in terms of whisky for their range of whisky liqueurs.

Plans for a new distillery in Falkirk were already announced in 2008 and in May 2010, the project got the final approval. Little news has come from the initiators, Fiona and Alan Stewart since then, but in June 2013 the owners were awarded a grant of £444,000 by the Scottish Government. The future distillery will be located at Cadgers Brae, Polmont and the projected cost is estimated at £5m.

Another project which we have heard about for some time now is a distillery on the Isle of Barra. Future casks have

been sold to the public since early 2008 and most of the plans regarding building and construction are ready. In July 2010, Peter Brown bought all the shares owned by Andrew Currie (of Arran Distillery fame) which had been part of the project since its inception. The original idea was to start building in autumn of 2009, but the economic climate has made funding difficult and the future for the distillery is now uncertain. The proposed site is Borve situated on the west side of the island.

In 2010, Alasdair Day launched a new blended Scotch whisky by the name of The Tweeddale. To be entirely accurate, it wasn´t a new whisky, but a recreation of a whisky manufactured in the 19th century. A company called J & A Davidson launched the brand and in 1895, Alasdair´s great grandfather, Richard Day, joined the company and eventually took over the business. When Richard Day retired after World War II, the business ceased to trade and the brand fell into oblivion. Alasdair inherited the old company´s cellar book and based on the recipes noted there, he was able to recreate the whisky. Alasdair Day soon began planning for a distillery of his own and decided it should be situated along the Scottish Borders south of Edinburgh. The funding, however, was the issue but that problem seems to have been solved when Alasdair teamed up with Bill Dobbie, entrepreneur and co-founder of online dating site, Cupid. A company, R&B Distillers, was formed, and yet another plan for a new distillery took shape. This will be built on the small Isle of Raasay, off the Isle of Skye. A derelict Victorian hotel will be turned into a distillery and a pre-planning application was submitted to Highland

Council in early 2015. If everything goes according to plan the Raasay distillery should be up and running by 2017. The owners also decided that the project along the Borders should be put on hold until the distillery on Raasay had been built. Meanwhile, R&B Distillers are asking whisky lovers around the world for their help to choose a location for the second distillery.

Apart from the R&B project, there are two proposed distillery projects for the Borders. No matter who will be the first to open up their doors, it will be the first distillery in the area since 1837. In February 2014, Mossburn Distillers (who are currently involved in building a whisky distillery on Skye), revealed their plans to build a considerably larger distillery south of Jedburgh. The Mossburn distillery will be a combined malt and grain distillery with a total capacity of no less than 2,5 million litres of alcohol. The budget for the construction amounts to a staggering £35-40m! The plans also involve a bottling hall, maturation facilities and the largest whisky shop in Scotland. Building could commence as soon as 2015 with an aim to start production in 2017. Finlay Calder, the Scottish rugby legend, was been appointed project manager.

The third proposed distillery in the area is one which could be built in Hawick by the Three Stills Company, a company incorporated in 2013 and spearheaded by John Fordyce. They expect to raise £10m in order to forge ahead with their plans. The company has already released a blended Scotch called Clan Fraser made from sourced whiskies.

Plans to build a distillery in the Shetlands were announced in 2002 by a company called Blackwood Distillers. After a few years, it became abundantly clear that the plans would never materialize. The issue regarding a distillery was resuscitated in 2013 by the whisky consultant, Stuart Nickerson, former distillery manager at Highland Park and Glenglassaugh. The idea was to build it at Saxa Vord on Unst, the most northerly of the islands. The first stage, a gin distillery producing Shetland Reel Gin, materialised in summer 2014. In September 2015, four cask strength whiskies, distilled at the Glenglassaugh distillery but bottled on Unst by The Shetland Distillery Co, were released.

The famous, first written record of whisky was a letter to Friar John Cor, a monk at the Abbey of Lindores, dated 1494 where, by order of King James IV, he was instructed to make "aqua vitae, VIII bolls of malt". Five hundred and twenty years later, there are plans to revive whisky production at Lindores Abbey. Drew McKenzie Smith, whose family has owned the land, on which the derelict abbey has stood, for a century, has attracted investors in order to build a distillery on farmland near the abbey. To assist him is the well-known whisky consultant, Dr. Jim Swan, and if everything goes according to plans, construction of the £5m distillery could start in early 2016 with a possible production start in 2017.

Eden Mill Brewery in Fife, was the first brewery in Scotland also to expand into distilling spirits. It now appears as if another brewery is to follow. In summer 2015, BrewDog announced that it intends to build a distillery called Granite City Distillery in its Ellon brew house in Aberdeenshire. The distillery will be equipped with two 3,000 litre copper pot stills, one 600 litre gin still and a 20 metre tall copper column still - all made by Arnold Holstein in Germany. The

wash will, of course, come from its own brewery and the plan is to start with whisky, gin and vodka. BrewDog was founded in 2007 and in just eight years it has reached annual sales of £33m. The company employs 360 people and its beers are sold in over 55 countries.

Other distillery projects in Scotland include one at Anniston Farms at Lunan, Angus where the owners were awarded a grant of £292,000 in summer 2013 and Drimnin distillery near Mull in Western Scotland.

Ireland

The Teeling family is currently involved in two distillery projects in Ireland. Jack and Stephen Teeling recently opened the first new distillery in Dublin for 125 years (more on page 226). The second project involves John Teeling (father of Jack and Stephen) and is of an even bigger magnitude. The Irish Whiskey Company (IWC) which has the Teeling family as majority owners signed an agreement with Diageo in August 2013, thus taking over the Great Northern Brewery in Dundalk. Diageo was about to move the brewing operation to Dublin and the IWC started restructuring the site into Great Northern Distillery at a cost of €35m. This will be the second biggest distillery in Ireland after Midleton, with the capacity to produce 3.6 million litres of pot still whiskey and 8 million litres of grain spirit. At the time of going to press, the grain part of the distillery has just started producing, while the pot still part is expected to come on stream at the beginning of September. At the same time, the Teelings are in discussions with investors to sell a 20% stake in the distillery for €5m. The main part of the business will be to supply whiskey to private label brands rather than under their own name.

Two distillery projects are under way in County Mayo in western Ireland. The Connacht Whiskey Company has made the greatest strides. They are setting up a distillery in an old bakery in Ballina and most of the equipment had arrived by the end of August 2015. A few miles to the south, in Lahardane, Paul Davis and Mark Quick are setting up the Nephin Distillery which will be fairly big (with a projected production of 500,000 litres per year) when it opens in December 2015. They will also open up a cooperage on site

Planning permission was granted in May 2013 for the Northern Irish producer of cream liqueurs, Niche Drinks to build a new distillery in Londonderry. Almost two years later it seems that construction work is about to start in Campsie on the outskirts of the city. The total investment will amount to £15m and the capacity will be 1 million litres per year. In anticipation of their own whiskey, the company will be releasing a new blend called The Quiet Man based on whiskies from other sources.

Slane Castle, north of Dublin, has been famous since the 1980s for hosting rock concerts with many of the biggest celebrities. But the Conyngham family, who owns the estate, has also in recent years established a whiskey brand, Slane Castle Whiskey, which has since become popular in the USA. Until recently, the whiskey has been produced at the Cooley Distillery but, when Beam Inc. took over the distillery in 2012, they also stopped selling whiskey to independent bottlers. This prompted the Conyngham family to plan for a distillery of their own on the estate. The plan-

Stills at Great Northern Distillery in Dundalk

ning application was approved in 2013 and the family part-nered with Camus Wine & Spirits to begin the construction. Eventually, Camus withdrew from the project but a new company showed an interest in June 2015. Brown-Forman, one of the biggest players in the industry, announced that they were buying Slane Castle Irish Whiskey Ltd. for $50m with the hope of having a distillery with a one million litre capacity ready by 2015/2016.

Dublin Whiskey Company is planning to rebuild a former mill in Dublin´s Liberties less than 500 metres from St Patrick´s Cathedral and converting it into a whiskey distillery. The project came to a temporary halt in April 2014 when the City Council asked the company to make a number of revisions to their plans but in summer 2014, DWC were finally given planning permission.

One of the more renowned entrepreneurs in the Irish whiskey business at the moment, is Mark Reynier, the former owner of Bruichladdich distillery on Islay. In 2014 he bought the former Guinness brewery in Waterford to rebuild it into a distillery. When it opens in 2016, it will certainly not be a small craft distillery. With two pot stills and one column still, Reynier will be able to produce 3 mil-lion litres, but he also has plans to double that in due time. The focus will be 100% on Irish whiskey and no white spirits will be distilled.

Peter Lavery, lottery millionaire and founder of The Bel-fast Distillery Company, has started to transform the former Crumlin Road jail into a whiskey distillery. Three stills have been ordered from Forsyth´s in Rothes and the capa-city will be 300,000 litres of alcohol. The whole investment is expected to be £6.8m and construction is now underway with a possible start of production in early 2016. Lavery is already selling whiskey produced by other distilleries under the brand names Danny Boy, Titanic and McConnell´s

Bernard Walsh, CEO and owner of The Irishman Brands with whiskies like The Irishman and Writer´s Tears (both produced at Midleton), announced in autumn 2013 that he planned to build a distillery at Royal Oak, Carlow. At the same time he also revealed that his company and the project was backed up by the major Italian drinks company Illva Saronno. In July 2014, planning permission was ob-tained and Walsh is hoping to start production in 2016. The distillery, with a capacity of producing 4 million bottles per year, will be equipped with both pot- and column stills and the total cost will amount to €25m.

Irish Fiddler Whiskey Company, with Colm Regan at the helm, has launched a range of whiskeys sourced from other distilleries. They are now looking to build their own distillery in County Galway. Sliabh Liag Distillery in south west Donegal, is the brainchild of Margaret Cunningham who has involved Oliver Hughes as its director. He is the co-founder, owner and CEO of Porterhouse Group, Ireland´s largest independent brewer with pubs and restaurants in Ireland, London and New York. Hughes was also involved in opening Dingle Distillery in 2012. Other ongoing distillery projects include Burren (Co. Clare), Dúchas (Co. Cork), Kilmacthomas (Co. Waterford), Tipperary (Co. Tip-perary) and Portaferry (Co. Down).

Bottling grapevine

Let´s start with the oldest whisky released this year. Unsurprisingly, it comes from the vast stocks of independent bottler, Gordon & MacPhail. A few years ago they started on a new range of bottlings called Generations. The first two expressions were Mortlach and Glenlivet – both 70 years old. In September 2015, it was time for what is generally believed to be the oldest single malt ever released – a Mortlach 75 year old from a first fill sherry cask filled in November 1939. Even if this whisky is out of reach for most of us (only 100 bottles with a price of £20,000 per bottle), it is still fascinating to know that this whisky was produced a couple of months after World War II started.

But this wasn´t, however, the only new bottling with a respectable age. From Tobermory came two 42 year olds – one Tobermory and one Ledaig and it almost seemed like Master Blender, Ian MacMillan wanted to end his days with Burn Stewart with a bang before leaving for Bladnoch to help the new owners resurrect the closed distillery. No less than four distilleries released their oldest bottlings so far; Tullibardine baffled aficionados with a 60 year old, the first in the new Custodian Collection. From Glen Ord came The Master´s Casks 40 years old, Knockdhu launched an anCnoc 1975 and also Glenglassaugh released their oldest whisky so far, 51 years old, together with a second batch of single casks. The other two distilleries in the group were also diligent. From BenRiach came three, new 18 year olds, each with a different finish; Dunder (dark rum), Albariza (PX sherry) and latada (madeira). To that you may add a number of single casks, as well as a new 10 year old. Glendronach presented a new 8 year old, Hielan, in the core range plus batch number 12 of single casks.

If we return to the subject of old bottlings, it is not surprising that Glenfarclas shows up. This year they not only spoiled their fans with a 1956 sherry cask, but also the release of a whisky with the peculiar name £511.19s.0d Family Reserve. It celebrates the 150th anniversary since the Grant

family took over Glenfarclas and the name refers to the amount that was initially paid for the distillery. There were plenty more anniversaries this year. One was Tomintoul, founded in 1965, and to mark the occasion, Five Decades was released, containing whiskies from 1965, 1975, 1985, 1995 and 2005. Kilchoman on Islay turned ten years and fittingly a 10 year old was released, together with new editions of 100% Islay and Loch Gorm as well as a madeira cask maturation. But the two jubilees that attracted the greatest interest was probably the 200th anniversary of both Laphroaig and Ardbeg respectively. The latter celebrated with Ardbeg Perpetuum released on Ardbeg Day and the fifth and final release of Supernova. The Laphroaig team decided to re-launch an old favourite, the 15 year old, in a temporary come-back and a 21 year old was released to celebrate the 21st anniversary of Friends of Laphroaig. The special bottling for this year´s Fesi Ile was a Cairdeas made from 100% floor malted barley back in 2003, but the real clou wasn't due until December by way of a 32 year old oloroso cask.

We have become accustomed to expressions without an age statement in recent years and 2014/2015 was no exception. Diageo was quite industrious with the release of Talisker Skye, Oban Little Bay and Dalwhinnie Winter´s Gold. The latter was sold with the recommendation of putting it in the freezer to enhance the honeyed notes of the whisky. From the hands of Chivas Brothers and Scapa distillery, we got Skiren, which was matured in first fill bourbon, while Glenmorangie delighted us with Duthac and Tusail. The latter had been made using floor malted Maris Otter barley. Glenlivet Founder´s Reserve was probably the one among NAS bottlings which received the most attention. It was launched to gradually replace the top selling Glenlivet 12 year old. At the same time, the owners launched two new duty free exclusives in the Master Distiller´s Reserve range – Solera vatted and Small Batch. At Ardmore, Legacy replaced Traditional in autumn 2014, but the latter made a come-back in 2015 as Tradition, and this time it was only available for duty free. Another new release from the distillery was Triple Wood in summer 2015 and with the new owners (Beam Suntory), there will probably be more available from Ardmore in the near future.

But there are still producers that stick to age statements on the labels and none more so than Dewar´s. Last year they started an impressive relaunch of whiskies from all five distilleries and the project got a not so modest name, The Last Great Malts of Scotland. And to top it all – all the whiskies carry an age statement. In 2014, Aberfeldy, Craigellachie and Aultmore received all the attention with a number of new expressions. These were followed up in 2015 with an 18 year old Aultmore and a 16 year old Aberfeldy. This year it was time to showcase Royal Brackla and Macduff distilleries. The first one was blessed with a new range made up of 12, 16 and 21 year olds, while the latter was renamed The Deveron (the distillery is still called Macduff) and saw the release of 10, 12 and 18 year olds.

For three distilleries, 2015 brought the second, completely new range in just a few years. Loch Lomond is now represented by Loch Lomond Original Single Malt and (to catch up on a new trend) Loch Lomond Single Grain. At the same time, the distillery offers Inchmurrin 12 and 18 year olds, as well as Madeira

Mortlach 75 years old and Glenglassaugh 51 years old

Wood and Glengarry NAS and 12 years old. Sharing the same owner as Loch Lomond, Glen Scotia launched their Double Cask, 15 year old and Victoriana. The third distillery with a new and expanded range is Speyside. Included in the core range are Tenné, 12 and 18 year old, and as limited releases we have Lord Byron and two special Michael Owen expressions (he´s the new face in marketing for the distillery) – Golden Choice and 1412. Finally, from Speyside, delighting all the fans of quaintly black whiskies (like Loch Dhu and Cu Duhb), the new Beinn Dubh was released.

The owners of Tomatin have been diligent in recent years when it comes to new releases and 2015 was no exception. The core range was expanded by a Tomatin Cask Strength and the separate range, Cu Bocan, was blessed with a Virgin Oak. Customers were also offered a mini tasting when Tomatin Contrast Duo Pack was launched. The two bottles (one matured in bourbon casks and the other in sherry butts) is a good opportunity to compare whisky, side by side, coming from the same distillery, but matured in different casks.

Benromach launched a new core 15 year old, as well as two finishes, Sassicaia and Hermitage and from Highland Park came Odin, the final instalment in the Valhalla Collection. For the first time, Glen Moray offered their smoky version with Glen Moray Classic Peated and a 25 year old port finish. In addition a 30 year old was also launched by the owners. Kilkerran from Glengyle distillery has almost come of age, which in this case means 12 years, and the two Kilkerran Work in Progress bottlings which were released this year will be the last before the core 12

year old is launched next year. Bowmore released more of both their prestigious 50 year old, as well as new editions of the Devil´s Casks and Tempest, but they also had a rare surprise up their sleeve – a Mizunara Cask Finish. With Japanese owners, Bowmore stands a better chance than most distilleries to come by this rare variety of Japanese oak.

Bruichladdich distillery and its thousands of fans around the world, said goodbye to the legendary Jim McEwan, who retired in June 2015. His farewell bottling was High Noon 134 where the number refers to the combined age of whiskies that were married to create it. From the distillery also came a PC12 and the heavily peated (208ppm) Octomore 7.1. In the meantime, Springbank went green with two, new organic expressions, a bourbon matured 12 year old, followed by a 13 year old sherry maturation. They also added a new edition of Longrow Red, but this time finished in pinot noir casks originating from New Zealand.

Tamdhu added a cask strength version to its 10 year old, called Tamdhu Batch Strength. From Glencadam came a 25 year old and Ben Nevis released a 12 year old white port finish. The owners of Knockdhu continue on its successful peated track with the release of Peatlands for select and domestic markets, as well as Barrow for duty free.

Finally, for the first time in Malt Whisky Yearbook history, the official word on this year´s Special Releases from Diageo hadn´t been revealed at the time of writing. Speculations are rife, as always, and I´ve decided to add my own "guesses" (probably safely substantiated thanks to the official registration of various labels) at the various distillery entries throughout the book.

BenRiach Dunder, Ledaig 42 years, Ardmore Triple Wood, Glen Scotia Victoriana and Royal Brackla 12 years

Independent
bottlers

The independent bottlers play an important role
in the whisky business. With their innovative bottlings, they increase
diversity. Single malts from distilleries where the owners' themselves
decide not to bottle also get a chance through the independents.
The following are a selection of the major companies.
All tasting notes have been prepared by Gavin Smith.

Gordon & MacPhail

www.gordonandmacphail.com

Established in 1895 the company, which is owned by the Urquhart family, still occupies the same premises in Elgin. Apart from being an independent bottler, there is also a legendary store in Elgin and, since 1993, an own distillery, Benromach. There is a wide variety of bottlings, for example Connoisseurs Choice (single malts bottled at either 43 or 46%), Private Collection (single malts, some dating back to the 1950s, usually bottled at 45%), MacPhail´s Collection (single malts bottled at 43%), Distillery Labels (a relic from a time when Gordon & MacPhail released more or less official bottlings for several producers. Currently 9 distilleries are represented in the range and the whisky is bottled at either 40 or 43%), Rare Old (exclusive whiskies from distilleries that are closed and sometimes even demolished, for example Glenugie, Glenury Royal, Coleburn, Glenesk and Glenlochy), Secret Stills (single malts from 1966 to 2000 where the distillery name in not disclosed, bottled at 45%), Cask Strength (a range of single malts bottled at cask strength), Rare Vintage (single malts, including several Glen Grant and Glenlivet bottlings going back to the 1940s) and Speymalt (a series of single malts from Macallan from 1938 and onwards).

In 2010, a new range was launched under the name Generations. To say that these are rare and old whiskies is an understatement. The first release was a Mortlach 70 year old, which was followed the year after by a Glenlivet 70 year old. In September 2015, it was time for the third instalment in the series - a Mortlach 75 years old. This is the oldest single malt ever bottled and only 100 bottles were released. Amazing bottles were also released in autumn 2014 when Private Collection Ultra was launched. To celebrate the transition of the family-owned business to the next generation, four extremely rare whiskies were presented; Linkwood 1953 61 years old, Glenlivet 1952 62 years old, Mortlach 1951 63 years old and Strathisla 1957 57 years old. The whiskies were chosen by members of the third and fourth generations of the Urquhart family.

Gordon & MacPhail rarely buy matured whisky from other producers. Instead, around 95% is bought as new make spirit and filled by the company. Some 7,000 casks are maturing in one racked and one dunnage warehouse in Elgin, another 7,000 casks are found at various distillers around Scotland and 20,000 casks are located in the warehouses at Benromach.

Glenturret 2000, 43%
Nose: Earthy, with treacle and dark berries, plus fruit pieces.
Palate: Full-bodied, with spicy, zesty plums, raisins, cocktail cherries and nutmeg.
Finish: Lengthy, with black treacle, cocoa powder, and an edge of slightly bitter oak.

Inchgower 2000, 46%
Nose: Fresh, with ripe apples, soft malt and vanilla.
Palate: Intensely fruity, with ginger and brittle toffee.
Finish: Medium to long, with persistent mango and cocoa notes.

Berry Bros. & Rudd

www.bbr.com

Britain´s oldest wine and spirit merchant, founded in 1698 has been selling their own world famous blend, Cutty Sark, since 1923. Berry Brothers had been offering their customers private bottlings of malt whisky for years, but it was not until 2002 that they launched Berry´s Own Selection of single malt whiskies. Under the supervision of Spirits Manager, Doug McIvor, some 30 expressions are on offer every year. Bottling is usually at 46% but expressions bottled at cask strength are also available. The super premium blended malt, Blue Hanger, is also included in the range. So far, eleven different releases have been made, each different from the other. The sixth edition sets itself apart from the rest as it combines both sherried malts from Speyside and peated Islay whisky. The ninth release, based on whiskies from Clynelish, Glen Elgin and Bunnahabhain, also showed some peaty notes as did the 11th and latest edition, which was released in late 2014. In autumn 2014 the Exceptional Casks Collection was launched. Handpicked by Doug McIvor a 50 year old North Britısh single grain, two single casks of Glenlivet 1972 and a Jamaican rum from 1977 were the first bottles in the new range. In 2010, BBR sold Cutty Sark blended Scotch to

Edrington and obtained The Glenrothes single malt in exchange.

A visit to Berry Bros. & Rudd at 3 St James's Street in London is an extraordinary experience. The business was established in 1698 by the Widow Bourne and the company has traded from the same shop for over 300 years! Originally selling coffee, the company soon expanded into wine and started supplying the Royal Family during the reign of King George III and still continues to do so today. An odd service which started in the mid 1700s was weighing the customers on the giant coffee scale. The results were entered into ledgers which have been maintained. Even today, it still happens that certain customers are offered this service.

Blue Hanger 11th release, 45.6%
Nose: Damp heather, lightly toasted oak, malt and old leather.
Palate: Smooth and sophisticated, yet robust.Cinnamon, sweet sherry and orchard fruits.
Finish. Long and spicy, drying slowly, with espresso coffee notes.

Berry's Own Bunnahabhain 1987, 46%
Nose: Sweet brine, warm leather and barleysugar.
Palate: Rounded, deeply fruity, with ginger, nutmeg and coffee.
Finish: Drying slightly, with soft spices, and tangy fruits. Lengthy.

Signatory Vintage Scotch Whisky

Founded in 1998 by Andrew and Brian Symington, Signatory lists at least 50 single malts at any one occasion. The most widely distributed range is Cask Strength Collection which sometimes contains spectacular bottlings from distilleries which have long since disappeared. One of the most recent additions to the range in 2014 was a rare Glencraig 1976, 38 years old. Another range is The Unchill Filtered Collection bottled at 46%. Some of the latest bottlings released are also spectacular; Craigduff 1973 (an extremely rare, peated Strathisla), Mosstowie 1979 and Glen Mhor 1982. Andrew Symington bought Edradour Distillery from Pernod Ricard in 2002

Ian Macleod Distillers

www.ianmacleod.com

The company was founded in 1933 and is one of the largest independent family-owned companies within the spirits industry. Gin, rum, vodka and liqueurs, apart from whisky, are found within the range and they also own Glengoyne and Tamdhu distilleries. In total 15 million bottles of spirit are sold per year. Their single malt ranges are single casks either bottled at cask strength or (more often) at reduced strength, always natural colour and un chill-filtered. The Chieftain's cover a range of whiskies from 10 to 50 years old while Dun Bheagan is divided into two series – Regional Malts, 8 year old single malts expressing the character from 4 whisky regions in Scotland and Rare Vintage Single Malts, a selection of single cask bottlings from various distilleries. There are two As We Get It single malt expressions – Highland and Islay, both 8 year olds and bottled at cask strength. The Six Isles blended malt contains whisky from all the whisky-producing islands and

is bottled at 43%. One of the top sellers is the blended malt Isle of Skye with five expressions – 8, 12, 18 (new since 2014), 21 and 50 years old. Finally, Smokehead, a heavily, peated single malt from Islay, was introduced in 2006. There is also a Smokehead Extra Black 18 years old and Smokehead Extra Rare (which basically is a 1 litre duty free bottling of the 12 year old). The company also has a blended Scotch portfolio which includes its biggest seller, King Robert II as well as Langs Supreme (5 years old) and Langs Select (12 years old).

Isle of Skye 18, 40%
Nose: Honey, sultanas, malt and spice, with a wisp of smoke.
Palate: Very smooth, with sherry and orchard fruits, backed up by spicy cocoa.
Finish: Medium in length, with lingering fruitiness.

Isle of Skye 21, 40%
Nose: Initially quite reticent, then melon and banana, with a hint of smoke.
Palate: Silky, with overt fruitiness and subtle, spicy smoke.
Finish: Relatively lengthy, with cinnamon and slight peat.

Blackadder International

www.blackadder.se

Blackadder is owned by Robin Tucek, one of the authors of The Malt Whisky File. Apart from the Blackadder and Blackadder Raw Cask, there are also a number of other ranges – Smoking Islay, Peat Reek, Aberdeen Distillers, Clydesdale Original and Caledonian Connections. One of the latest brands in the Blackadder family is Riverstown which is especially earmarked for the Asian market. The company has also been known for bottling unusual expressions of Amrut single malt. All bottlings are single cask, uncoloured and un chill-filtered. Most of the bottlings are diluted to 43-46% but Raw Cask is always bottled at cask strength. Around 100 different bottlings are launched each year.

Duncan Taylor

www.duncantaylor.com

Duncan Taylor was founded in Glasgow in 1938 as a cask broker and trading company. Over the decades, the company built strong ties with distillers over Scotland, with the company bringing their own casks to the distilleries to be filled with new make spirit. This resulted in a collection of exceptionally rare casks, many from distilleries which are now closed. Duncan Taylor was acquired by Euan Shand in 2001 and operations were moved to Huntly.

Duncan Taylor's flagship brand is the blended Scotch Black Bull, a brand with a history going back to 1864. The brand was trademarked in the US on the repeal of prohibition in 1933 and was re-branded in 2009 by Duncan Taylor. The range consists of three core releases – Kyloe, a 12 year old and a 21 year old. There are also three limited versions, 30 year old, 40 year old and Special Reserve. The Black Bull brand is complimented by Smokin' which is a blend of peated Speyside, Islay and grain whisky from the Lowands.

The portfolio also includes Rarest of the Rare (single cask, cask strength whiskies of great age from demolished distilleries), Battlehill (younger malts at 43%), Dimensions (a collection of single malts and single grains aged up to 39 years and bottled either at cask strength or at 46%) and The Octave (single malt whiskies matured or 'Octavised' for a further period in small, 50 litre ex-sherry octave casks). New additions to the Duncan Taylor single malt range are The Rarest, which included a Macallan 1969 as the first release, The Tantalus, which offers a selection of whiskies all aged in their 40s and The Duncan Taylor Single Range including whiskies aged 30 years or more from closed distilleries. The blended malt category is represented by Auld Reekie, a 10 year old from Islay, which is similar to Big Smoke, although the latter is younger, more peated and available in two strengths, 40% and 60%.

In 2014, a new bottling plant with additional warehousing was opened in Huntly. Since 2007, the owners have also had plans to build their own distillery. Construction work has began but no final date for the production start has been given.

Black Bull 12 year old, 50%

Nose: Malt, toffee, sherry, sultanas and spice. Toasted coconut and raisins.

Palate: Rich, sweet and fruity, with soft, sherried spices and glace cherries.

Finish: Relatively long, with chilli and ginger. Drying to aniseed.

Big Smoke 60, 60%

Nose: Sweet bonfire smoke, barley and vanilla.

Palate: Oily and sweet, with big notes of peat, bright spices, and ripe apples.

Finish: Long and oily, with bonfire embers.

Scotch Malt Whisky Society

www.smws.com

The Scotch Malt Whisky Society, established in the mid 1980s and owned by Glenmorangie Co since 2003, has more than 20,000 members worldwide and apart from UK, there are 17 chapters around the world. The idea from the very beginning was to buy casks of single malts from the producers and bottle them at cask strength without colouring or chill filtration. The labels do not reveal the name of the distillery. Instead there is a number but also a short description which will give you a clue to which distillery it is. A Tasting Panel selects and bottles around twenty new casks the first Friday of every month. The SMWS also arranges tastings at their different venues but also at other locations. The society produces an excellent, award winning members magazine called Unfiltered.

In March 2015, Glenmorangie Company announced that SMWS had been sold to a group of private investors. The reason for the sale was that Glenmorangie wished to focus on building their two single malt brands - Glenmorangie and Ardbeg.

Wm Cadenhead & Co

www.wmcadenhead.com

This company was established in 1842 and is owned by J & A Mitchell (who also owns Springbank) since 1972. The single malts from Cadenheads are neither chill filtered nor coloured. The last couple of years, there have been three different ranges of whisky; Authentic Collection (cask strength), Duthie's (diluted to 46%) and Chairman's Stock (older and rarer whiskies). Authentic is still bottled for the company's own shops while the latter two have been phased out to make way for four new ranges developed by Mark Watt who joined the company in 2012 after having worked for Duncan Taylor for several years.

The first is Creations which will be small batch blended malts or blended Scotch. The first two releases were Robust Smoky Ember, a 21 year old blended malt consisting of Ardbeg, Bowmore and Caol Ila and a 20 year old sherry cask blended Scotch called Rich Fruit Sherry with Bruichladdich, Mortlach, Cameronbridge and Invergordon in the bottle. This was followed up in 2014 with the 17 year old Light Creamy Vanilla, a vatting of Ardmore, Auchroisk, Caperdonich, Clynelish and Invergordon grain.

The second new range is Small Batch which are single malt (or single grain) vattings of two to three casks and bottled at 46%. Recent releases include Cragganmore 1999 and Glengoyne 2001. The

third range is similar but the whiskies are bottled at cask strength. Bottlings in June 2015 included Glenlivet 1973, Littlemill 1990 and Invergordon 1991. The fourth range, the Single Cask Range, has included releases such as Glen Keith 1973, Mannochmore 1977 and Miltonduff 1978.

A chain of ten whisky shops working under the name Cadenhead's can be found in the UK, Denmark, The Netherlands, Germany, Poland, Italy and Switzerland.

BenRiach 1996 19 years old, 47.1%

Nose: Soft and floral, with honey-coated malt, ripe bananas and milk chocolate.

Palate: Creamy and fruity, with more banana and honey, plus nougat and coffee.

Finish: Citrus fruits linger long with cocoa powder.

Glenfarclas 1990 25 years old, 52.6%

Nose: Floral, with heather in bloom and a hint of new leather. Background malt and honey.

Palate: Mouth-coating, with vibrant ginger, vanilla, ripe apricots and cocoa.

Finish: Lengthy, fruity instant coffee and shortcake.

Compass Box Whisky Co

www.compassboxwhisky.com

Most people within the whisky industry acknowledge the fact that the cask has the greatest influence on the flavour of whisky, but none more so than the founder and owner of Compass Box, John Glaser. His philosophy is strongly influenced by meticulous selection of oak for the casks, clearly inspired by his time in the wine business. But he also has a lust for experimenting and innovation to test the limits, which was clearly shown when Spice Tree was launched in 2005. For an additional maturation, Glaser filled malt whisky in casks prepared with extra staves of toasted French oak suspended within the cask.

The company divides its ranges into a Signature Range and a Limited Range. Spice Tree (a blended malt), The Peat Monster (a combination of peated islay whiskies and Highland malts), Oak Cross (American oak casks fitted with heads of French oak), Asyla (a blended whisky matured in first-fill ex-bourbon American oak) and Hedonism (a vatted grain whisky) are included in the former.

In the Limited range, whiskies are regularly replaced and at times only to resurface a couple of years later in new variations. Among the recent releases are Hedonism Maximus, Peat Monster Reserve and Flaming Heart. In 2011 it became illegal to use the term vatted malt. Instead the term blended malt must be used. To mark this change of terminology, which was heavily debated, Compass Box released two limited expressions – The Last Vatted Malt (a vatting of 36 year old Glenallachie and 26 year old Caol Ila) and The Last Vatted Grain (with whiskies from Invergordon, Cameron Bridge and the two closed distilleries Carsebridge and Port Dundas).

A third range was added in summer of 2011 when Great King Street was launched. The range will offer blended Scotch with a 50% proportion of malt whisky and using new French oak for complexity. The first expression was called Artist's Blend and in autumn 2014 Glasgow Blend was released. In August 2013, the blended Scotch Delilah's was released to commemorate the 20th anniversary of Delilah's bar in Chicago. A few months later, The General was released. This was a vatting of two old (30-40 years) blended whiskies, both of which had been maturing in casks as blends. In 2014, the company made an attempt to re-create one of their very first bottlings, the blended malt Eleuthera, which was discontinued in 2004. The new expression, named The Lost Blend, was a vatting of Clynelish, Allt-a-Bhainne and Caol Ila. The first new bottling in 2015 was a cask strength version of Peat Monster which was sold in magnum bottles. This was followed up by a 15th anniversary expression of Hedonism called Quindecimus. Bottled at 46%, it is a vatting of five different grain whiskies, both young and old. Finally, autumn 2015 saw the launch of a 15th anniversary bottling of Flaming Heart, first released in 2006. It combines peaty smokiness with spicy notes from new, French oak.

In autumn 2014, Compass Box made a long-term agreement with John Dewar & Sons where the Bacardi-owned company would supply Compass Box with stocks of whisky for future bottlings. In spring 2015 it was further announced that Bacardi had acquired a minority share of the independent bottler. Both parties confirm that Compass Box will be run as a separate entity from Bacardi also in the future with John Glaser continuing to lead and run the business.

Hedonism Quindecimus, 46%
Nose: Stewed fruits, ginger, linseed and a hint of new leather.
Palate: Sweet and smooth,with coconut, toffee, vanilla and orchard fruits, becoming increasingly peppery.
Finish: Warming, with ginger and light oak.

Flaming Heart 15th Anniversary
Nose: Sweet and malty, with a slight coastal edge. Subtle, fragrant smoke and a hint of sherry.
Palate: Relatively full-bodied, with cinnamon, heather, Brazil nuts, spicy peat and brine.
Finish: Cocoa, black tea, lingering wood spice.

Creative Whisky Company
www.creativewhisky.co.uk

David Stirk, who worked in the whisky industry for many years, started the Creative Whisky Co in 2005. Creative Whisky exclusively bottles single casks, divided into three series: The Exclusive Malts are bottled at cask strength and vary in age between 8 and 40 years. Around 20 bottlings are made annually. This is followed by the Exclusive Range which comprises of somewhat younger whiskies, between 8 and 16 years, bottled at either 45% or 45.8%. Finally, Exclusive Casks are single casks, which have been 'finished' for three months in another type of cask, e. g. Madeira, Sherry, Port or different kinds of virgin oak.

In 2015, David celebrated the company´s 10th anniversary with a range of 7 new bottlings; Tamdhu 1980 (35 years), Invergordon 1984 (30), Glen Garioch 1990 (25), North Highland 1995 (20), Bowmore 1999 (15), Laphroaig 2005 (10) and Ardmore 2009 (5 years old). This was followed up by The Exclusive Malts Ireland – a 13 year old single cask from Cooley. Previously, David has also released a blended Scotch called The Exclusive Blend made up of 80% malt whiskies and 20% grain. All the whiskies were distilled in 1991 and matured in ex-sherry casks for 21 years.

Master of Malt
www.masterofmalt.com

Master of Malt is one of the biggest and most innovative whisky retailers in the UK. The company also has ranges of its own bottled single malts. One range is called the Secret Bottlings Series, where no distillery names appear on the labels. It's split between editions of well aged releases (30, 40, 50 & 60 year olds from Speyside) and a range of no age statement regional bottlings at 40% (Highland,

Speyside, Lowland, Islay and Island). The bottlings are very competitively priced, not least the older ones. Master of Malt also bottles single casks at natural cask strength from various distilleries in their Single Cask Series. Some of the latest are a 24 year old Highland Park, a 23 year old Ardbeg and a 6 year old Paul John. In 2011, Master of Malt invited whisky bloggers to create their own whisky blends. The winning contribution was released in autumn 2011 under the name St Isidore.

Ardbeg 23 years old, 50.6%
Nose: New leather, sweet, warm peat, grapefruit and wood shavings
Palate: Very sweet, honey, fruity peat and ginger.
Finish: Long and warming, with toffee and persistent peat embers.

Macallan 40 years old, 43%
Nose: Rich and warm, with figs, sherry, marzipan, ginger and orange zest.
Palate: Jaffa oranges, plain chocolate, pepper and oak.
Finish: Lengthy and mouth-drying, with prickly spices.

Atom Brands
www.masterofmalt.com

Part of the same ATOM group as retailer Master of Malt, ATOM Brands includes a number of independent bottlers. They're distributed by Maverick Drinks who are also part of the group and were awarded the IWSC Spirit Distributor of the Year Trophy in 2014, having only been established the year before. Maverick also import many American craft whiskeys from the likes of St. George Spirits and FEW Spirits.

That Boutique-y Whisky Company was launched in September 2012, with the idea being to produce single malt bottlings by blending whisky from the same distillery but at different ages. Sometimes the age difference between these whiskies can be as much as 30 years and hence there is no age statement on the bottlings. The whiskies turned out to be an instant success and over 100 different bottlings have now been released from over 50 different distilleries including Springbank, Highland Park, Glenallachie and Arran. The range also includes a handful of blended malts, blends, single grains and even a bourbon.

The first whiskies from Darkness! were released in spring 2014 and the key words for these expressions are dark and heavily sherried. To create the character, single malts are filled into specially commissioned 50 litre first fill Sherry casks where they are finished for more than 3 months. Pedro Ximénez and Oloroso Sherry casks were initially used, but more recent releases include a 17 year old Ben Nevis finished in a Palo Cortado cask and a 23 year old Invergordon grain finished in a hybrid PX and Oloroso cask made up with staves from each.

Darkness Invergordon 23 years old, 50.8%
Nose: Lanolin, vanilla custard, sweet sherry, toffee bonbons and contrasting blackberry notes.
Palate: Silky in the mouth, with sultanas, cocoa powder and maraschino cherries.
Finish: Medium in length, slight fruitiness and milk choclate caramel.

The Blended Whisky Company produces the Lost Distilleries Blend. As the name implies, every batch is made exclusively from whiskies produced at now closed distilleries. The sixth batch contains whiskies from Caperdonich, Imperial, Brora, Port Ellen, Glenisla, Glen Mhor and Mosstowie while the grain part comes from Port Dundas.

Drinks by the Dram is another, the brilliant idea here being to sell single malts (and other spirits) in 30ml bottles. They're now available in a range of themed Tasting Sets at a variety of price levels as well as in the Whisky Advent Calendar and spirits-filled Christmas Crackers.

There's also Reference Series, a range of educational blended malts that goes vertically from I, to II, to III, each made with the same four whiskies but with III being made with a far greater proportion of the older, more complex malts compared to I, which is more youthful and clean. The range then continues horizontally, if you will, with .1 (additional PX finish), .2 (includes 10% heavily peated Islay malt) and .3 (containing for e150a caramel colouring) editions. Comparisons can then be made throughout the range to help drinkers understand how composition and maturity can affect the final whisky.

Reference Series II, 47.5%
Nose: Soft, creamy and fragrant, with heather, pine, ginger and almonds.
Palate: Luscious in the mouth, with sherry, malt and developing chilli notes.
Finish: Drying, with oak, dark choco-late and spicy liquorice.

The Whisky Agency

www.whiskyagency.de

The man behind this company is Carsten Ehrlich, to many whisky aficionados known as one of the founders of the annual Whisky Fair in Limburg, Germany. His experience from sourcing casks for limited Whisky Fair bottlings led him to start as an independent bottler in 2008 under the name The Whisky Agency. He is currently working with three ranges; The Whisky Agency with 12 series of whiskies released so far – from Butterflies to Bugs – the names alluding to the motif on the labels, The Perfect Dram (at least 30 expressions so far) and Specials with some un-usual bottlings, for example a Tomatin 1967 sherry butt. A recent collaboration between The Whisky Agency and The Whisky Exchange in London resulted in the bottling of a 48 year old Girvan single grain, distilled in 1964.

A Dewar Rattray Ltd

www.adrattray.com

This company was founded by Andrew Dewar Rattray in 1868. In 2004 the company was revived by Tim Morrison, previously of Morrison Bowmore Distilleries and fourth generation descendent of Andrew Dewar, with a view to bottling single cask malts from different regions in Scotland. One of its best-sellers is the 12 year old single malt named Stronachie. It is named after a distillery that closed in the 1930s. Tim Morrison bought one of the few remaining bottles of Stronachie and found a Highland distillery that could reproduce its character. The distillery was shrouded in secrecy until 2010, when it was revealed as being Benrinnes in Speyside. Each Stronachie bottling is a batch of 6-10 casks. The 12 year old Stronachie was bottled in 2010 by another expression, 18 years old and in 2014, the 12 year old was replaced by a new 10 year old. A peated, blended malt, Cask Islay, became available in 2011 and released again in 2013 but this time as a single malt. In 2012 a new, 5 year old blend was launched under the name Bank Note.

The AD Rattray's Cask Collection is a range of single cask whiskies bottled at cask strength and without colouring or chill-filtration. This range was recently complemented by Vintage Cask Collection, including rare and older whiskies.

In 2011, the company opened A Dewar Rattray's Whisky Experience & Shop in Kirkoswald, South Ayrshire. Apart from having a large choice of whiskies for sale, there is a sample room, as well as a cask room. The Whisky Experience also provides organised tastings for individuals and groups, from half an hour speed tasting of three whiskies, to a master class of five very old and rare single casks. Exclusively available in the shop, there is a specially selected single cask. The most recent, Malt 4, is a Tamdhu 2008 first fill sherry cask.

Girvan 1964 50 years old, 43%
Nose: Plums soaked in linseed and vanilla essence and sprinkled with ginger.
Palate: Very spicy, with fleeting soft fruit notes and caramel.
Finish: Dries in a relatively short finish to quite bitter oak.

A D Rattrays's Whisky Experience - Malt 4, 40%
Nose: Nutty malt, peaches and toffee.
Palate: Lively fresh fruits, ginger, nutmeg and milky coffee.
Finish: Medium in length, cinnamon and tangy citrus fruits.

Samaroli Srl

www.samaroli.it

The company was founded by Silvano Samaroli in 1968 and at that time he was the only person from outside the UK who was working as an independent bottler of Scotch whisky. Over the years, Samaroli has built up a reputation as a discerning bottler, selecting only the best casks available.

Apart from single malts, Samaroli also produces a blended malt every year known as Evolution (previously called No Age). The 2013 release was a vatting of single malts from 15 to 50 years. The whisky is diluted to 45% and that process takes a year to complete. Small amounts of water are added repeatedly and in between the whisky rests, in order to let the water and the whisky merge. There are also blended malts called Samaroli Blend, based on single malts from Speyside and Islay. Two of the most recent are 18 and 33 years old respectively.

Glenkeir Treasures

www.whiskyshop.com

The Whisky Shop is the biggest whisky retail chain in the UK with 22 shops – the latest, a flagship store, having been established in Piccadilly opposite the Ritz Hotel in London just before Christmas 2012. The company was founded in 1992 and was bought by the current owner, Ian Bankier, in 2004. Apart from having an extensive range of malt (and other) whiskies, they also select and bottle their own range of single malts called Glenkeir Treasures. Once a cask has been chosen it is re-racked into smaller oak casks which are then put out for display in each store. The whisky is bottled to order and the customer can also try the whisky in the shop before buying. Glenkeir Treasures comes in three bottle sizes – 10, 20 and 50 cl and is bottled at 40%. The current range consists of Craigellachie 6 year old, Inchgower 6 years old, Fettercairn 6 years old and Speyside 13 years old. To celebrate the chain's 10th anniversary, the first blended Scotch under the Glenkeir Treasures label was launched in 2014.

Adelphi Distillery

www.adelphidistillery.com

Adelphi Distillery is named after a distillery which closed in 1902. The company is owned by Keith Falconer and Donald Houston, who recruited Alex Bruce from the wine trade to act as Marketing Director. Their whiskies are always bottled at cask strength, uncoloured and non chill-filtered. Adelphi bottles around 50 casks a year. Two of their recurrent brands are Fascadale (a Highland Park) and Liddesdale (a Bunnahabhain) which are in batches of approximately 1,500 bottles. They also have their own blended Scotch, Adelphi Private Stock, which is bottled at 40%. In 2012, Adelphi Distillery received planning permission to build their own distillery in Glenbeg on the Ardnamurchan peninsula, a couple of miles from the company's office. Most of the buildings were complete by August 2013, the equipment started to arrive in the autumn and, finally on 11th July 2014, the distillery came on stream. The official opening was two weeks later on the 25th July. It is now the most westerly distillery in mainland Scotland and the goal is to produce 100,000 litres per year. The owners are planning for two different styles of whisky; peated and unpeated.

Fascadale 12 year old batch 5, 46%
Nose: Icing sugar, vanilla, milk chocolate, apricots, plus mildly smoky malt.
Palate: Full-bodied and fruity, with toffee, black pepper and a hint of peat in the background.
Finish: Medium in length, slowly drying.

Liddesdale 21 year old batch 4, 46%
Nose: A whiff of warm balloons, hand-rolling tobacco, raisins, sweet new leather and pencil shavings.
Palate: Rich, sweet, spicy notes, soft peat, sherry and fruit malt loaf.
Finish: Bold, with lingering citrus fruit.

Douglas Laing & Co

www.douglaslaing.com

Established in 1948 by Douglas Laing, this firm was run for many years by his two sons, Fred and Stewart. In May 2013, the brothers decided to go their separate ways. Douglas Laing & Co is now run by Fred Laing and his daughter, Cara. The other side of the business, run by Stewart and his sons, Scott and Andrew, is called Hunter Laing. Douglas Laing now has the following brands in their portfolio; Provenance (single casks typically aged between 8 and 20 years and bottled at 46%), Director's Cut (old and rare single malts bottled at cask strength), Premier Barrel (single malts in ceramic decanters bottled at 46%), Clan Denny (two blended malts and a selection of old single grains), Double Barrel (two malts vatted together and bottled at 46%), Big Peat (a vatting of selected Islay malts) and Old Particular, a range of single malts which in 2015 was complemented by four single grains; North British 21, Cameronbridge 25, Girvan 25 and Strathclyde 27 years old.

New releases since 2013 also include Scallywag, a blended malt influenced by sherried whiskies from Speyside and recently complemented by the limited

Scallywag Cask Strength Edition, Timorous Beastie, also a vatted malt where the character relies on malts like Dalmore, Glengoyne and Glen Garioch, Syndicate, a 12 year old blended Scotch bottled at 40% and Rock Oyster, a vatted malt bottled at 46.8% and combining malts from Islay, Arran, Orkney and Jura to create a coastal flavour profile.

Rock Oyster, 46.8%
Nose: Damp beaches, brine and ginger. A hint of peat.
Palate: Initial ozone, with vanilla, cereal and peppery peat.
Finish: Quite dry, with salt and spice.

Timorous Beastie, 46.8%
Nose: Soft and fragrant, with malt, sweet dried fruits, sherry and heather honey.
Palate: Full and sweet, with oranges, apples and hints of aniseed and cinnamon.
Finish: Medium in length, early orange, then drying with spice.

Hunter Laing & Co

www.hunterlaing.com

This company was formed after the demerger between Fred and Stewart Laing in 2013 (see below). It is run by Stewart Laing and his two sons, Scott and Andrew. The relatively new company Edition Spirits, founded by the sons has also been absorbed into Hunter Laing with the range of single malts called The First Editions.

From the demerger, the following ranges and brands ended up in the Hunter Laing portfolio; The Old Malt Cask (rare and old malts, bottled at 50%), The Old and Rare Selection (an exclusive range of old malts offered at cask strength), Douglas of Drumlanrig (single casks bottled at 46%) and Sovereign (a range of grain whiskies). The Sovereign range was re-launched end of 2014 with four extraordinary bottlings; Port Dundas 25 and 36 years old, Strathclyde 25 years old and a North British 52 years old!, The portfolio also includes blended Scotch such as John Player Special, House of Peers and Langside. A new range bearing the name Hepburn's Choice was launched in spring 2014. These single malts are younger than The Old Malt Cask expressions (often 8 to 12 years) and bottled at 46%. A little later in the year, the blended malt Highland Journey was released.

OMC Laphroaig 12 years old, 50%
Nose: Sweet, fruity peat, vanilla, brine and spice.
Palate: Viscous; iniitally sweet and spicy, with Germolene notes, then darker smoky peat and black currants emerge.
Finish: Liquorice and drying peat. Medium in length.

Macallan 30 years old, 48.3%
Nose: Musty apples, with walnuts, malt and a suggestion of mint.
Palate: Big fresh fruit notes, with honey, caramel and nutmeg.
Finish: Becoming dusty, with drying oak.

Malts of Scotland

www.malts-of-scotland.com

Thomas Ewers from Germany, bought casks from Scottish distilleries and decided in the spring of 2009 to start releasing them as single casks bottled at cask strength and with no colouring or chill filtration. At the moment he has released circa 90 bottlings. The most recent include Arran 1996, Tullibardine 1980, Glenrothes

1996, Port Charlotte 2002 and a rare Strathmill 1975. He also has two expressions called Glen First Class (a Glenfarclas) and Glen Peat Class (a vatting of Ardbeg, Laphroaig and Bowmore), both bottled at 50%. Another new series is Amazing Casks, dedicated to very special and superior casks.

Wemyss Malts

www.wemyssmalts.com

This family-owned company, a relatively newcomer to the whisky world, was founded in 2005. The family owns another three companies in the field of wine and gin and in November 2014 they opened up their own whisky distillery at Kingsbarns in Fife. Based in Edinburgh, Wemyss Malts takes advantage of Charles MacLean´s experienced nose when choosing their casks. There are two ranges; one of which consists of single casks bottled at 46% or the occasional cask strength. The names of the whiskies reflect what they taste like although for some time now, the distillery name is also printed on the label. For instance, some of the most recent releases are called Foraged Fruit Fool (1991 Blair Athol), Kumquat Cluster (1993 Glenrothes) and Stem Ginger Preserve (Mortlach 1995). All whiskies are un chill-filtered and without colouring. The other range is made up of blended malts of which there are three at the moment – Spice King, Peat Chimney and The Hive. When first launched in 2005 they were bottled at the age of 5. Four years later the range was expanded with 8 year olds and in 2010 with 12 year olds. In spring 2015, the 8 year old versions were replaced by bottlings without age statements, un chill-filtered and bottled at 40%. A limited addition to the range of blended malts, appeared in September 2015. Kiln Embers is a smokier version of Peat Chimney, bottled at 46% and without age statement. In 2012, the company released its first premium blended whisky based on a selection of malt and grain whiskies aged a minimum of 15 years. The whisky is named Lord Elcho after the eldest son of the 5th Earl of Wemyss. The Lord Elcho range has since then been expanded with a no age statement version.

The Hive NAS, 46%
Nose: Milk chocolate-covered honeycomb, bananas and malt.
Palate: Spicy malt, sweet orange and barley sugar.
Finish: Drying slightly, warming, with ginger and lingering orange.

Spice King NAS, 46%
Nose: Sweet spice, toffee and honey.
Palate: Oily, with lively spices, citrus fruit, dark berries.
Finish: Drying, with liquorice and soft oak.

Among the latest releases are a 2 year old rye from Few Spirits, a 5 year old from Amrut, a 13 year old Arran and an 18 year old Glen Elgin. Since 2013 the company also arranged whisky events in New York state called Whisky Jewbilee. The fourth Whisky Jewbilee was held in June 2015 and in October 2015 it is time for the inaugural Whisky Jewbilee in Chicago.

Speciality Drinks

www.specialitydrinks.com

Sukhinder Singh, known by most for his very well-stocked shop in London, The Whisky Exchange, is behind this company. In the beginning of October every year, he is also hosting The Whisky Show in London, one of the best whisky festivals in the world. In 2005 he started as an independent bottler of malt whiskies operating under the brand name The Single Malts of Scotland. There are around 50 bottlings on offer at any time, either as single casks or as batches bottled at cask strength or at 46%. In 2009 a new range of Islay malts under the name Port Askaig was introduced, starting with a cask strength, a 17 year old and a 25 year old. In 2011 the 17 year old was replaced by a 19 year old and a 30 year old was also added to the range. In 2012, a 12 year old was introduced and in 2013, cask strength versions of the 19 year old and the 30 year old were released. The new version for 2015 is Port Askaig 100 proof, bottled without age statement. Elements of Islay, a series in which all Islay distilleries are, or will be, represented was introduced around the same time. The list of the product range is cleverly constructed with periodical tables in mind in which each distillery has a two-letter acronym followed by a batch number, for example Pe_4 (Port Ellen), Lg_3 (Lagavulin) or Br_3 (the third edition from Bruichladdich).

Elements of Islay Br_3, 51.6%
Nose: Mildly coastal with almond oil, apricots, vanilla and salted caramel.
Palate: Zesty spices, toffee, oranges and lemons.
Finish: Spices continue through the relatively long, fruity finish, with development of slightly drying oak.

Port Askaig 100 proof, 57.1%
Nose: Soft and sweet, with barley and fragrant peat. Mild antispectic and emerging oranges and lemons.
Palate: Smooth and engaging, with big notes of earthy peat, seaweed and citrus fruit.
Finish: Long, with oily peat and lingering citrus.

Jewish Whisky Company

www.singlecasknation.com

Jason Johnstone-Yellin and Joshua Hatton, two well-known whisky bloggers have, in alliance with Seth Klaskin, taken their first step into the world of whisky bottling. The idea with Single Cask Nation somewhat reminds you of Scotch Malt Whisky Society in the sense that you have to become a member of the nation in order to buy the bottlings. You can choose between three different cost levels which will give you various benefits, including one or more bottlings from the current range of whiskies. The first four bottlings that were released in 2012 were Arran 12 years, a peated BenRiach 17 years, a Glen Moray 12 years and a 4 year old Kilchoman. Recently they have also included American and Indian whiskies in the range.

Meadowside Blending

www.meadowsideblending.com

The company may be a newcomer to the family of independent bottlers but the founder certainly isn´t. Donald Hart, a Keeper of the Quaich and co-founder of the well-known bottler Hart Brothers, runs the Glasgow company together with his son, Andrew. There are two sides to the business – blends sold under the name The Royal Thistle where the core expression is a 3 year old, as well as single malts labelled The Maltman. Some of the most recent single malts include Glenturret 30 year old, Glenrothes 18 year old and Arran 16 year old.

Mackillop´s Choice

www.mackillopschoice.com

Mackillop's Choice, founded in 1996, is an independent bottler owned by Angus Dundee Distillers. The brand is named after Lorne McKillop who selects the casks. The whole range is single casks with no colouring or chill filtration. Some of the bottlings are at cask strength, while others are diluted to 40 or 43%. Among the latest new bottlings are Caol Ila 1980, Highland Park 1986, Scapa 19991, Ardbeg 1993 and Mortlach 1991.

The Vintage Malt Whisky Company

www.vintagemaltwhisky.com

The Vintage Malt Whisky Co. was founded in 1992 by Brian Crook who previously had twenty years experience in the malt whisky industry. In recent years, Brian has been joined in the company by his son Andrew. The company also owns and operates a sister company called The Highlands & Islands Scotch Whisky Co. The most famous brands in the range are undoubtedly two single Islay malts called Finlaggan and The Ileach. The latter comes in two versions, bottled at 40% and 58%. The Finlaggan range was recently repackaged and now consists of Old Reserve (40%), Eilean Mor (46%), Port Finish (46%), Sherry Finish (46%) and Cask Strength (58%). Other expressions include two blended malts, Glenalmond and Black Cuillin and, not least, a wide range of single cask single malts under the name The Cooper´s Choice. They are bottled at 46% or at cask strength and are all non coloured and non chillfiltered. In 2012, the company launched a range extension called Cooper´s Choice Golden Grains with a selection of old single grain whiskies from closed distilleries.

Invergordon 1984 30 years old, 57%
Nose: Toffee bonbons, linseed, vanilla and white pepper.
Palate: Rich and creamy, with a hint of pear, creme brulee and ginger.
Finish: Persistently peppery, with continued creaminess.

Ben Nevis 1996 18 years old, 46%
Nose: Savoury, with ripe figs, treacle and a hint of tar.
Palate: Viscous, with toffee spices, red cherries and developing citrus fruit.
Finish: Lengthy and warming, with drying oak and black coffee.

Svenska Eldvatten

www.eldvatten.se

This new, Swedish independent bottler, founded in 2011, is run by Tommy Andersen and Peter Sjögren. They both have extensive experience from whisky and other spirits, which they have gained from arranging tastings for many years. Through the contacts that they had established in Scotland, they managed to get hold of three high quality casks of Bunnahabhain 1991 and 1997 and these were the first to be bottled in 2011. Since then, more than 40 single casks, bottled at cask strength, have been released. Among the most recent bottlings are a Ledaig 1997, a sherrymatured Tobermory from 1994 and a Glendullan 2001 with a maturation in an ex-Bowmore cask. The owners have also released their own blended malt, bottled at 50% under the name Glenn (a humorous tip of the hat to their home town Gothenburg where Glenn is one of the most common names). In their range of spirits they also have aged tequila and rum from the famous and sadly closed Trinidadian distillery Caroni.

They have also recently launched their own gin and aquavit and summer 2015 saw the release of their own rum, WeiRon Super Premium Aged Carribean Rum. The name of their company, Svenska Eldvatten, can be translated to Swedish Firewater.

Deerstalker Whisky Co

www.deerstalkerwhisky.com

The Deerstalker brand, which dates from 1880 was originally owned by J.G. Thomson & Co of Leith and subsequently Tennent Caledonian Breweries. It was purchased by Glasgow based Aberko Ltd in 1994 and is managed by former Tennent's Export Director Paul Aston. The Deer-stalker range covers single cask as well as blended malt whiskies. The 12 year old single malt (46%, un chill-filtered, natural colour) is the best known and has sourced its malt from Balmenach distillery for over 35 years. More recent additions are 'Limited Release' single cask bottlings of Allt-a-Bhainne (18 year old), Braeval (19 year old) and Auchroisk (16 year old) at 48%. A Deerstalker Blended Malt (Highland Edition) was launched in 2014.

Deerstalker Blended Malt, 43%
Nose: Initially slightly vegetal, with potato skins and unsalted peanuts, becoming more floral, with mild spice.
Palate: Sweet and savoury notes, with a sprinkling of salt and a splash of soy sauce.
Finish: Tangy, with lingering fruity spice.

Deerstalker 12 year old, 46%
Nose: Sweet and fruity, with sherry and chilli. Faintly savoury.
Palate: Fruity and very spicy, with black pepper and hints of sherry.
Finish: More chill in the finish, plus plain chocolate-coated raisins.

The Ultimate Whisky Company

www.ultimatewhisky.com

Founded in 1994 by Han van Wees and his son Maurice, this Dutch independent bottler has until now bottled more than 500 single malts. All whiskies are un chill-filtered, without colouring and bottled at either 46% or cask strength. Recent bottlings include Benrinnes 1997, Glen Grant 1992, Glen Keith 1991 and Glentauchers 1996. Younger malts can also be found; Tamdhu 2006 and Ledaig 2008. The van Wees family also operate one of the finest spirits shops in Europe - Van Wees Whisky World in Amersfoort - with i.a. more than 1,000 different whiskies including more than 500 single malts.

The Rest & Be Thankful Whisky Co.

www.foxfitzgerald.co.uk

One of the newest independent bottlers, started by Eamonn Jones who already in 2010 founded Fox Fitzgerald Whisky Trading Co. Two of that company´s products are Peat´s Beast, a heavily peated single malt, and Glen Stuart blended Scotch. As an independent bottler, Eamonn Jones and his partner Aidan Smith, have already launched a couple of rare bottlings - a sauternes-matured Octomore from 2007 and a Port Charlotte 2001 matured in a syrah cask. Even stranger though, is the Strathmore single grain 1972. Strathmore is also known under the name North of Scotland, a grain distillery opened and operated by George Christie, the founder of Speyside distillery, until 1980 when it was closed.

Whisky
shops

AUSTRALIA

The Odd Whisky Coy
PO Box 2045
Glynde, SA, 5070
Phone: +61 (0)8 8365 4722
www.theoddwhiskycoy.com.au
Founded and owned by Graham Wright,
this on-line whisky specialist has an
impressive range. They are agents for
famous brands such as Springbank,
Benromach and Berry Brothers and
arrange recurrent seminars on the subject.

World of Whisky
Shop G12, Cosmopolitan Centre
2-22 Knox Street
Double Bay NSW 2028
Phone: +61 (0)2 9363 4212
www.worldofwhisky.com.au
A whisky specialist which offers a range
of 300 different expressions, most of them
single malts. The shop is also organising
and hosting regular tastings.

AUSTRIA

Potstill
Strozzigasse 37
1080 Wien
Phone: +43 (0)664 118 85 41
www.potstill.org
Austria's premier whisky shop with over
1100 kinds of which c 900 are malts,
including some real rarities. Arranges
tastings and seminars and ships to several
European countries. On-line ordering.

Cadenhead Austria
Alter Markt 1
5020 Salzburg
Phone: +43 (0)662 84 53 05
www.cadenhead.at
Number 8 in the famous Cadenhead´s
chain of whisky shops. At the moment
they offer 350 different whiskies, mostly
single malts and they also arrange monthly
tastings.

BELGIUM

Whiskycorner
Kraaistraat 16
3530 Houthalen
Phone: +32 (0)89 386233
www.whiskycorner.be
A very large selection of single malts,
no less than 1100 different! Also other
whiskies, calvados and grappas. The site is
in both French and English. Mail ordering,
but not on-line. Shipping worldwide.

Jurgen´s Whiskyhuis
Gaverland 70
9620 Zottegem
Phone: +32 (0)9 336 51 06
www.whiskyhuis.be
An absolutely huge assortment of more
than 2,000 different single malts with 700
in stock and the rest delivered within the
week. Also 40 different grain whiskies and
120 bourbons. Worldwide shipping

Huis Crombé
Doenaertstraat 20
8510 Marke
Phone: +32 (0)56 21 19 87
www.crombewines.com
A wine retailer which also covers all kinds
of spirits. The whisky range is very nice
where a large assortment of Scotch is
supplemented with whiskies from Japan,
the USA and Ireland to mention a few.

Anverness
Grote Steenweg 74
2600 Berchem – Antwerpen
Phone: +32 (0)3 218 55 90
www.anverness.be
Peter de Decker has established himself
as one of the best Belgian whisky retailers
where, apart from an impressive range of
whiskies, recurrent tastings and whisky
dinners play an important role.

We Are Whisky
Avenue Rodolphe Gossia 33
1350 Orp-Jauche (Jauche)
Phone: +32 (0)471 134556
www.wearewhisky.com
A fairly new shop and on-line retailer
with a range of more than 400 different
whiskies. They also arrange 3-4 tasting
every month.

Dram 242
Opwijksestraat 242
9280 Lebbeke
Phone: +32 (0)477 260993
www.dram242.be
Started in 2012 by Dirk Verleysen, this
shop has a wide range of whiskies. Apart
from the core official bottlings, Dirk has
focused on rare, old expressions as well as
whiskies from small, independent bottlers.

BRAZIL

Single Malt Brasil
Phone: +55 (21) 3566-0158
www.lojadewhisky.com.br
The biggest whisky specialist In Brazil
(and one of few in the country) with a nice
range of other spirits as well, especially
cachaça. The sister site, singlemalt.com.

br, is a great, educational site about single
malt whisky.

CANADA

Kensington Wine Market
1257 Kensington Road NW
Calgary
Alberta T2N 3P8
Phone: +1 403 283 8000
www.kensingtonwinemarket.com
With 400 different bottlings this is the
largest single malt assortment in Canada.
Also 2,500 different wines. Regular
tastings in the shop.

DENMARK

Juul´s Vin & Spiritus
Værnedamsvej 15
1819 Frederiksberg
Phone: +45 33 31 13 29
www.juuls.dk
A very large range of wines, fortified
wines and spirits. Around 500 single malts.
Also a good selection of drinking glasses.
On-line ordering.

Cadenhead´s WhiskyShop Denmark
Kongensgade 69 F
5000 Odense C
Phone: +45 66 13 95 05
www.cadenheads.dk
Whisky specialist with a very good range,
not least from Cadenhead's. Nice range
of champagne, cognac and rum. Arranges
whisky and beer tastings. On-line ordering
with worldwide shipping.

Whisky.dk
Sjølund Gade 12
6093 Sjølund
Phone: +45 2081 3743
www.whisky.dk
Henrik Olsen and Ulrik Bertelsen are
well-known in Denmark for their whisky
shows but they also run an on-line spirits
shop with an emphasis on whisky but also
including an impressive stock of rums.

Kokkens Vinhus
Hovedvejen 102
2600 Glostrup
Phone: +45 44 97 02 30
Peter Bangs Vej 74
2000 Frederiksberg
Phone: +45 38 87 86 70
www.kokkensvinhus.dk
A shop with a complete assortment of
wine, spirit, coffee, tea and delicatessen.
More than 500 whiskies are in stock,
mostly single malts. They are specialists in

independent bottlings. On-line ordering for shipments within Denmark.

ENGLAND

The Whisky Exchange
Vinopolis, 1 Bank End
London SE1 9BU
Phone: +44 (0)207 403 8688
www.thewhiskyexchange.com
This is an excellent whisky shop owned by Sukhinder Singh. Started off as a mail order business which was run from a showroom in Hanwell, but since some years back there is also an excellent shop at Vinopolis in downtown London. The assortment is huge with well over 1000 single malts to choose from. Some rarities which can hardly be found anywhere else are offered much thanks to Singh's great interest for antique whisky. There are also other types of whisky and cognac, calvados, rum etc. On-line ordering and ships all over the world.

The Whisky Shop
(See also Scotland, The Whisky Shop)
Unit 1.09 MetroCentre
Gateshead NE11 9YG
Phone: +44 (0)191 460 3777

11 Coppergate Walk
York YO1 9NT
Phone: +44 (0)1904 640300

510 Brompton Walk
Lakeside Shopping Centre
Grays, Essex RM20 2ZL
Phone: +44 (0)1708 866255

7 Turl Street
Oxford OX1 3DQ
Phone: +44 (0)1865 202279

3 Swan Lane
Norwich NR2 1HZ
Phone: +44 (0)1603 618284

70 Piccadilly
London W1J 8HP
Phone: +44 (0)20 7499 6649

Unit 7 Queens Head Passage
Paternoster
London EC4M 7DZ
Phone: +44 (0)207 329 5117

3 Exchange St
Manchester M2 7EE
Phone: +44 (0)161 832 6110

25 Chapel Street
Guildford GU1 3UL
Phone: +44 (0)1483 450900

Unit 35 Great Western Arcade
Birmingham B2 5HU
Phone: +44 (0)121 212 1815

64 East Street
Brighton BN1 1HQ
Phone: +44 (0)1273 327 962

3 Cheapside
Nottingham NG1 2HU
Phone: +44 (0)115 958 7080

Trentham Shopping Village
Trentham, Stoke on Trent
Staffordshire ST4 8JG
Phone: +44 (0)178 264 4483
www.whiskyshop.com

The first shop opened in 1992 in Edinburgh and this is now the UK's largest specialist retailer of whiskies with 22 outlets. The two most recent to open up were flagship stores in Piccadilly, London and in Manchester. A large product range with over 700 kinds, including 400 malt whiskies and 140 miniature bottles, as well as accessories and books. They also run The W Club, the leading whisky club in the UK where the excellent Whiskeria magazine is one of the member's benefits. On-line ordering and shipping all over the world except to the USA.

Royal Mile Whiskies
3 Bloomsbury Street
London WC1B 3QE
Phone: +44 (0)20 7436 4763
www.royalmilewhiskies.com
The London branch of Royal Mile Whiskies. See also Scotland, Royal Mile Whiskies.

Berry Bros. & Rudd
3 St James' Street
London SW1A 1EG
Phone: +44 (0)800 280 2440

The Warehouse Shop
Hamilton Close, Houndmills
Basingstoke RG21 6YB
Phone: +44 (0)800 280 2440
www.bbr.com/whisky
A legendary shop that has been situated in the same place since 1698. One of the world's most reputable wine shops but with an exclusive selection of malt whiskies. There are also shops in Dublin and Hong Kong.

The Wright Wine and Whisky Company
The Old Smithy, Raikes Road, Skipton, North Yorkshire BD23 1NP
Phone: +44 (0)1756 700886
www.wineandwhisky.co.uk
An eclectic selection of near to 1000 different whiskies to choose from. 'Tasting Cupboard' of nearly 100 opened bottles for sampling with hosted tasting evenings held on a regular basis. Great 'Collector to Collector' selection of old and rare whiskies plus a fantastic choice of 1200+ wines, premium spirits and liqueurs. International mail order.

Master of Malt
2 Leylands Manor
Tubwell Lane
Crowborough TN6 3RH
Phone: +44 (0)1892 888 376
www.masterofmalt.com
Independent bottler and online retailer since 1985. A very impressive range of more than 1,000 Scotch whiskies of which 800 are single malts. In addition to whisky from other continents there is a wide selection of rum, cognac, Armagnac and tequila. The website is redesigned and contains a wealth of information on the distilleries. They have also launched "Drinks by the Dram" where you can order 3cl samples of more than 500 different whiskies to try before you buy a full bottle.

Whiskys.co.uk
The Square, Stamford Bridge
York YO4 11AG
Phone: +44 (0)1759 371356
www.whiskys.co.uk
Good assortment with more than 600 different whiskies. Also a nice range of armagnac, rum, calvados etc. On-line ordering, ships outside of the UK. The owners also have another website, www.whiskymerchants.co.uk with a huge amount of information on just about every whisky distillery in the world.

The Wee Dram
5 Portland Square, Bakewell
Derbyshire DE45 1HA
Phone: +44 (0)1629 812235
www.weedram.co.uk
Large range of Scotch single malts (c 450) with whiskies from other parts of the world and a good range of whisky books. Run 'The Wee Drammers Whisky Club' with tastings and seminars. End of October they arrange the yearly Wee Dram Fest whisky festival.

Hard To Find Whisky
1 Spencer Street
Birmingham B18 6DD
Phone: +44 (0)8456 803 489
www.htfw.com
As the name says, this family owned shop specialises in rare, collectable and new releases of single malt whisky. The range is astounding - almost 3,000 different bottlings including no less than 263 different Macallan. World wide shipping.

Mainly Wine and Whisky
3-4 The Courtyard, Bawtry
Doncaster DN10 6JG
Phone: +44 (0)1302 714 700
www.whisky-malts-shop.com
A good range with c 400 different whiskies of which 300 are single malts. Arranges tastings and seminars. On-line ordering with shipping also outside the UK.

Nickolls & Perks
37 High Street, Stourbridge
West Midlands DY8 1TA
Phone: +44 (0)1384 394518
www.nickollsandperks.co.uk
Mostly known as wine merchants but also has a good range of whiskies with c 300 different kinds including 200 single malts. On-line ordering with shipping also outside of UK. Since 2011, Nickolls & Perks also organize the acclaimed Midlands Whisky Festival, see www.whiskyfest.co.uk

Gauntleys of Nottingham
4 High Street
Nottingham NG1 2ET
Phone: +44 (0)115 9110555
www.gauntley-wine.co.uk
A fine wine merchant established in 1880. The range of wines are among the best in the UK. All kinds of spirits, not least whisky, are taking up more and more space and several rare malts can be found. The monthly whisky newsletter by Chris Goodrum makes good reading and there is also a mail order service available.

Hedonism Wines
3-7 Davies St.
London W1K 3LD
Phone: +44 (020) 729 078 70
www.hedonism.co.uk
Located in the heart of London´s Mayfair,
this is a new temple for wine lovers but
also with an impressive range of whiskies
and other spirits. They have over 1,200
different bottlings from Scotland and the
rest of the world! The very elegant shop is
in itself well worth a visit.

The Wine Shop
22 Russell Street, Leek
Staffordshire ST13 5JF
Phone: +44 (0)1538 382408
www.wineandwhisky.com
In addition to wine there is a good
range of 300 whiskies and also calvados,
cognac, rum etc. They also stock a range
of their own single malt bottlings under the
name of 'The Queen of the Moorlands'.
Mailorder within the UK.

The Lincoln Whisky Shop
87 Bailgate
Lincoln LN1 3AR
Phone: +44 (0)1522 537834
www.lincolnwhiskyshop.co.uk
Mainly specialising in whisky with more
than 400 different whiskies but also 500
spirits and liqueurs and some 100 wines.
Mailorder only within UK.

Milroys of Soho
3 Greek Street
London W1D 4NX
Phone: +44 (0)207 734 2277
www.milroys.co.uk
A classic whisky shop in Soho with a very
good range with over 700 malts and a
wide selection of whiskies from around the
world. On-line ordering.

Arkwrights
114 The Dormers
Highworth
Wiltshire SN6 7PE
Phone: +44 (0)1793 765071
www.whiskyandwines.com
A good range of whiskies (over 700 in
stock) as well as wine and other spirits.
Regular tastings in the shop. On-line
ordering with shipping all over the world
except USA and Canada.

Edencroft Fine Wines
8-10 Hospital Street, Nantwich
Cheshire, CW5 5RJ
Phone: +44 (0)1270 629975
www.edencroft.co.uk
Family owned wine and spirits shop since
1994. Around 250 whiskies and also a
nice range of gin, cognac and other spirits
including cigars. Worldwide shipping.

Cadenhead´s Whisky Shop
26 Chiltern Street
London W1U 7QF
Phone: +44 (0)20 7935 6999
www.whiskytastingroom.com
One in a chain of shops owned by
independent bottlers Cadenhead. Sells
Cadenhead's product range and c. 200
other whiskies. Regular tastings and on-
line ordering.

Constantine Stores
30 Fore Street
Constantine, Falmouth
Cornwall TR11 5AB
Phone: +44 (0)1326 340226
www.drinkfinder.co.uk
A full-range wine and spirits dealer with a
good selection of whiskies from the whole
world (around 800 different, of which
600 are single malts). Worldwide shipping
except for USA and Canada.

The Vintage House
42 Old Compton Street
London W1D 4LR
Phone: +44 (0)20 7437 5112
www.sohowhisky.com
A huge range of 1400 kinds of malt
whisky, many of them rare. Supplementing this is also a selection of fine wines.
On-line ordering.

Whisky On-line
Units 1-3 Concorde House, Charnley
Road, Blackpool, Lancashire FY1 4PE
Phone: +44 (0)1253 620376
www.whisky-online.com
A good selection of whisky and also
cognac, rum, port etc. On-line ordering
with shipping all over the world.

FRANCE

La Maison du Whisky
20 rue d´Anjou
75008 Paris
Phone: +33 (0)1 42 65 03 16

6 carrefour d l´Odéon
75006 Paris
Phone: +33 (0)1 46 34 70 20

(2 shops outside France)
47 rue Jean Chatel
97400 Saint-Denis, La Réunion
Phone: +33 (0)2 62 21 31 19

The Pier at Robertson Quay
80 Mohamed Sultan Road, #01-10
Singapore 239013
Phone: +65 6733 0059
www.whisky.fr
France's largest whisky specialist with
over 1200 whiskies in stock. Also a
number of own-bottled single malts. La
Maison du Whisky acts as a EU distributor
for many whisky producers around the
world. Four shops and on-line ordering.
Ships to some 20 countries.

The Whisky Lodge
7 rue Ferrandière
69002 Lyon
Phone: +33 (0)4 78 42 48 22
Located in the heart of Lyon and run by
Pierre Tissandier, son of the founder, this
shop carries more than 1,200 different
whiskies!

GERMANY

Celtic Whisk(e)y & Versand
Otto Steudel
Bulmannstrasse 26
90459 Nürnberg
Phone: +49 (0)911 45097430
www.whiskymania.de/celtic
A very impressive single malt range with

well over 1000 different single malts and
a good selection from other parts of the
world. On-line ordering with shipping also
outside Germany.

SCOMA
Am Bullhamm 17
26441 Jever
Phone: +49 (0)4461 912237
www.scoma.de
Very large range of c 750 Scottish malts
and many from other countries. Holds
regular seminars and tastings. The
excellent, monthly whisky newsletter
SCOMA News is produced and can be
downloaded as a pdf-file from the website.
On-line ordering.

The Whisky Store
Am Grundwassersee 4
82402 Seeshaupt
Phone: +49 (0)8801 30 20 000
www.whisky.de
A very large range comprising c 700
kinds of whisky of which 550 are malts.
Also sells whisky liqueurs, books and
accessories. The website is a goldmine of
information. On-line ordering.

Cadenhead´s Whisky Market
Luxemburger Strasse 257
50939 Köln
Phone: +49 (0)221-2831834
www.cadenheads.de
Good range of malt whiskies (c 350
different kinds) with emphasis on
Cadenhead's own bottlings. Other
products include wine, cognac and rum
etc. Arranges recurring tastings and also
has an on-line shop.

Cadenhead´s Whisky Market
Mainzer Strasse 20
10247 Berlin-Friedrichshain
Phone: +49 (0)30-30831444
www.cadenhead-berlin.de
Excellent product range with more than
700 different kinds of whiskies with
emphasis on Cadenhead's own bottlings
as well as cognac and rum. Arranges
recurrent tastings.

Malts and More
Hosegstieg 11
22880 Wedel
Phone: +49 (0)40-23620770
www.maltsandmore.de
Large assortment with over 800 different
single malts as well as whiskies from
many other countries. Also a nice selection
of cognac, rum etc. On-line ordering.

Reifferscheid
Mainzer Strasse 186
53179 Bonn / Mehlem
Phone: +49 (0)228 9 53 80 70
www.whisky-bonn.de
A well-stocked shop often listed as one of
the best in Germany. Aside from a large
range of whiskies, wine, spirit, cigars
and a delicatessen can be found. Regular
tastings.

Whisky-Doris
Germanenstrasse 38
14612 Falkensee
Phone: +49 (0)3322-219784
www.whisky-doris.de

Large range of over 300 whiskies and also sells own special bottlings. Orders via email. Shipping also outside Germany.

Finlays Whisky Shop
Friedrichstrasse 3
65779 Kelkheim
Phone: +49 (0)6195 9699510
www.finlayswhiskyshop.de
Whisky specialists with a large range of over 1,400 whiskies. Finlays also work as the importer to Germany of Douglas laing, James MacArthur and Wilson & Morgan. On-line ordering.

Weinquelle Lühmann
Lübeckerstrasse 145
22087 Hamburg
Phone: +49 (0)40-25 63 91
www.weinquelle.com
An impressive selection of both wines and spirits with over 1000 different whiskies of which 850 are malt whiskies. Also an impressive range of rums. On-line ordering.

The Whisky-Corner
Reichertsfeld 2
92278 Illschwang
Phone: +49 (0)9666-951213
www.whisky-corner.de
A small shop but large on mail order. A very large assortment of over 1600 whiskies. Also sells blended and American whiskies. The website is very informative with features on, among others, whisky making, tasting and independent bottlers. On-line ordering.

World Wide Spirits
Hauptstrasse 12
84576 Teising
Phone: +49 (0)8633 50 87 93
www.worldwidespirits.de
A nice range of c 500 whiskies with some rarities from the twenties. Also large selection of other spirits.

Whisk(e)y Shop Tara
Rindermarkt 16
80331 München
Phone: +49 (0)89-26 51 18
www.whiskyversand.de
Whisky specialists with a very broad range of, for example, 800 different single malts. On-line ordering.

WhiskyKoch
Weinbergstrasse 2
64285 Darmstadt
Phone: +49 (0)6151 99 27 105
www.whiskykoch.de
A combination of a whisky shop and restaurant. The shop has a nice selection of single malts as well as other Scottish products and the restaurant has specialised in whisky dinners and tastings.

Kierzek
Weitlingstrasse 17
10317 Berlin
Phone: +49 (0)30 525 11 08
www.kierzek-berlin.de
Over 400 different whiskies in stock. In the product range 50 kinds of rum and 450 wines from all over the world are found among other products. Mail order is available.

House of Whisky
Ackerbeeke 6
31683 Obernkirchen
Phone: +49 (0)5724-399420
www.houseofwhisky.de
Aside from over 1,200 different malts also sells a large range of other spirits (including over 100 kinds of rum). On-line ordering.

Whiskyworld
Ziegelfeld 6
94481 Grafenau / Haus i. Wald
Phone: +49 (0)8555-406 320
www.whiskyworld.de
A very good assortment of more than 1,000 malt whiskies. Also has a good range of wines, other spirits, cigars and books. On-line ordering.

World Wide Whisky (2 shops)
Eisenacher Strasse 64
10823 Berlin-Schöneberg
Phone: +49 (0)30-7845010
Hauptstrasse 58
10823 Berlin-Schöneberg
www.world-wide-whisky.de
Large range of 1,500 different whiskies. Arranges tastings and seminars. Has a large number of rarities. Orders via email.

HUNGARY
Whisky Net / Whisky Shop
Kovács Làszlò Street 21
2000 Szentendre

Veres Pálné utca 8.
1053 Budapest
Phone: +36 1 267-1588
www.whiskynet.hu
www.whiskyshop.hu
The largest selction of whisky in Hungary. Agents for Arran, Benriach, Glenfarclas, Gordon & MacPhail, Kilchoman among others. Also mailorder.

INDIA
The Vault
World Whiskies & Fine Spirits
Mumbai
Phone: +91 22-22028811/22
www.vaultfinespirits.com
India's first curated fine spirits platform, opened in October 2013. An interesting concept where personal assistance in choosing a gift or planning an event is also part of the service. Expect many more brands to be added in coming years.

IRELAND
Celtic Whiskey Shop
27-28 Dawson Street
Dublin 2
Phone: +353 (0)1 675 9744
www.celticwhiskeyshop.com
More than 70 kinds of Irish whiskeys but also a good selection of Scotch, wines and other spirits. World wide shipping.

ITALY
Whisky Shop
by Milano Whisky Festival
Via Cavaleri 6, Milano
Phone: +39 (0)2 48753039
www.whiskyshop.it
The team behind the excellent Milano Whisky Festival also have an on-line whiskyshop with almost 500 different single malts including several special festival bottlings.

Cadenhead's Whisky Bar
Via Poliziano, 3
20154 Milano
Phone: +39 (0)2 336 055 92
www.cadenhead.it
This is the newest addition in the Cadenhead's chain of shops. Concentrating mostly on the Cadenhead's range but they also stock whiskies from other producers.

THE NETHERLANDS
Whiskyslijterij De Koning
Hinthamereinde 41
5211 PM 's Hertogenbosch
Phone: +31 (0)73-6143547
www.whiskykoning.nl
An enormous assortment with more than 1400 kinds of whisky including c 800 single malts. Arranges recurring tastings. On-line ordering. Shipping all over the world.

Van Wees - Whiskyworld.nl
Leusderweg 260
3817 KH Amersfoort
Phone: +31 (0)33-461 53 19
www.whiskyworld.nl
A very large range of 1000 whiskies including over 500 single malts. Also have their own range of bottlings (The Ultimate Whisky Company). On-line ordering.

Wijnhandel van Zuylen
Loosduinse Hoofdplein 201
2553 CP Loosduinen (Den Haag)
Phone: +31 (0)70-397 1400
www.whiskyvanzuylen.nl
Excellent range of whiskies (circa 1100) and wines. Email orders with shipping to some ten European countries.

Wijnwinkel-Slijterij
Ton Overmars
Hoofddorpplein 11
1059 CV Amsterdam
Phone: +31 (0)20-615 71 42
www.tonovermars.nl
A very large assortment of wines, spirits and beer which includes more than 400 single malts. Arranges recurring tastings. Orders via email.

Wijn & Whisky Schuur
Blankendalwei 4
8629 EH Scharnegoutem
Phone: +31 (0)515-520706
www.wijnwhiskyschuur.nl
Large assortment with 1000 different whiskies and a good range of other spirits as well. Arranges recurring tastings. On-line ordering.

Versailles Dranken
Lange Hezelstraat 83
6511 Cl Nijmegen
Phone: +31 (0)24-3232008
www.versaillesdranken.nl
A very impressive range with more than
1500 different whiskies, most of them
from Scotland but also a surprisingly good
selection (more than 60) of Bourbon. Arranges recurring tastings. On-line ordering.

Alba Malts
Kloosterstraat 15
6981 CC Doesburg
Phone: +31 (0)65-4295905
www.albamalts.com
A new whisky shop situated in an old
chapel dating back to 1441. Marnix Okel
has a passion for Scotland and will focus
on Scotch single malt only, with a range of
400 whiskies to start with.

NEW ZEALAND
Whisky Galore
66 Victoria Street
Christchurch 8013
Phone: +64 (3) 377 6824
www.whiskygalore.co.nz
The best whisky shop in New Zealand with
550 different whiskies, approximately
350 which are single malts. There is also
online mail-order with shipping all over
the world except USA and Canada.

POLAND
George Ballantine´s
Krucza str 47 A, Warsaw
Phone: +48 22 625 48 32

Pulawska str 22, Warsaw
Phone: +48 22 542 86 22

Marynarska str 15, Warsaw
Phone: +48 22 395 51 60

Francuska str 27, Warsaw
Phone: +48 22 810 32 22
www.sklep-ballantines.pl
The biggest assortment in Poland with
more than 360 different single malts.
Apart from whisky there is a full range of
spirits and wines from all over the world.
Recurrent tastings and mailorder.

RUSSIA
Whisky World Shop
9, Tverskoy Boulevard
123104 Moscow
Phone: +7 495 787 9150
www.whiskyworld.ru
Huge assortment with more than 1,000
different single malts. The range is
supplemented with a nice range of cognac,
armagnac, calvados, grappa and wines.
Tastings are also arranged.

SCOTLAND
Gordon & MacPhail
58 - 60 South Street, Elgin
Moray IV30 1JY
Phone: +44 (0)1343 545110
www.gordonandmacphail.com
This legendary shop opened already in
1895 in Elgin. The owners are perhaps

the most well-known among independent
bottlers. The shop stocks more than 800
bottlings of whisky and more than 600
wines and there is also a delicatessen
counter with high-quality products.
Tastings are arranged in the shop and there
are shipping services within the UK and
overseas. The shop attracts visitors from
all over the world.

Royal Mile Whiskies (2 shops)
379 High Street, The Royal Mile
Edinburgh EH1 1PW
Phone: +44 (0)131 2253383

3 Bloomsbury Street
London WC1B 3QE
Phone: +44 (0)20 7436 4763
www.royalmilewhiskies.com
Royal Mile Whiskies is one of the most
well-known whisky retailers in the UK.
It was established in Edinburgh in 1991.
There is also a shop in London since 2002
and a cigar shop close to the Edinburgh
shop. The whisky range is outstanding
with many difficult to find elsewhere.
They have a comprehensive site regarding
information on regions, distilleries,
production, tasting etc. Royal Mile
Whiskies also arranges 'Whisky Fringe'
in Edinburgh, a two-day whisky festival
which takes place annually in mid August.
On-line ordering with worldwide shipping.

The Whisky Shop
(See also England, The Whisky Shop)
Buchanan Galleries
Buchanan Street
Glasgow G1 2GF
Phone: +44 (0)141 331 0022

17 Bridge Street
Inverness IV1 1HD
Phone: +44 (0)1463 710525

11 Main Street
Callander FK17 8DU
Phone: +44 (0)1877 331936

93 High Street
Fort William PH33 6DG
Phone: +44 (0)1397 706164

52 George Street
Oban PA34 5SD
Phone: +44 (0)1631 570896

Unit 14
Gretna Gateway Outlet Village
Gretna DG16 5GG
Phone: +44 (0)1461338004

Unit RU58B, Ocean Terminal
Edinburgh EH6 6JJ
Phone: +44 (0)131 554 8211

Unit 23
Princes Mall
Edinburgh EH1 1BQ
Phone: +44 (0)131 558 7563

28 Victoria Street
Edinburgh EH1 2JW
Phone: +44 (0)131 225 4666
www.whiskyshop.com
The first shop opened in 1992 in
Edinburgh and this is now the United
Kingdom's largest specialist retailer of
whiskies with 22 outlets. The two most
recent to open up were flagship stores in
Piccadilly, London and in Manchester. A

large product range with over 700 kinds,
including 400 malt whiskies and 140
miniature bottles, as well as accessories
and books. The own range 'Glenkeir
Treasures' is a special assortment of
selected malt whiskies. The also run
The W Club, the leading whisky club in
the UK where the excellent Whiskeria
magazine is one of the member´s benefits.
On-line ordering and shipping all over the
world except to the USA.

Loch Fyne Whiskies
Inveraray
Argyll PA32 8UD
Phone: +44 (0)1499 302 219
www.lfw.co.uk
A legendary shop! The range of malt
whiskies is large and they have their own
house blend, the prize-awarded Loch Fyne,
as well as their 'The Loch Fyne Whisky
Liqueur'. There is also a range of house
malts called 'The Inverarity'. On-line
ordering with worldwide shipping.

Single Malts Direct
36 Gordon Street
Huntly
Aberdeenshire AB54 8EQ
Phone: +44 (0) 845 606 6145
www.singlemaltsdirect.com
Owned by independent bottler Duncan
Taylor. In the assortment is of course the
whole Duncan Taylor range but also a
selection of their own single malt bottlings
called Whiskies of Scotland. A total of
almost 700 different expressions. On-line
shop with shipping worldwide. The website has information on whisky production
and a glossary of whisky terms.

The Whisky Shop Dufftown
1 Fife Street, Dufftown
Moray AB55 4AL
Phone: +44 (0)1340 821097
www.whiskyshopdufftown.co.uk
Whisky specialist in Dufftown in the heart
of Speyside, wellknown to many of the
Speyside festival visitors. More than 500
single malts as well as other whiskies.
Arranges tastings as well as special events
during the Festivals. On-line ordering.

Cadenhead's Whisky Shop
(Eaglesome)
30-32 Union Street
Campbeltown
Argyll PA28 6JA
Phone: +44 (0)1586 551710
www.wmcadenhead.com
One in a chain of shops owned by
independent bottlers Cadenhead. Sells
Cadenhead's products and other whiskies
with a good range of Springbank. On-line
ordering.

Cadenhead´s Whisky Shop
172 Canongate, Royal Mile
Edinburgh EH8 8BN
Phone: +44 (0)131 556 5864
www.wmcadenhead.com
The oldest shop in the chain owned by Cadenhead. Sells Cadenhead's product range
and a good selection of other whiskies
and spirits. Recurrent tastings. On-line
ordering.

The Good Spirits Co.
23 Bath Street,
Glasgow G2 1HW
Phone: +44 (0)141 258 8427
www.thegoodspiritsco.com
A newly opened specialist spirits store selling whisky, bourbon, rum, vodka, tequila, gin, cognac and armagnac, liqueurs and other spirits. They also stock quality champagne, fortified wines and cigars.

The Scotch Whisky Experience
354 Castlehill, Royal Mile
Edinburgh EH1 2NE
Phone: +44 (0)131 220 0441
www.scotchwhiskyexperience.co.uk
The Scotch Whisky Experience is a must for whisky devotees visiting Edinburgh. An interactive visitor centre dedicated to the history of Scotch whisky. This five-star visitor attraction has an excellent whisky shop with almost 300 different whiskies in stock. Reccently, after extensive refurbishment, a brand new and interactive shop was opened.

Whiski Shop
4 North Bank Street
Edinburgh EH1 2LP
Phone: +44 (0)131 225 1532
www.whiskishop.com
www.whiskirooms.co.uk
A new concept located near Edin-burgh Castle, combining a shop, a tasting room and a bistro. Also regular whisky tastings. Online mail order with worldwide delivery.

Robbie's Drams
3 Sandgate, Ayr
South Ayrshire KA7 1BG
Phone: +44 (0)1292 262 135
www.robbiesdrams.com
Over 600 whiskies available in store and over 900 available from their on-line shop. Specialists in single cask bottlings, closed distillery bottlings, rare malts, limited edition whisky and a nice range of their own bottlings. Worldwide shipping.

The Whisky Barrel
PO Box 23803, Edinburgh, EH6 7WW
Phone: +44 (0)845 2248 156
www.thewhiskybarrel.com
Online specialist whisky shop based in Edinburgh. They stock over 1,000 single malt and blended whiskies including Scotch, Japanese, Irish, Indian, Swedish and their own casks. Worldwide shipping.

The Scotch Malt Whisky Society
www.smws.com
A society with more than 20 000 members worldwide, specialised in own bottlings of single casks and release between 150 and 200 bottlings a year. Orders on-line for members only. Shipping only within UK.

Drinkmonger
100 Atholl Road
Pitlochry PH16 5BL
Phone: +44 (0)1796 470133

11 Bruntsfield Place
Edinburgh EH10 4HN
Phone: +44 (0)131 229 2205
www.drinkmonger.com
Two new shops opened in 2011 by the well-known Royal Mile Whiskies. The idea is to have a 50:50 split between wine and specialist spiritswith the addition of a cigar assortment. The whisky range is a good cross-section with some rarities and a focus on local distilleries.

A.D. Rattray´s Whisky Experience & Whisky Shop
32 Main Road
Kirkoswald
Ayrshire KA19 8HY
Phone: +44 (0) 1655 760308
www.adrattray.com
A combination of whisky shop, sample room and educational center owned by the independent bottler A D Rattray. Tasting menus with different themes are available.

Robert Graham Ltd (3 shops)
194 Rose Street
Edinburgh EH2 4AZ
Phone: +44 (0)131 226 1874

Robert Graham´s Global Whisky Shop
111 West George Street
Glasgow G2 1QX
Phone: +44 (0)141 248 7283

Robert Graham's Treasurer 1874
254 Canongate
Royal Mile
Edinburgh EH8 8AA
Phone: +44 (0)131 556 2791
www.whisky-cigars.co.uk
Established in 1874 this company specialises in Scotch whisky and cigars. They have a nice assortment of malt whiskies and their range of cigars is impressive.

SOUTH AFRICA
Aficionados Premium Spirits Online
M5 Freeway Park
Cape Town
Phone: +27 21 511 7337
www.aficionados.co.za
An online liquor retailer specialising in single malt whisky. They claim to offer the widest of range of whiskies available in South Africa and hold regular tastings around the country. Shipping only within South Africa.

SWITZERLAND
P. Ullrich AG
Schneidergasse 27
4051 Basel
Phone: +41 (0)61 338 90 91
Another two shops in Basel:
Laufenstrasse 16 & Unt. Rebgasse 18
and one in Talacker 30 in Zürich
www.ullrich.ch
A very large range of wines, spirits, beers, accessories and books. Over 800 kinds of whisky with almost 600 single malt. Online ordering. Recently, they also founded a whisky club with regular tastings (www.whiskysinn.ch).

Eddie's Whiskies
Dorfgasse 27
8810 Horgen
Phone: +41 (0)43 244 63 00
www.eddies.ch
A whisky specialist with more than 700 different whiskies in stock with emphasis on single malts (more than 500 different). Also arranges tastings.

World of Whisky
Via dim Lej 6
7500 St. Moritz
Phone: +41 (0)81 852 33 77
www.world-of-whisky.ch
A legendary shop situated in the Hotel Waldhaus Am See which has an also legendary whisky bar, the Devil´s Place. The shop stocks almost 1,000 different whiskies and has a good range of other spirits such as rum, cognac and armagnac.

Scot & Scotch
Wohllebgasse 7
8001 Zürich
Phone: +41 44 211 90 60
www.scotandscotch.ch
A whisky specialist with a great selection including c 560 single malts. Mail orders, but no on-line ordering.

Angels Share Shop
Unterdorfstrasse 15
5036 Oberentfelden
Phone: +41 (0)62 724 83 74
www.angelsshare.ch
A combined restaurant and whisky shop. More than 400 different kinds of whisky as well as a good range of cigars. Scores extra points for short information and photos of all distilleries. On-line ordering.

UKRAINE
Good Wine
Illi Mechnykova 9
Kiev 01133
Phone: +38 (0)44 4911075
www.goodwine.ua
Wine Bureau Company is the largest importer of wine and spirits in Ukraine, with sales of 4 million bottles per year. They also operate a flagship store in Kiev called Good Wine. The impressive and large shop (2,500 sqm) has a range of more than 1000 different spirits including 400 single malts. The owners arrange weekly tastings and also produce their own magazine.

USA
Binny´s Beverage Depot
5100 W. Dempster (Head Office)
Skokie, IL 60077
Phone:
Internet orders, 888-942-9463 (toll free)
Whiskey Hotline, 888-817-5898 (toll free)
www.binnys.com
A chain of no less than 31 stores in the Chicago area, covering every-thing within wine and spirits. Some of the stores also have a gourmet grocery, cheese shop and, for cigar lovers, a walk-in humidor. The range is impressive with more than 1500 whisk(e)y (740 single malts, 200 bourbons, 60 Irish whiskeys) and more. Among other products almost 340 kinds of tequila, more than 400 vodkas and almost 300 rums should be mentioned. Online mail order service.

Statistics

The following pages have been made possible,
first and foremost thanks to kind cooperation from The IWSR.
Data has also been provided by Drinks International, The Scotch Whisky
Industry Review and the Scotch Whisky Association.

Whisk(e)y forecast (volume)
by region and category 2014-2019

■ = positive volume growth   ■ = negative volume growth

SW=Scotch Whisky, IW=Irish Whiskey, UW=US Whiskey, CW=Canadian Whisky, OW=Other Whisky, TOT=Total.
The figures show CAGR% (Compound Annual Growth Rate) i. e. year-over-year growth rate.

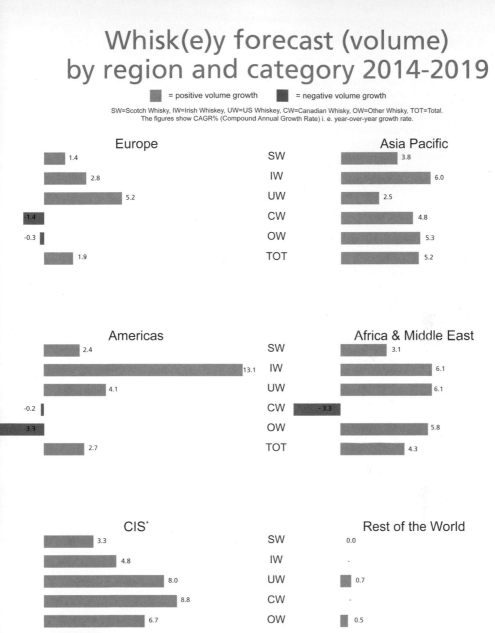

Europe

SW	1.4
IW	2.8
UW	5.2
CW	-1.4
OW	-0.3
TOT	1.9

Asia Pacific

SW	3.8
IW	6.0
UW	2.5
CW	4.8
OW	5.3
TOT	5.2

Americas

SW	2.4
IW	13.1
UW	4.1
CW	-0.2
OW	3.3
TOT	2.7

Africa & Middle East

SW	3.1
IW	6.1
UW	6.1
CW	-3.3
OW	5.8
TOT	4.3

CIS*

SW	3.3
IW	4.8
UW	8.0
CW	8.8
OW	6.7
TOT	4.0

* Russia and other former Soviet Socialist Republic states

Rest of the World

SW	0.0
IW	-
UW	0.7
CW	-
OW	0.5
TOT	0.0

The World

SW	2.3
IW	8.2
UW	4.2
CW	-0.2
OW	5.2
TOT	4.2

Source: © The IWSR 2015

The Top 30 Whiskies of the World
Sales figures for 2014 (units in million 9-litre cases)

Officer's Choice (Allied Blenders & Distillers), Indian whisky — 30,2
McDowell's No. 1 (United Spirits), Indian whisky — 25,0
Johnnie Walker (Diageo), Scotch whisky — 17,9
Royal Stag (Pernod Ricard), Indian whisky — 15,7
Imperial Blue (Pernod Ricard), Indian whisky — 14,0
Jack Daniel's (Brown-Forman), Tennessee whiskey — 11,7
Old Tavern (United Spirits), Indian whisky — 11,2
Original Choice (John Distilleries), Indian whisky — 10,5
Hayward's Fine (United Spirits), Indian whisky — 10,4
Bagpiper (United Spirits), Indian whisky — 9,5
Jim Beam (Beam Suntory), Bourbon — 7,4
Ballantine's (Pernod Ricard), Scotch whisky — 5,9
Crown Royal (Diageo), Canadian whisky — 5,3
Jameson (Pernod Ricard), Irish whiskey — 4,7
Chivas Regal (Pernod Ricard), Scotch whisky — 4,6
Blenders Pride (Pernod Ricard), Indian whisky — 4,4
William Grant's (William Grant & Sons), Scotch whisky — 4,4
8PM (Radico Khaitan), Indian whisky — 4,3
J&B Rare (Diageo), Scotch whisky — 3,7
Director's Special (United Spirits), Indian whisky — 3,6
Kakubin (Suntory), Japanese whisky — 3,2
William Lawson's (Bacardi), Scotch whisky — 3,1
William Peel (Bélvedère), Scotch whisky — 2,8
Royal Challenge (United Spirits), Indian whisky — 2,7
Dewar's (Bacardi), Scotch whisky — 2,7
Label 5 (La Martiniquaise), Scotch whisky — 2,6
Bell's (Diageo), Scotch whisky — 2,5
Seagram's 7 Crown (Diageo), American blended — 2,4
Black Velvet (Constellation Brands), Canadian whisky — 2,3
Director's Special Black (United Spirits), Indian whisky — 2,3

Source: Drinks International, The Millionaires Club 2015

Global Exports of Scotch by Region

Volume (litres of pure alcohol) Region	2014	2013	chg %	Value (£ Sterling) Region	2014	2013	chg %
Africa	22,713,779	23,248,240	-2	Africa	219,338,371	245,198,819	-11
Asia	67,056,990	66,885,104	0	Asia	795,445,614	848,620,987	- 6
Australasia	8,520,307	9,327,531	-9	Australasia	92,416,542	92,451,929	0
C&S America	39,715,835	44,804,019	-11	C&S America	361,399,671	457,153,767	- 21
Eastern Europe	1,242,474	3,877,037	-68	Eastern Europe	14,754,642	38,826,012	-62
Europe (other)	6,106,119	6,128,353	0	Europe (other)	95,439,307	93,572,349	2
European Union	125,108,785	126,816,793	-1	European Union	1,243,088,413	1,318,765,906	- 6
Middle East	14,976,159	12,868,936	16	Middle East	209,700,635	171,842,952	22
North America	48,363,451	50,242,875	-4	North America	913,583,180	994,212,224	-8
Total	**333,803,899**	**344,198,888**	**-3**	**Total**	**3,945,166,375**	**4,260,644,945**	**- 7**

Source: Scotch Whisky Association

World Consumption of Blended Scotch

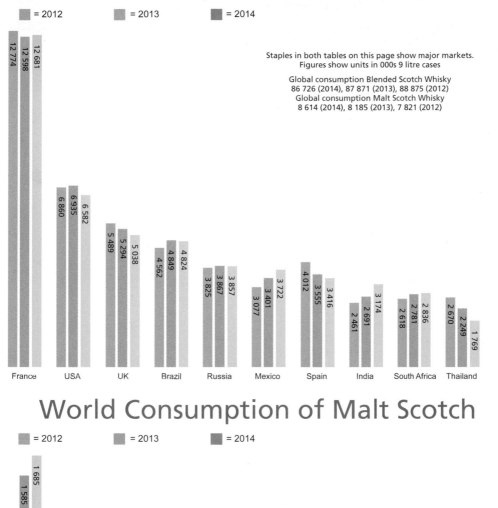

■ = 2012   ■ = 2013   ■ = 2014

Staples in both tables on this page show major markets.
Figures show units in 000s 9 litre cases

Global consumption Blended Scotch Whisky
86 726 (2014), 87 871 (2013), 88 875 (2012)
Global consumption Malt Scotch Whisky
8 614 (2014), 8 185 (2013), 7 821 (2012)

France: 12 774 / 12 598 / 12 681
USA: 6 860 / 6 935 / 6 582
UK: 5 489 / 5 294 / 5 038
Brazil: 4 562 / 4 849 / 4 824
Russia: 3 825 / 3 867 / 3 857
Mexico: 3 077 / 3 401 / 3 722
Spain: 4 012 / 3 555 / 3 416
India: 2 461 / 2 691 / 3 174
South Africa: 2 618 / 2 781 / 2 836
Thailand: 2 670 / 2 249 / 1 769

Source: © The IWSR 2015

World Consumption of Malt Scotch

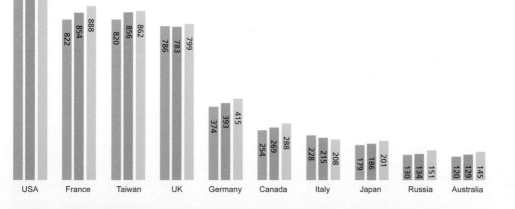

■ = 2012   ■ = 2013   ■ = 2014

USA: 1 404 / 1 585 / 1 685
France: 822 / 854 / 888
Taiwan: 820 / 856 / 862
UK: 786 / 783 / 799
Germany: 374 / 393 / 415
Canada: 254 / 269 / 288
Italy: 228 / 215 / 208
Japan: 179 / 186 / 201
Russia: 130 / 134 / 151
Australia: 120 / 129 / 145

Source: © The IWSR 2015

Top 10 Scotch Malt Whisky brands - world market share %

The Glenlivet	2014	12,4
	2013	11,9
	2012	11,4
Glenfiddich	2014	12,2
	2013	12,8
	2012	13,7
The Macallan	2014	9,6
	2013	9,7
	2012	9,2
Glenmorangie	2014	5,7
	2013	5,6
	2012	5,7
The Singleton Dufftown, Glendullan, Glen Ord	2014	4,8
	2013	4,2
	2012	3,5
Aberlour	2014	3,4
	2013	3,1
	2012	3,0
Glen Grant	2014	3,2
	2013	3,3
	2012	3,4
Laphroaig	2014	3,1
	2013	2,7
	2012	2,7
Balvenie	2014	2,8
	2013	2,8
	2012	2,8
Talisker	2014	2,2
	2013	2,1
	2012	1,8

Top 10 Scotch Blended Whisky brands - world market share %

Johnnie Walker	2014	21,6
	2013	21,6
	2012	20,9
Ballantine's	2014	6,8
	2013	6,7
	2012	6,6
Chivas Regal	2014	5,4
	2013	5,4
	2012	5,5
Grant's	2014	5,1
	2013	5,3
	2012	5,3
J&B	2014	4,1
	2013	4,2
	2012	4,6
Famous Grouse	2014	3,6
	2013	3,5
	2012	3,6
William Lawson's	2014	3,3
	2013	3,0
	2012	2,9
Dewar's	2014	3,1
	2013	3,4
	2012	3,4
William Peel	2014	3,1
	2013	3,0
	2012	3,0
Label 5	2014	2,7
	2013	2,5
	2012	2,6

Source: © The IWSR 2015

Distillery Capacity

Litres of pure alcohol - Scottish, active distilleries only

Distillery	Capacity	Distillery	Capacity	Distillery	Capacity
Glenfiddich	14 000 000	Bowmore	2 000 000	Springbank	750 000
Roseisle	12 500 000	Knockdhu	2 000 000	Kingsbarns	600 000
Ailsa Bay	12 000 000	Balblair	1 800 000	Speyside	600 000
Glen Ord	11 000 000	Pulteney	1 800 000	Annandale	500 000
Macallan	11 000 000	Bruichladdich	1 500 000	Ardnamurchan	500 000
Glenlivet	10 500 000	Glendronach	1 400 000	Benromach	500 000
Dalmunach	10 000 000	Glen Spey	1 400 000	Royal Lochnagar	500 000
Teaninich	9 800 000	Knockando	1 400 000	Glenturret	340 000
Balvenie	6 800 000	Glen Garioch	1 370 000	Harris	230 000
Caol Ila	6 500 000	Ardbeg	1 300 000	Kilchoman	200 000
Glen Grant	6 200 000	Glencadam	1 300 000	Edradour	130 000
Dufftown	6 000 000	Scapa	1 300 000	Wolfburn	125 000
Glen Keith	6 000 000	Glenglassaugh	1 100 000	Ballindalloch	100 000
Glenmorangie	6 000 000	Glengoyne	1 100 000	Glasgow	100 000
Mannochmore	6 000 000	Tobermory	1 000 000	Eden Mill	80 000
Auchroisk	5 900 000	Oban	870 000	Daftmill	65 000
Miltonduff	5 800 000	Glen Scotia	800 000	Strathearn	30 000
Glenrothes	5 600 000	Arran	750 000	Abhainn Dearg	20 000
Linkwood	5 600 000	Glengyle	750 000		
Ardmore	5 550 000				
Dailuaine	5 200 000				
Glendullan	5 000 000				
Loch Lomond	5 000 000				
Tomatin	5 000 000				
Clynelish	4 800 000				
Kininvie	4 800 000				
Longmorn	4 500 000				
Tormore	4 400 000				
Glenburgie	4 200 000				
Glentauchers	4 200 000				
Speyburn	4 200 000				
Craigellachie	4 100 000				
Allt-a-Bhainne	4 000 000				
Braeval	4 000 000				
Dalmore	4 000 000				
Glenallachie	4 000 000				
Royal Brackla	4 000 000				
Tamdhu	4 000 000				
Tamnavulin	4 000 000				
Aberlour	3 800 000				
Mortlach	3 800 000				
Glenlossie	3 700 000				
Aberfeldy	3 500 000				
Benrinnes	3 500 000				
Glenfarclas	3 500 000				
Cardhu	3 400 000				
Macduff	3 340 000				
Glen Moray	3 300 000				
Laphroaig	3 300 000				
Tomintoul	3 300 000				
Aultmore	3 200 000				
Fettercairn	3 200 000				
Inchgower	3 200 000				
Deanston	3 000 000				
Tullibardine	3 000 000				
Balmenach	2 800 000				
Benriach	2 800 000				
Blair Athol	2 800 000				
Bunnahabhain	2 700 000				
Glen Elgin	2 700 000				
Talisker	2 700 000				
Strathmill	2 600 000				
Glenkinchie	2 500 000				
Highland Park	2 500 000				
Lagavulin	2 450 000				
Strathisla	2 450 000				
Cragganmore	2 200 000				
Dalwhinnie	2 200 000				
Jura	2 200 000				
Auchentoshan	2 000 000				
Ben Nevis	2 000 000				

Summary of Malt Distillery Capacity by Category

Category	Litres of alcohol	% of Industry	Average capacity
Speyside (48)	226 150 000	61,1	4 711 000
Islands (8)	10 700 000	2,9	1 338 000
Highlands (33)	93 145 000	25,2	2 822 000
Islay (8)	19 950 000	5,4	2 494 000
Lowlands (8)	17 845 000	4,8	2 231 000
Campbeltown (3)	2 300 000	0,6	767 000
Total (108)	**370 090 000**	**100**	**3 427 000**

Summary of Malt Distillery Capacity by Owner

Owner (number of distilleries)	Litres of alcohol	% of Industry
Diageo (28)	120 210 000	32,5
Pernod Ricard (14)	69 150 000	18,7
William Grant (4)	37 600 000	10,2
Edrington Group (4)	19 440 000	5,2
Bacardi (John Dewar & Sons) (5)	18 140 000	4,9
Emperador Inc (Whyte & Mackay) (4)	13 400 000	3,6
Pacific Spirits (Inver House) (5)	12 600 000	3,4
Beam Suntory (2)	8 850 000	2,4
Moët Hennessy (Glenmorangie) (2)	7 300 000	2,0
Distell (Burn Stewart) (3)	6 700 000	1,8
Campari (Glen Grant) (1)	6 200 000	1,7
Loch Lomond Group (2)	5 800 000	1,6
Suntory (Morrison Bowmore) (3)	5 370 000	1,5
Benriach Distillery Co (3)	5 300 000	1,4
Ian Macleod Distillers (2)	5 100 000	1,4
Tomatin Distillery Co (1)	5 000 000	1,4
Angus Dundee (2)	4 600 000	1,2
J & G Grant (Glenfarclas) (1)	3 500 000	0,9
La Martiniquaise (Glen Moray) (1)	3 300 000	0,9
Picard (Tullibardine) (1)	3 000 000	0,8
Nikka (Ben Nevis Distillery) (1)	2 000 000	0,5
Rémy Cointreau (Bruichladdich) (1)	1 500 000	< 0,5
J & A Mitchell (2)	1 500 000	< 0,5
Isle of Arran Distillers (1)	750 000	< 0,5
Wemyss Malts (Kingsbarns) (1)	600 000	< 0,5
Harvey´s of Edinburgh (Speyside) (1)	600 000	< 0,5
Adelphi Distillery (Ardnamurchan) (1)	500 000	< 0,5
Annandale Distillery Co. (1)	500 000	< 0,5
Gordon & MacPhail (Benromach) (1)	500 000	< 0,5
Isle of Harris Distillers (1)	230 000	< 0,5
Kilchoman Distillery Co (1)	200 000	< 0,5
Signatory Vintage (Edradour) (1)	130 000	< 0,5
Wolfburn Distillery (1)	125 000	< 0,5
Ballindalloch Estate (1)	100 000	< 0,5
Glasgow Distillery Company (1)	100 000	< 0,5
Paul Miller (Eden Mill) (1)	80 000	< 0,5
Francis Cuthbert (Daftmill) (1)	65 000	< 0,5
Strathearn Distillery (1)	30 000	< 0,5
Mark Thayburn (Abhainn Dearg) (1)	20 000	< 0,5

↑ Shetland
151

2 3

ORKNEY ISLANDS

132

4 Wick

NORTH HIGHLANDS

Isle of Lewis
129
Isle of Harris
149

5 6

7

8

139 10 9

11

12 13 16 Inverness

SKYE

14

15

SPEYSIDE

18

19 20

21

23

22

Barra
1

73

Kyle of Lockalsh

17

24

Loch Ness

25

Aberdeen

148

26

27

28

30 31

29

CENTRAL HIGHLANDS

32

Fort William

EAST HIGHLANDS

134

Pitlochry 37

36

34

35

72

WEST HIGHLANDS

39

33

36

Oban

Loch Tay

143

40

Dundee

41 133

MULL

44 Perth

145 42

St. Andrews

Loch Lomond

144

137

142 43

45 Stirling

JURA

46

50

68 128 70

130

47 48

53

71

49 Glasgow

54 Edinburgh

67 66

51 52

55

ISLAY 138

150

69

62

152

136

65 63 64

58

ARRAN

THE LOWLANDS

146

59

Campbeltown Ayr

60 61

147

56 127

Dumfries

131

Stranraer

57

Do you want to find out more in detail where the different distilleries are situated? We suggest that you pay a visit to **www.maltmadness.com/whisky/map/Scotland/** where you will find a very nice, interactive map made by Johannes van den Heuvel. Another favourite is found at **bit.ly/daNJMP** where Steffen Bräuner has plotted not only all the Scottish and Irish distilleries but there are also maps for the Americas and for distilleries from the rest of the world.

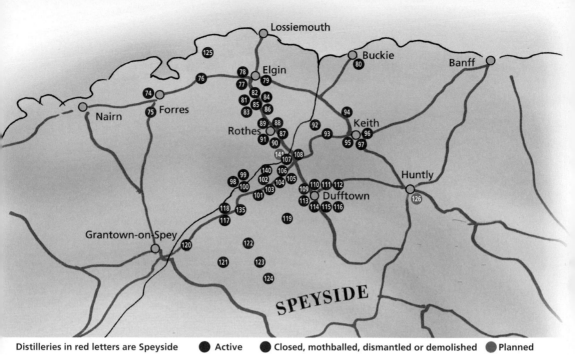

Distilleries in red letters are Speyside ● **Active** ● **Closed, mothballed, dismantled or demolished** ● **Planned**

c = Closed, m = Mothballed, dm = Dismantled, d = Demolished

39 Aberfeldy	38 Edradour	21 Knockdhu	1 Barra	52 Glen Flagler (d)	103 Dailuaine
106 Aberlour	130 Falkirk	56 Ladyburn (dm)	2 Highland Park	53 Rosebank (c)	104 Benrinnes
129 Abhainn Dearg	32 Fettercairn	63 Lagavulin	3 Scapa	54 St Magdalene (dm)	105 Glenallachie
127 Ailsa Bay	138 Gartbreck	64 Laphroaig	4 Pulteney	55 Glenkinchie	106 Aberlour
119 Allt-a-Bhainne	150 Glasgow Dist.	145 Lindores Abbey	5 Brora (c)	56 Ladyburn (dm)	107 Macallan
143 Aniston Farms	13 Glen Albyn (d)	79 Linkwood	6 Clynelish	57 Bladnoch	108 Craigellachie
131 Annandale	105 Glenallachie	48 Littlemill (d)	7 Balblair	58 Arran	109 Convalmore (dm)
62 Ardbeg	76 Glenburgie	46 Loch Lomond	8 Glenmorangie	59 Springbank	110 Dufftown
25 Ardmore	34 Glencadam	36 Lochside (d)	9 Ben Wyvis (c)	60 Glengyle	111 Pittyvaich (d)
134 Ardnamurchan	23 Glendronach	84 Longmorn	10 Teaninich	61 Glen Scotia	112 Glenfiddich
58 Arran	116 Glendullan	107 Macallan	11 Dalmore	62 Ardbeg	113 Balvenie
49 Auchentoshan	85 Glen Elgin	141 Macallan II	12 Glen Ord	63 Lagavulin	114 Kininvie
92 Auchroisk	35 Glenesk (dm)	20 Macduff	13 Glen Albyn (d)	64 Laphroaig	115 Mortlach
94 Aultmore	101 Glenfarclas	81 Mannochmore	14 Glen Mhor (d)	65 Port Ellen (dm)	116 Glendullan
7 Balblair	112 Glenfiddich	15 Millburn (dm)	15 Millburn (dm)	66 Bowmore	117 Tormore
135 Ballindalloch	52 Glen Flagler (d)	77 Miltonduff	16 Royal Brackla	67 Bruichladdich	118 Cragganmore
120 Balmenach	24 Glen Garioch	144 Morrison family	17 Tomatin	68 Kilchoman	119 Allt-a-Bhainne
113 Balvenie	18 Glenglassaugh	115 Mortlach	18 Glenglassaugh	69 Caol Ila	120 Balmenach
19 Banff (d)	50 Glengoyne	33 North Port (d)	19 Banff (d)	70 Bunnahabhain	121 Tomintoul
1 Barra	87 Glen Grant	40 Oban	20 Macduff	71 Jura	122 Glenlivet
30 Ben Nevis	60 Glengyle	111 Pittyvaich (d)	21 Knockdhu	72 Tobermory	123 Tamnavulin
82 Benriach	96 Glen Keith	128 Port Charlotte	22 Glenugie (dm)	73 Talisker	124 Braeval
104 Benrinnes	55 Glenkinchie	65 Port Ellen (dm)	23 Glendronach	74 Benromach	125 Roseisle
74 Benromach	122 Glenlivet	4 Pulteney	24 Glen Garioch	75 Dallas Dhu (c)	126 Duncan Taylor
9 Ben Wyvis (c)	31 Glenlochy (d)	53 Rosebank (c)	25 Ardmore	76 Glenburgie	127 Ailsa Bay
57 Bladnoch	83 Glenlossie	125 Roseisle	26 Speyside	77 Miltonduff	128 Port Charlotte
37 Blair Athol	14 Glen Mhor (d)	16 Royal Brackla	27 Royal Lochnagar	78 Glen Moray	129 Abhainn Dearg
66 Bowmore	8 Glenmorangie	27 Royal Lochnagar	28 Glenury Royal (d)	79 Linkwood	130 Falkirk
124 Braeval	78 Glen Moray	54 St Magdalene (dm)	29 Dalwhinnie	80 Inchgower	131 Annandale
5 Brora (c)	12 Glen Ord	3 Scapa	30 Ben Nevis	81 Mannochmore	132 Wolfburn
67 Bruichladdich	89 Glenrothes	151 Shetland Dist.	31 Glenlochy (d)	82 Benriach	133 Strathearn
70 Bunnahabhain	61 Glen Scotia	88 Speyburn	32 Fettercairn	83 Glenlossie	134 Ardnamurchan
69 Caol Ila	91 Glenspey	26 Speyside	33 North Port (d)	84 Longmorn	135 Ballindalloch
90 Caperdonich (c)	93 Glentauchers	59 Springbank	34 Glencadam	85 Glen Elgin	136 Tweeddale
99 Cardhu	41 Glenturret	133 Strathearn	35 Glenesk (dm)	86 Coleburn (dm)	137 Inchdairnie
152 Clydeside Distillery	22 Glenugie (dm)	97 Strathisla	36 Lochside (d)	87 Glen Grant	138 Gartbreck
6 Clynelish	28 Glenury Royal (d)	95 Strathmill	37 Blair Athol	88 Speyburn	139 "Teaninich 2"
86 Coleburn (dm)	149 Harris	73 Talisker	38 Edradour	89 Glenrothes	140 Dalmunach
109 Convalmore (dm)	147 Hawick	98 Tamdhu	39 Aberfeldy	90 Caperdonich (c)	141 Macallan II
118 Cragganmore	2 Highland Park	123 Tamnavulin	40 Oban	91 Glenspey	142 Eden Mill
108 Craigellachie	137 Inchdairnie	139 "Teaninich 2"	41 Glenturret	92 Auchroisk	143 Aniston Farms
42 Daftmill	102 Imperial (d)	72 Tobermory	42 Daftmill	93 Glentauchers	144 Morrison
103 Dailuaine	80 Inchgower	17 Tomatin	43 Kingsbarns	94 Aultmore	145 Lindores Abbey
75 Dallas Dhu (c)	47 Inverleven (d)	121 Tomintoul	44 Tullibardine	95 Strathmill	146 Jedburgh
11 Dalmore	146 Jedburgh	148 Torabhaig	45 Deanston	96 Glen Keith	147 Hawick
140 Dalmunach	71 Jura	117 Tormore	46 Loch Lomond	97 Strathisla	148 Torabhaig
29 Dalwhinnie	68 Kilchoman	44 Tullibardine	47 Inverleven (d)	98 Tamdhu	149 Harris
45 Deanston	51 Kinclaith (d)	136 Tweeddale	48 Littlemill (d)	99 Cardhu	150 Glasgow Distillery
110 Dufftown	43 Kingsbarns	132 Wolfburn	49 Auchentoshan	100 Knockando	151 Shetland
126 Duncan Taylor	114 Kininvie		50 Glengoyne	101 Glenfarclas	152 Clydeside Distillery
142 Eden Mill	100 Knockando		51 Kinclaith (d)	102 Imperial (d)	

Index

Bold figures refer to the main entry in the distillery directory.

Index

Bold figures refer to the main entry in the distillery directory.

Index

Bold figures refer to the main entry in the distillery directory.